Fundamental concepts
of topology

Fundamental concepts of topology

PETER V. O'NEIL
Department of Mathematics
College of William and Mary in Virginia

GORDON AND BREACH SCIENCE PUBLISHERS
New York   London   Paris

Copyright © 1972 by Gordon and Breach, Science Publishers, Inc.

Gordon and Breach Science Publishers Ltd.
42 William IV Street
London WC2N 4DE

Gordon and Breach, Science Publishers, Inc.
One Park Avenue
New York, NY 10016

Gordon & Breach
7-9 rue Emile Dubois
Paris F-75014

First Published June 1972
Second Printing April 1981

Library of Congress catalog card number 75-140134. ISBN 0 677 03420 2. All rights reserved. No part of this book may be reproduced or utilized in any form or by any means, electronic or mechanical, including photocopying, recording, or by any information storage and retrieval system, without permission in writing from the publishers. Printed in the United States of America.

To my family

Preface

THIS BOOK IS INTENDED as a rigorous first course in set topology, with an introduction to topological groups, homotopy theory and homotopy groups. It should also serve as a foundation for more advanced work in topology and for modern analysis.

The first seven chapters cover basic set topology, with Chapter Eight devoted to the particular properties of metric spaces. The final three chapters are largely independent, and may be read or omitted individually, with the exception that the definition of homotopy given in section one of Chapter Ten is used in Chapter Eleven. Experience has shown that the whole book can be completed at a comfortable pace in two semesters.

My chief aim in writing the book was to present a balanced blend of theory and techniques of proof, examples and counterexamples, and applications. The importance of a feeling for good examples and counterexamples cannot be overemphasized: they serve not only to illustrate the theory, but, perhaps of greater value, to clarify its limitations. For this reason I have developed a number of the classical examples of topology with the beginner in mind. These should be read as carefully as the theorems and proofs.

Applications to other areas of mathematics, particularly analysis, have also been included to indicate some of the more far reaching implications of the theory. Among these are the theorems of Baire and Ascoli for metric spaces, a homotopy proof of the Fundamental Theorem of Algebra, and Andre Weil's proof of the existence of the Haar integral as an introduction to harmonic analysis.

The main factor in choosing the order of presentation of the material was motivation. Wherever practical I have tried to introduce new concepts with some stated purpose, either as solutions to specific problems, or as natural extensions of previously developed theory. For example, the limitations inherent in sequences necessitate more general theories of convergence, which in turn suggest a need for the Hausdorff axiom. Questions involving continuous extensions lead to regular and normal spaces, and completely regular spaces arise in considering the separation properties of products and subspaces of normal spaces.

In several places I have given personal taste preference over custom. Traditionally beginning students are first introduced to metric spaces, then to general topological spaces. My feeling is that, while Euclidean n-space certainly holds a good deal of intuitive appeal, metric spaces, in general, exhibit properties sufficiently unlike those of Euclidean space to make this approach potentially misleading. To cite one instance of this, the student is usually very surprised to learn (if he ever does) that the closure of an open sphere of radius r need not consist of all points at a distance less than or equal to r from the center, as is the case in E^n. For this reason, I place metric spaces at the end of Chapter One in order to have them available for discussion and examples, but postpone a treatment of their particular topological properties until Chapter Eight, by which time a good deal of general theory has been developed. I also prefer filterbases to filters as providing a neater theory of convergence. However, filter convergence is developed in parallel with the filterbase theory in the exercises.

The problems were written to complement the text. Some are routine calculations for practice. However, most of the problems suggest additional concepts and approaches, examples and counterexamples, applications, and, occasionally, a line of research to pursue in the literature. Several important topics not included in the text, among them paracompactness, function spaces, and uniformities, are developed in some depth in series of exercises. The more difficult problems are marked with an asterisk. No proof in the text depends upon any problem.

A short (and necessarily incomplete) history of the development of topology is given following the introductory remarks on prerequisites. More detailed historical notes are given from time to time to help the student achieve a perspective in the subject.

Each chapter is divided into sections, which are numbered from one on. Theorems, definitions, figures and examples are numbered consecutively by section. Theorem 3.4 is the fourth theorem of the third section of the chapter in which the reference occurs. A reference to another chapter is accompanied by the chapter number. For example, Definition 2.1.IV is the first definition, second section of Chapter IV. Corollaries and Lemmas are simply numbered consecutively within each section. The end of a proof or example is marked by the symbol ∎.

The bibliography lists books and articles dealing with various aspects of topology and related analysis, as well as background material in history, set theory and algebra. The level of the references ranges from expository to light technical to research oriented books and papers. Each book is coded to indicate its relevance as a reference for this book.

Following the bibliography are summary charts on separation axioms, compactness and covering theorems, metric spaces and hereditary and productive properties, an index of counterexamples, and an index of symbols and abbreviations.

Prerequisites The reader should have had the usual college background in calculus. Usually elementary calculus is sufficient.

The reader must also be fully conversant with basic set notation and identities. Mathematical induction, construction by induction, Zorn's Lemma, the Axiom of Choice, ordinals, and transfinite induction are discussed informally in Appendix B. The use of ordinals is restricted to examples. A purely intuitive grasp of these concepts is all that is assumed.

Equivalence relations are used in connection with quotient spaces and topological groups. These are the subject of Appendix C.

Finally, a very elementary knowledge of group theory, at a level generally taught in the second or third year of college, is needed in Chapters Nine and Eleven. The basic ideas are sketched in Appendix D.

A historical note Modern topology developed from geometry, analysis, and a deepening questioning and understanding of the nature of mathematics.

Probably the first example of reasoning in a way which might now be called topological was Leonhard Euler's solution of the famous Königsberg Bridge Problem in the eighteenth century. The problem was to determine whether or not one could plan a walk in such a way as to cross each of the seven bridges (Figure 1) exactly once. The critical feature in Euler's solution was his abstraction of the essential part of the problem, the diagram of the proposed walk, which he drew as in Figure 2. Distances and other geometrical features, such as the shape of the paths, were correctly judged irrelevant. Euler characterized his solution as belonging to the "geometry of position", depending as it did upon position alone.

Bernhard Riemann's studies in analysis and geometry in the nineteenth century stimulated the first vague notions of a topological space and topological invariants. Laying the foundations along with Riemann were Cauchy, Abel, Bolzano and Weierstrass, who probed the nature of limits and continuity, and Gauss, who investigated surfaces and non-Euclidean geometries. Further impetus was provided in the second half of the century by the studies of sets and the real number system by Dedekind and Cantor. Cantor's work led him to an understanding of much of the topology of Euclidean n-space as we know it today.

At the end of the nineteenth century, Henri Poincaré continued the development of the theory of surfaces and manifolds initiated by Riemann. Poincaré's results belong to what we now call combinatorial topology, and were motivated by his interest in differential equations.

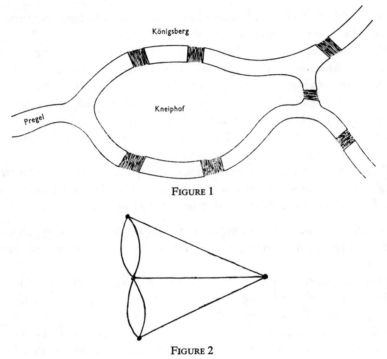

FIGURE 1

FIGURE 2

In 1906 Maurice Fréchet demonstrated how strongly function theory had come to depend upon set theory, previewing the central role topology would soon play in analysis. Around 1911 L. E. J. Brouwer announced his important invariance theorems. Opinions still differ as to whether topology dates from Euler, Riemann, Cantor, Poincaré, or Brouwer (with an occasional vote for Weierstrass).

Poincaré's contemporary, David Hilbert, influenced the development of topology along axiomatic lines. About 1914, Hausdorff, who is usually recognized as the father or general topology, wrote the axioms for modern Hausdorff spaces, working from Hilbert's notion of a neighborhood. A deeper appreciation of the separation axioms soon followed, due largely to P. S. Urysohn and the Russian school. In the 1920's Tychonov developed his notion of compactness. The Moore-Smith theory of convergence (1922) was followed by Henri Cartan's theory of filters (1937).

Algebraic methods were brought to bear on topological problems with the development of algebraic topology. The fundamental group of a space was introduced by Poincaré in 1895. Witold Hurewicz defined the higher homotopy groups in 1935–36. Homology groups are also attributed to Poincaré, with today's axiomatic approach due to Samuel Eilenberg and Norman Steenrod (1952).

Algebraic topology in turn exerted its own influence. A new subject called homological algebra appeared, with much of the early work due to Cartan and Eilenberg. In addition, differential geometry and algebraic geometry have been revitalized by cohomology and the theory of fibre bundles.

In recent years a number of classical problems have been solved, or appear accessible for the first time, and the subject continues to be one of the most vigorous branches of modern mathematics.

Peter V. O'Neil

Contents

	Preface	vii
CHAPTER I	**Topologies**	1
	1.1 Definition of a topology and examples 1	
	1.2 Basic concepts 5	
	1.3 Bases and subbases 12	
	1.4 Metric spaces 20	
	Problems 22	
CHAPTER II	**Continuity and homeomorphism**	28
	2.1 Continuity 28	
	2.2 Homeomorphism 34	
	Problems 37	
CHAPTER III	**Relative, product, identification aud quotient topology**	40
	3.1 Relative topology 40	
	3.2 Product topology 45	
	3.3 Identification and quotient topology 56	
	Problems 61	
CHAPTER IV	**Convergence**	66
	4.1 Sequential convergence 66	
	4.2 Nets 69	
	4.3 Filterbases 75	
	4.4 Equivalence of net and filterbase convergence 79	
	4.5 Ultrafilterbases 80	
	Problems 83	
CHAPTER V	**Separation axioms**	87
	5.1 Hausdorff spaces 87	
	5.2 Regular spaces 92	

5.3 Normal spaces 97
5.4 Completely regular spaces 109
  Problems 116

CHAPTER VI **Connectivity** 119
6.1 Connectedness 119
6.2 Path connectedness 130
  Problems 135

CHAPTER VII **Compactness and covering theorems** . . . 138
7.1 Compactness 138
7.2 Local compactness 154
7.3 Countable compactness 161
7.4 Other kinds of compactness and 1°-countable spaces 164
7.5 2°-countable, separable and lindelöf spaces 167
  Problems 171

CHAPTER VIII **A closer look at metric spaces** 176
8.1 Invariance and hereditary properties 176
8.2 Equivalent metrics 177
8.3 Continuity and uniform continuity 178
8.4 Isometries 181
8.5 Adequacy of sequences 182
8.6 Compactness and covering theorems in metric spaces 183
8.7 Complete metric spaces 188
8.8 The Baire category theorem 196
8.9 Ascoli's theorem 202
8.10 Metrization 204
  Problems 209

CHAPTER IX **Topological groups** 214
9.1 Basic terminology, notation and examples 214
9.2 Bases and separation 219
9.3 Subgroups and connectedness 221
9.4 Compactness, local compactness and quotient groups 223
9.5 Uniform continuity on topological groups 225
9.6 Application: analysis on topological groups 228
  Problems 231

Chapter X	**The concept of homotopy** 235
	10.1 Introduction 235
	10.2 Homotopy 235
	10.3 Nullhomotopy 241
	10.4 Brouwer's theorem for S^1 247
	10.5 Applications 257
	Problems 260
Chapter XI	**Homotopy groups** 262
	11.1 The fundamental group of a space 262
	11.2 $\pi(S^1, (1, 0))$ 268
	11.3 Dependence upon the base point 273
	11.4 The fundamental group of a product space 275
	11.5 Maps and fundamental groups 276
	11.6 Homotopy type 278
	11.7 Higher homotopy groups 281
	Problems 282

Appendix A: logical preliminaries 285

Appendix B: set theory 287

Appendix C: equivalence relations . . . 302

Appendix D: groups 303

Bibliography 306

Summary charts 310

Index of counterexamples 313

Symbols and abbreviations 314

Index 317

CHAPTER I

Topologies

1.1 Definition of a topology and examples

LET X BE A SET. A topology on X is a collection of subsets of X which contains the empty set and X itself, and is closed under the operations of taking arbitrary unions and finite intersections.

1.1 DEFINITION T *is a topology on* X
if and only if
(1) $T \subset \mathscr{P}(X)$.
(2) $\phi \in T$ *and* $X \in T$.
(3) If $A \subset T$, *then* $\cup A \in T$.
(4) If $\phi \neq A \subset T$ *and* A *is finite, then* $\cap A \in T$.

When T is a topology on X, the pair (X, T) is called a topological space. To simplify notation, we usually refer to X as a topological space, or just a space, whenever a specific topology is understood to have been defined on X. In those instances where two or more topologies are being considered on the same set, then of course it is necessary to specify in each statement the topology relevant to that statement.

1.1 EXAMPLE $\mathscr{P}(X)$ is a topology on X for any set X. ∎

1.2 EXAMPLE $\{\phi, X\}$ is a topology on X. ∎

The topology of Example 1.1 is called the discrete topology on X, and that of Example 1.2, the indiscrete topology. These two topologies, which always differ if X has more than one element, are at opposite extremes—one contains as many, the other as few, as possible subsets of X consistent with the definition. Such topologies are often useful in constructing counterexamples.

1.3 EXAMPLE Let $X = \{a, b\}$ and $T = \{\phi, \{a\}, X\}$. Then T is the Sierpinski topology on X. ∎

1.4 EXAMPLE Let X consist of all n-tuples of real numbers (n is some positive integer). Traditionally, this set is denoted E^n. E^1 may be identified geometrically with the real line, E^2 with the plane, and so on.

If $(x_1, \ldots, x_n) \in E^n$ and $(y_1, \ldots, y_n) \in E^n$, let

$$\varrho_n((x_1, \ldots, x_n), (y_1, \ldots, y_n)) = \left\{\sum_{i=1}^{n} (x_i - y_i)^2\right\}^{1/2}.$$

It is not difficult to verify that ϱ_n enjoys the following properties:

$\varrho_n(a, b) \geq 0$,

$\varrho_n(a, b) = 0 \Leftrightarrow a = b$,

$\varrho_n(a, b) = \varrho_n(b, a)$,

$\varrho_n(a, b) \leq \varrho_n(a, c) + \varrho_n(c, b)$,

whenever $a, b, c \in E^n$.

Define a set T of subsets of E^n by specifying that, if $M \subset E^n$, then $M \in T$ if, given $x \in M$, there is some positive number ε (probably dependent upon x) such that $E^n \cap \{y | \varrho_n(x, y) < \varepsilon\} \subset M$.

Geometrically, a subset M of E^2 is in T if, given $x \in M$, some circle (or sphere in the case of E^3), of radius, say, ε, can be drawn about x so that all points inside the circle (or sphere) are in M. For example, in E^2, $\{(x, 0) | x \text{ is real}\} \notin T$, but $\{(x, y) \mid |x - 1| < 3 \text{ and } |y - 4| < 2\} \in T$.

We shall verify that T is a topology on E^n. Clearly $T \subset \mathscr{P}(E^n)$; $\phi \in T$ is vacuously true, and $E^n \in T$ is immediate (if $x \in E^n$, choose $\varepsilon = 1$).

To verify Def. 1.1 (3), let $A \subset T$. We must show that $\cup A \in T$. To do this, let $x \in \cup A$. Then, $x \in a$ for some $a \in A$. Since $A \subset T$, then $a \in T$. Then, for some $\varepsilon > 0$, $E^n \cap \{y | \varrho_n(x, y) < \varepsilon\} \subset a \subset \cup A$.

Finally, consider Def. 1.1(4). Let $\phi \neq A \subset T$ and let A be finite. For concreteness, suppose $A = \{a_1, \ldots, a_s\}$. To show that $\bigcap_{i=1}^{s} a_i \in T$, let $x \in \bigcap_{i=1}^{s} a_i$. Given i, $1 \leq i \leq s$, then $x \in a_i \in T$, so for some $\varepsilon_i > 0$, $E^n \cap \{y | \varrho_n(x, y) < \varepsilon_i\} \subset a_i$. Let $\varepsilon = \min_{1 \leq i \leq s} \varepsilon_i$. Then $\varepsilon > 0$ and $\varepsilon \leq \varepsilon_i$ for $i = 1, \ldots, s$, so that $E^n \cap \{y | \varrho_n(x, y) < \varepsilon\} \subset a_i$.

It is interesting to note that the argument just given for finite intersections may fail if A is infinite. If, say, $A = \{a_1, a_2, \ldots\}$, with infinitely many elements, then we may have infinitely many numbers ε_i, and it may not be possible to choose a smallest one. For example, if $\varepsilon_i = 1/i$, there is no positive number smaller than each ε_i, and, in fact, $\inf\{1/i | i \in Z^+\} = 0$.

As a concrete example of this, note that, for $n = 1$, $]-1/j, 1/j[\in T$ for each positive integer j. But $\bigcap_{j=1}^{\infty}]-1/j, 1/j[= \{0\} \notin T$. ∎

The topology just described is called the Euclidean topology on E^n, and (E^n, T) is called Euclidean n-space. Henceforth, whenever we write E^n, we mean Euclidean n-space.

1.5 EXAMPLE Let X be any set, and $T = (\mathscr{P}(X) \cap \{A | X - A \text{ is finite}\}) \cup \{\phi\}$. Then T is the cofinal topology on X (sometimes T is also known as the topology of finite complements). Note that, if X is finite, then T is the discrete topology. ∎

1.6 EXAMPLE Let a and b be real numbers, with $a < b$.

Let $X = \{f | f \text{ is continuous on } [a, b] \text{ into the reals}\}$. This set is customarily written $C([a, b])$.

If f and g are in X, let
$$d(f, g) = \sup \{|f(x) - g(x)| \,|\, a \leq x \leq b\}.$$
Note that:
$$d(f, g) \geq 0,$$
$$d(f, g) = 0 \Leftrightarrow f = g,$$
$$d(f, g) = d(g, f),$$
$$d(f, g) \leq d(f, h) + d(h, g)$$
for f, g, h in $C([a, b])$.

Define $T \subset \mathscr{P}(X)$ by specifying, if $A \subset X$, then $A \in T$ if, given $f \in A$, there is some $\varepsilon > 0$ such that $X \cap \{g | d(f, g) < \varepsilon\} \subset A$.

The argument used in Example 1.4 can be adapted almost verbatim (replace E^n by $C([a, b])$ and ϱ_n by d) to show that T is a topology on $C([a, b])$. T is called the sup norm topology on $C([a, b])$, for reasons which will become clear in Chapter IV (see Example 1.4 in that chapter). ∎

1.7 EXAMPLE Let $X = \{f | f: [0, 1] \to R\}$, where R denotes the set of real numbers. Then X consists of all real-valued functions on the unit interval and contains $C([0, 1])$ as a subset.

If n is a positive integer and $x_1, \ldots, x_n$ are points in $[0, 1]$, and $\varepsilon > 0$ and $f \in X$, let
$$U(f; x_1, \ldots, x_n; \varepsilon) = X \cap \{g \,|\, |g(x_i) - f(x_i)| < \varepsilon \text{ for } i = 1, \ldots, n\}.$$

If $A \subset X$, define $A \in T$ to mean:

If $f \in A$, then there is some $\varepsilon > 0$ and there are points $x_1, \ldots, x_n$ in $[0, 1]$ such that $U(f; x_1, \ldots, x_n; \varepsilon) \subset A$.

Then, T is a topology on X. Verification of Definition 1.1, (1) through (3), is straightforward. To verify Def. 1.1(4), suppose first that $a \in T$ and $b \in T$. Let $f \in a \cap b$. Since $f \in a$, then some $U(f; x_1, \ldots, x_n; \varepsilon) \subset a$. Since

$f \in b$, then some $U(f; y_1, ..., y_m; \delta) \subset b$. Choose $\xi = \min(\varepsilon, \delta)$. Then $U(f; x_1, ..., x_n, y_1, ..., y_m; \xi) \subset a \cap b$, implying that $a \cap b \in T$. Now Def. 1.1(4) follows by an easy induction argument. ∎

These seven examples will suffice for now. As an exercise, the reader might try constructing all the topologies on the set $\{a, b, c\}$ (there are twenty-nine of them).

Elements of a topology T on X are customarily called T-open sets, or, informally, open sets. It is important to note that the word open is meaningful only in reference to a given topology. Thus, the "open interval" $]0, 1[$ is open in the Euclidean topology on E^1, but is not open as a subset of R in the cofinal or indiscrete topologies on R. Similarly, $\{2\}$ is open in the discrete, but not in the Euclidean, indiscrete or cofinal topologies on R.

A subset A of X is a T-neighborhood (T-nghd., or just nghd.) of x if A is T-open and $x \in A$.

Finally, a subset A of X is T-closed if $X - A$ is T-open (i.e., $X - A \in T$).

Note that open is not the opposite of closed, nor are the terms mutually exclusive. In the discrete topology on X, every subset of X is both open and closed. In the Euclidean topology on E^1, $]0, 1]$ is neither open nor closed.

For completeness, we list these notions as Definition 1.2. The obvious fact that ϕ and X are always both open and closed is given as Theorem 1.1.

1.2 DEFINITION *Let T be a topology on X.*

(1) If $A \subset X$, then, A is T-open $\Leftrightarrow A \in T$.
(2) If $x \in X$, then, A is a T-nghd. of $x \Leftrightarrow x \in A \in T$.
(3) If $A \subset X$, then, A is T-closed $\Leftrightarrow X - A \in T$.

1.1 THEOREM *Let T be a topology on X. Then,*

(1) ϕ is T-open and T-closed.
(2) X is T-open and T-closed.

Proof ϕ and X are both T-open, since $\phi \in T$ and $X \in T$ by definition. Since ϕ is T-open, then $X = X - \phi$ is T-closed. Then also ϕ is T-closed, since $X - X = \phi$. ∎

Definition 1.1 (3) and (4) may be rephrased: arbitrary unions, and finite intersections, of open sets are open. As we saw in Example 1.4, an arbitrary intersection of open sets need not be open ($\{0\}$ is closed and not open in the Euclidean topology on E^1). The reader can construct an example of a union of closed sets which is not closed.

Theorem 1.2 says that arbitrary intersections and finite unions of closed sets are closed. These statements are the duals of Definition 1.1, (3) and (4), through DeMorgan's Laws.

1.2 Theorem *Let T be a topology on X. Let $A \subset \mathscr{P}(X)$ such that, if $a \in A$, then a is T-closed. Then,*

(1) If $A \neq \phi$, then $\cap A$ is T-closed.
(2) If A is finite, then $\cup A$ is T-closed.

Proof of (1) Suppose $A \neq \phi$.
Note that $X - \cap A = \bigcup_{a \in A} (X - a)$. Now, if $a \in A$, then $X - a \in T$, so $\bigcup_{a \in A} (X - a) \in T$ by Definition 1.1 (3). Hence $X - \cap A \in T$, and $\cap A$ is T-closed.

Proof of (2) If $A = \phi$, then $\cup A = \phi$ is T-closed by Theorem 1.1 (1).
If $A \neq \phi$, then $X - \cup A = \bigcap_{a \in A} (X - a)$. But, $X - a \in T$ if $a \in A$. Further, A is finite, so $\bigcap_{a \in A} (X - a) \in T$ by Definition 1.1 (4). Hence $X - \cup A \in T$, implying that $\cup A$ is T-closed. ∎

1.2 Basic concepts

Given a subset A of X, and a topology T on X, there are certain kinds of points in X which bear a particularly important relationship to A relative to T. These are the interior and exterior points, cluster points, boundary points, and isolated points, of A. The names are motivated by the geometry of E^n, as will be seen shortly in the examples, but be careful not to carry geometric intuition into arbitrary topological spaces which might look quite different from E^n.

For the remainder of this section (X, T) is a topological space.

2.1 Definition *Let $A \subset X$. Then,*

*(1) x is T-interior to A
if and only if
for some a, $a \in T$ and $x \in a \subset A$.*
*(2) x is T-exterior to A
if and only if
for some a, $a \in T$ and $x \in a \subset X - A$.*

Put another way, x is interior to A (relative to T) if x has a neighborhood wich is wholly inside A; x is exterior to A if some neighborhood of x is wholly outside of A.

2.1 Example In the Sierpinski topology of Example 1.3, a is an interior point of $\{a\}$ and of X, but b is an interior point only of X. Note that a is an exterior point of $\{b\}$, but b is not an exterior point of $\{a\}$, as the only neighborhood of b is X, and X is not a subset of $X - \{a\}$. ∎

2.2 Example Let X be the set of real numbers and A the interval $]0, 1]$. Then, $\frac{1}{2}$ is an interior point of A relative to the Euclidean topology on E^1, since $\frac{1}{2} \in]0, \frac{3}{4}[\subset]0, 1]$, and $]0, \frac{3}{4}[$ is Euclidean open. But $\frac{1}{2}$ is not interior to A relative to the cofinal topology on X. For, if $\frac{1}{2} \in a \subset]0, 1]$, then $X - a$ is infinite, so a cannot be cofinal-open.

In the Euclidean topology, 3 is an exterior point of $]0, 1]$, but 0 and 1 are neither interior nor exterior. ∎

Strictly speaking, we never have to mention exterior points, since x is exterior to A exactly when x is interior to $X - A$.

2.1 Theorem *Let $x \in X$ and $A \subset X$.*
Then,
x is T-exterior to A
if and only if
x is T-interior to $X - A$.

Proof Immediate by Definition 2.1. ∎

For convenience, we denote the set of T-interior points of A by $\text{Int}_T(A)$, called the T-interior (or just interior) of A. In view of Theorem 2.1, there is no need to define a set $\text{Ext}_T(A)$.

2.2 Definition $\text{Int}_T(A) = \{x | x \text{ is } T\text{-interior to } A\}$.

The interior points of A constitute the largest open subset of A. As a result, a set coincides with its interior exactly when it is open.

2.2 Theorem *Let $A \subset X$.*
Then,

(1) $\text{Int}_T(A) \subset A$.
(2) $\text{Int}_T(A) = \cup \{a | a \in T \text{ and } a \subset A\}$.
(3) A *is* T-*open* $\Leftrightarrow \text{Int}_T(A) = A$.
(4) *If* $A \subset B \subset X$, *then* $\text{Int}_T(A) \subset \text{Int}_T(B)$.
(5) *If* $B \subset X$, *then* $\text{Int}_T(A) \cup \text{Int}_T(B) \subset \text{Int}_T(A \cup B)$.

Proof of (1) Let $x \in \text{Int}_T (A)$. For some a, $x \in a \subset A$ and $a \in T$. Then, $x \in A$.

Proof of (2) Let $x \in \text{Int}_T (A)$. Then, for some a, $x \in a \subset A$ and $a \in T$. Then, $x \in a \in \{b | b \subset A \text{ and } b \in T\}$, implying that $x \in \cup \{b | b \subset A \text{ and } b \in T\}$. Then, $\text{Int}_T (A) \subset \cup \{b | b \subset A \text{ and } b \in T\}$.

Conversely, suppose that $y \in \cup \{b | b \subset A \text{ and } b \in T\}$. For some a, $y \in a \subset A$ and $a \in T$. But then $y \in \text{Int}_T (A)$.

Proof of (3) If A is T-open, then $A \in \{a | a \subset A \text{ and } a \in T\}$, implying that $A \subset \cup \{a | a \subset A \text{ and } a \in T\} = \text{Int}_T (A)$. Then, $\text{Int}_T (A) = A$ by (1).

Conversely, suppose that $\text{Int}_T (A) = A$. By (2) and Definition 1.1 (3), $\text{Int}_T (A)$ is T-open.

Proof of (4) Let $x \in \text{Int}_T (A)$. For some a, $x \in a \subset A$ and $a \in T$. Then, $a \subset B$, so that $x \in \text{Int}_T (B)$.

Proof of (5) By (4), $\text{Int}_T (A) \subset \text{Int}_T (A \cup B)$. Similarly, $\text{Int}_T (B) \subset \text{Int}_T (A \cup B)$. Hence $\text{Int}_T (A) \cup \text{Int}_T (B) \subset \text{Int}_T (A \cup B)$. ∎

It is not in general true that $\text{Int}_T (A) \cup \text{Int}_T (B) = \text{Int}_T (A \cup B)$. For example, let $A = \{x | x \text{ is irrational and } 0 < x < 1\}$, and $B = \{x | x \text{ is rational and } 0 < x < 1\}$. In the Euclidean topology on E^1, $\text{Int}_T (A) = \text{Int}_T (B) = \phi$, since any open interval contains both rationals and irrationals. But $A \cup B =]0, 1[= \text{Int}_T (A \cup B)$.

A boundary point of A is a point which is neither interior nor exterior to A. That is, every neighborhood of the point contains at least one point in A and one point not in A.

A cluster point of A is one with the property that every neighborhood of the point, with the point removed, contains at least one element of A.

Finally, x is an isolated point of A if some neighborhood of x has only x in common with A.

2.3 Definition Let $x \in X$ and $A \subset X$.

(1) x is a T-boundary point of A
  if and only if
  if $x \in t \in T$, then $t \cap A \neq \phi$ and $t \cap (X - A) \neq \phi$.
(2) x is a T-cluster point of A
  if and only if
  if $x \in t \in T$, then $(t - \{x\}) \cap A \neq \phi$.
(3) x is a T-isolated point of A
  if and only if
  for some t, $x \in t \in T$ and $t \cap A = \{x\}$.

We shall make some observations and then give examples. Note that a boundary point of A may or may not belong to A. Also, the boundary points of A and of $X - A$ are the same, since the definition is unchanged if A is replaced by $X - A$.

The condition for x to be a cluster point of A is sometimes rephrased: every punctured neighborhood of x intersects A. Cluster points of A may also be in A or in $X - A$, and boundary points may be cluster points (and vice versa).

An isolated point of A is automatically in A, and is never a cluster point of A.

2.3 Example Let $X = \{a, b\}$, with the Sierpinski topology. Then, a is an isolated point of $\{a\}$, since $a \in \{a\}$ and $\{a\} \cap \{a\} \in T$. But a is not a cluster point of $\{a\}$, since $\{a\} \in T$ and $(\{a\} - \{a\}) \cap \{a\} = \phi$. a is not a boundary point of $\{a\}$, since $\{a\} \cap (X - \{a\}) = \phi$. a is an interior point of $\{a\}$, since $\{a\}$ is open.

b is a boundary point of $\{a\}$, since the only neighborhood of b is X, and $X \cap \{a\} \neq \phi$ and $X \cap (X - \{a\}) \neq \phi$. Hence b is also a boundary point of $X - \{a\} = \{b\}$. b is not an interior point of $\{b\}$; b is a cluster point of $\{a\}$, but not of $\{b\}$, and is an isolated point of $\{b\}$. Finally, a is not a cluster point of $\{b\}$. ∎

2.4 Example Let S be the subset of E^2 given by $S = \{(x, y) | x < y\} \cup \{(n, -n) | n$ is a positive integer$\}$. See Figure 2.1.

The isolated points of S are exactly the points $(n, -n)$ with n a positive integer. About each such point can be drawn a circular neighborhood containing no other points of S.

The boundary points of S are exactly the points $(n, -n)$, $n \in Z^+$ (these belong to S), and the points (x, x) with x real (these do not belong to S). Some, but not all, of the boundary points are isolated points.

The interior points of S are exactly the points (x, y) with x and y real and $x < y$.

The cluster points of S are the points (x, y) with $x \leq y$. Thus, some, but not all, of the cluster points are boundary points, and some, but not all, of the cluster points are interior points. ∎

2.5 Example Let $X = C([a, b])$, with T the sup norm topology of Example 1.6.

Let A be the subset of $C([a, b])$ consisting of the polynomials on $[a, b]$.

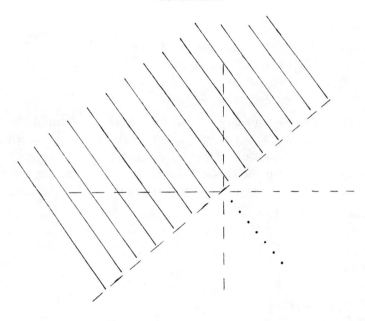

$\{(x,y) | x < y\} \cup \{(m,-m) | m \in Z^+\}$

FIGURE 2.1

Let E be the exponential function restricted to $[a, b]$. That is, $E(x) = e^x$ for $a \leq x \leq b$. Then, $E \in C([a, b])$ and $E \notin A$. We shall show that E is a cluster point of A.

Let t be any neighborhood of E. We must produce a polynomial lying in t.

By definition of the sup norm topology, there is some $\varepsilon > 0$ such that $g \in t$ whenever $g \in C([a, b])$ and $d(E, g) = \sup\{|E(x) - g(x)| \mid a \leq x \leq b\} < \varepsilon$. Thus, it suffices to produce a polynomial p which uniformly approximates E on $[a, b]$ to within an error of ε.

If the reader is familiar with the Weierstrass Approximation Theorem (see Cullen, Dugundji or Simmons), then we are through. If not, let

$$p_n(x) = \sum_{j=0}^{n} \frac{x^j}{j!} \text{ for } a \leq x \leq b \text{ and } n = 1, 2, \ldots$$

The sequence $\{p_n\}_{n=1}^{\infty}$ converges uniformly to E on $[a, b]$, so for some positive integer N, $d(E, p_N) < \varepsilon$.

It takes very little extra work to conclude that E is also a boundary point of A. If t is any neighborhood of E, then $t \cap A \neq \phi$ by the preceding argument. But $E \notin A$, so $t \cap (C([a, b]) - A) \neq \phi$ also.

The reader might try showing that A has no interior or isolated points. ∎

2.4 DEFINITION

(*1*) $\text{Bdry}_T(A) = \{x | x \text{ is a } T\text{-boundary point of } A\}$.
(*2*) $\text{Der}_T(A) = \{x | x \text{ is a } T\text{-cluster point of } A\}$.
(*3*) $\text{Cl}_T(A) = A \cup \text{Der}_T(A)$.
(*4*) $\text{Is}_T(A) = \{x | x \text{ is a } T\text{-isolated point of } A\}$.

$\text{Bdry}_T(A)$ is called the boundary (or sometimes frontier) of A, and $\text{Der}_T(A)$ the derived set of A. $\text{Cl}_T(A)$ consists of all points of A, together with all cluster points of A, and is called the closure of A. As we shall see, $\text{Cl}_T(A)$ is always closed, and is the smallest closed set containing A as a subset. Customarily, $\text{Cl}_T(A)$ is denoted $\bar{A}$, when the topology T is understood.

2.3 THEOREM *Let* $A \subset X$.
Then,

(*1*) $A \subset \bar{A}$.
(*2*) *Let* $x \in X$. *Then,*
$$x \in \bar{A} \Leftrightarrow t \cap A \neq \phi \text{ for each nghd. } t \text{ of } x.$$
(*3*) *If* $A \subset B \subset X$, *then* $\bar{A} \subset \bar{B}$.
(*4*) $\bar{A}$ *is T-closed.*
(*5*) $A = \bar{A} \Leftrightarrow A$ *is T-closed.*
(*6*) $\bar{\bar{A}} = \bar{A}$.
(*7*) $\bar{A} = \cap\{C | A \subset C \text{ and } C \text{ is } T\text{-closed}\}$.
(*8*) *If* $A \subset B$ *and* B *is T-closed, then* $\bar{A} \subset B$.
(*9*) *If* $B \subset X$, *then* $\overline{A \cup B} = \bar{A} \cup \bar{B}$.

Proof of (*1*)   Immediate from Definition 2.4 (3).

Proof of (*2*)   Suppose first that $x \in \bar{A}$. If $x \in A$, then $x \in t \cap A$ for each neighborhood t of x. If $x \in \text{Der}_T(A)$, then $(t - \{x\}) \cap A \neq \phi$ for each nghd. t of x, so also $t \cap A \neq \phi$.

Conversely, suppose that $t \cap A \neq \phi$ for each nghd. t of x. If $x \in t \cap A$ for some nghd. t of x, then $x \in A$, so that $x \in \bar{A}$. If $x \notin t \cap A$ for each nghd. t of x, then $(t - \{x\}) \cap A = t \cap A \neq \phi$, so $x \in \text{Der}_T(A)$, hence $x \in \bar{A}$.

Proof of (*3*)   Let $A \subset B \subset X$. Let $x \in \bar{A}$. If t is a nghd. of x, then $t \cap A \neq \phi$ by (2), so also $t \cap B \neq \phi$. Then, by (2), $x \in \bar{B}$.

Proof of (*4*)   We shall show that $X - \bar{A}$ is open. Let $y \in X - \bar{A}$. Then, $y \notin \text{Der}_T(A)$. Then, there is some nghd. t of y such that $(t - \{y\}) \cap A = \phi$. But, $y \notin A$, so $t \cap A = \phi$. Then $t \subset x - \bar{A}$, so $y \in \text{Int}_T(X - \bar{A})$, so $X - \bar{A}$ is open by Theorem 2.2 (3).

Proof of (5) If $A = \bar{A}$, then A is closed by (4).

Conversely, suppose that A is closed. To show that $A = \bar{A}$, it suffices to show that $\text{Der}_T(A) \subset A$. Let $x \in \text{Der}_T(A)$, and suppose $x \notin A$. Then, $x \in X - A \in T$, so $((X - A) - \{x\}) \cap A \neq \phi$, which is absurd.

Proof of (6) By (4), $\bar{A}$ is closed. Then, by (5), $\bar{\bar{A}} = \bar{A}$.

Proof of (7) By (1) and (4), $\bar{A} \in \{C | A \subset C \text{ and } C \text{ is closed}\}$. Then, $\cap \{C | A \subset C \text{ and } C \text{ is closed}\} \subset \bar{A}$.

Conversely, let $x \in \bar{A}$. Suppose that $A \subset C$ and C is closed. Then, by (3) and (5), $\bar{A} \subset \bar{C} = C$, so $x \in C$. Then, $x \in \cap \{C | A \subset C \text{ and } C \text{ is closed}\}$.

Proof of (8) Immediate by (7).

Proof of (9) By (3), $\bar{A} \subset \overline{A \cup B}$ and $\bar{B} \subset \overline{A \cup B}$, so $\bar{A} \cup \bar{B} \subset \overline{A \cup B}$.

To go the other way, note that, by (1) and (4), $\bar{A} \cup \bar{B} \in \{C | A \cup B \subset C$ and C is closed$\}$. Then, by (7), $\cap \{C | A \cup B \subset C$ and C is closed$\} = \overline{A \cup B} \subset \bar{A} \cup \bar{B}$. ∎

2.4 Theorem *Let $A \subset X$. Then,*

(1) $\text{Bdry}_T(A) = \bar{A} \cap \overline{(X - A)}$.
(2) $\text{Bdry}_T(A) = \text{Bdry}_T(X - A)$.
(3) $\text{Bdry}_T(A) = \bar{A} - \text{Int}_T(A)$.
(4) $\bar{A} = \text{Int}_T(A) \cup \text{Bdry}_T(A)$.
(5) The sets $\text{Int}_T(A)$, $\text{Bdry}_T(A)$ and $\text{Int}_T(A)$ are pairwise disjoint, and $X = \text{Int}_T(A) \cup \text{Bdry}_T(A) \cup \text{Int}_T(A)$.
(6) $\text{Is}_T(A) \cap \text{Der}_T(A) = \phi$.

Proof Left as an exercise. ∎

We conclude this section with a theorem which will be of use later.

2.5 Theorem *Let A be T-open. Let B be T-closed. Then,*

(1) $B - A$ is T-closed.
(2) $A - B$ is T-open.

Proof of (1) Note that $X - (B - A) = (X - B) \cup A$. Since B is closed, $X - B$ is open. Since A is open, $(X - B) \cup A$ is open.

Proof of (2) Left as an exercise. ∎

1.3 Bases and subbases

It is in general very cumbersome to specify a topology by explicitly describing each open set. Often, however, it is possible to specify T in terms of just some of its elements.

Consider, for example, the Euclidean topology on E^1. If A is open and $x \in A$, then $]x - \varepsilon_x, x + \varepsilon_x[\subset A$ for some $\varepsilon_x > 0$. Then $A = \bigcup_{x \in A}]x - \varepsilon_x, x + \varepsilon_x[$. The open intervals form a base for the Euclidean topology on E^1 in the sense that each open set is a union of open intervals. The base completely determines the topology (which consists of all possible unions of basic open sets) and has the advantage that it is often easier to work with.

3.1 Definition $\mathscr{B}$ *is a base for* T
if and only if
$\mathscr{B} \subset T$ *and* $T = \{\cup P | P \subset \mathscr{B}\}$.

3.1 Example Let $\mathscr{B}$ consist of all sets $\{g | d(f, g) < \varepsilon\}$, where $\varepsilon > 0$ and $f \in C([a, b])$. Then $\mathscr{B}$ is a base for the sup norm topology of Example 1.6. ∎

3.2 Example If T is a topology on X, then T is a base for T. ∎

For the remainder of this section, let T be a topology on X. It is often useful to know when a given subset $\mathscr{B}$ of T is a base for T. A necessary and sufficient condition is that one be able to fit inside each nghd. of x, a set from $\mathscr{B}$ containing x, for each point x in the space.

3.1 Theorem *Let* $\mathscr{B} \subset T$.
Then,
$\mathscr{B}$ *is a base for* T
if and only if
if $x \in t \in T$, *then there is some* $b \in \mathscr{B}$ *such that* $x \in b \subset t$.

Proof Suppose first that $\mathscr{B}$ is a base for T, and let $x \in t \in T$. For some P, $P \subset \mathscr{B}$ and $t = \cup P$. Then, $x \in b$ for some $b \in P$. Then $x \in b \subset t$ and $b \in \mathscr{B}$.

Conversely, suppose, if $x \in t \in T$, then $x \in b \subset t$ for some $b \in \mathscr{B}$. Let $A \in T$. If $a \in A$, choose some $b_a \in \mathscr{B}$ such that $x \in b_a \subset A$. Then, $\{b_a | a \in A\} \subset \mathscr{B}$ and $A = \bigcup_{a \in A} b_a$. ∎

Note that, in Examples 1.4, 1.6 and 1.7, we actually gave the topology in terms of a base, although the term base was not used at the time.

In view of Theorem 3.1, it is easy to test openness of a given set A by looking at just the basic open sets. A set A is open exactly when each point of A has a basic nghd. inside A. It is further possible to distinguish points of $\bar{A}$, as well as cluster points, by using just the basic neighborhoods in the criteria of Theorem 2.3 (2) and Definition 2.3 (2) respectively.

3.2 Theorem *Let $\mathscr{B}$ be a base for T. Let $A \subset X$. Then,*

(1) $A \in T$
 if and only if
 if $x \in A$, then there is some $b \in \mathscr{B}$ such that $x \in b \subset A$.

(2) Let $x \in X$.
 Then,
 $x \in \bar{A} \Leftrightarrow$ if $x \in b \in \mathscr{B}$, then $b \cap A \neq \phi$.

(3) Let $x \in X$.
 Then,
 x is a T-cluster point of A
 if and only if
 if $x \in b \in \mathscr{B}$, then $(b - \{x\}) \cap A \neq \phi$.

Proof of (1) Left as an exercise.

Proof of (2) Let $x \in \bar{A}$, and let $x \in b \in \mathscr{B}$. Since $\mathscr{B} \subset T$, then $b \in T$, so $b \cap A \neq \phi$ by Theorem 2.3 (2).

Conversely, suppose that $b \cap A \neq \phi$ whenever $x \in b \in \mathscr{B}$. Let $x \in t \in T$. By (1), for some b, $b \in \mathscr{B}$ and $x \in b \subset t$. Then $b \cap A \neq \phi$, so $t \cap A \neq \phi$, and $x \in \bar{A}$ by Theorem 2.3 (2).

Proof of (3) Left as an exercise. ∎

Note that a topology will in general have many bases. In practice, one attempts to choose as simple a base as possible to work with.

It is natural at this point to ask the following question. If $\mathscr{B}$ is a set of subsets of X, is $\mathscr{B}$ a base for some topology on X?

The answer is obviously no: $\{\phi\}$ is not a base for any topology on X if $X \neq \phi$. Even if $\cup \mathscr{B} = X$, however, $\mathscr{B}$ still need not be a topological base. For example, let $X = \{a, b, c\}$ and let $\mathscr{B}$ consist of ϕ, $\{a, b\}$, and $\{b, c\}$. If $\mathscr{B}$ were a base for a topology S on X, then S would have to consist of ϕ, X, $\{a, b\}$ and $\{b, c\}$, together will all possible unions of sets in $\mathscr{B}$. But S is not a topology, since $\{a, b\} \cap \{b, c\} = \{b\} \notin S$.

Sufficient conditions for a set of subsets of X to be a base for a topology on X are given in the next theorem.

3.3 Theorem Let $\mathscr{B} \subset \mathscr{P}(X)$ and $\cup \mathscr{B} = X$.
Then,
$\mathscr{B}$ is a base for a topology on X
if and only if
if $a \in \mathscr{B}$ and $b \in \mathscr{B}$ and $x \in a \cap b$, then $x \in c \subset a \cap b$ for some $c \in \mathscr{B}$.

Proof Necessity is a simple consequence of Theorem 3.1, and is left to the reader.

To prove sufficiency, suppose, if $a \in \mathscr{B}$ and $b \in \mathscr{B}$ and $x \in a \cap b$, then $x \in c \subset a \cap b$ for some $c \in \mathscr{B}$.

Let $T = \{\cup A | A \subset \mathscr{B}\}$. We shall show that T is a topology on X.

Note that $\phi \subset \mathscr{B}$, so $\cup \phi = \phi \in T$. Also, $\cup \mathscr{B} = X$ by hypothesis.

To verify Definition 1.1 (3), let $P = \{p_\alpha | \alpha \in \mathscr{A}\} \subset T$. Now, if $\alpha \in \mathscr{A}$, then $p_\alpha \in T$, so for some $A_\alpha \subset \mathscr{B}$, $p_\alpha = \cup A_\alpha$. For convenience, write $A_\alpha = \{z_{\alpha x} | x \in C_\alpha\}$. Then, $p_\alpha = \bigcup_{x \in C_\alpha} z_{\alpha x}$, and $\cup P = \bigcup_{\alpha \in \mathscr{A}} p_\alpha = \bigcup_{\alpha \in \mathscr{A}} \left(\bigcup_{x \in C_\alpha} z_{\alpha x} \right) = \bigcup_{(\alpha, x) \in \bigcup_{\alpha \in \mathscr{A}} (\{\alpha\} \times C_\alpha)} z_{\alpha x}$.

But, for each $(\alpha, x) \in \bigcup_{\alpha \in \mathscr{A}} (\{\alpha\} \times C_\alpha)$, we have $z_{\alpha x} \in \mathscr{B}$, so $\bigcup_{(\alpha, x) \in \bigcup_{\alpha \in \mathscr{A}} (\{\alpha\} \times C_\alpha)} z_{\alpha x} \in T$.

Finally, to verify Definition 1.1 (4), let $A \in T$ and $B \in T$. Write $A = \bigcup_{\alpha \in M} a_\alpha$ and $B = \bigcup_{\beta \in N} b_\beta$, where $a_\alpha \in \mathscr{B}$ for each $\alpha \in M$ and $b_\beta \in \mathscr{B}$ for each $\beta \in N$. If $x \in A \cap B$, then, for some $\alpha_x \in M$ and $\beta_x \in N$, we have $x \in a_{\alpha_x}$ and $x \in b_{\beta_x}$. Then, $x \in a_{\alpha_x} \cap b_{\beta_x}$, so by hypothesis there is some $c_x \in \mathscr{B}$ such that $x \in c_x \subset a_{\alpha_x} \cap b_{\beta_x}$.

It is easy now to verify that $A \cap B = \bigcup_{x \in A \cap B} c_x$, hence $A \cap B \in T$. By an obvious induction, T satisfies Definition 1.1 (4). ∎

As an application, the reader should show that the sets $U(f; x_1, \ldots, x_n; \varepsilon)$ of Example 1.7 form a base for a topology on the set of real valued functions on $[0, 1]$.

It is sometimes useful to be able to compare topologies in terms of their bases. Theorem 3.4 is a test for this.

3.4 THEOREM *Let $\mathscr{B}$ be a base for T on X. Let $\mathscr{B}'$ be a base for T' on X. Then,*
$$T \subset T'$$
if and only if
if $x \in A \in \mathscr{B}$, then, for some A', $A' \in \mathscr{B}'$ and $x \in A' \subset A$.

Proof Left as an exercise. ∎

The theorem of course applies when $\mathscr{B} = T$ and/or $\mathscr{B}' = T'$. If $T \subset T'$, we say that T' is finer than T (since T' has more open sets). Not every pair of topologies on X is comparable in this way—possibly neither $T \subset T'$ nor $T' \subset T$ (see Example 3.7).

Suppose now that S is any set of subsets of X. There are topologies on X containing S as a subset—$\mathscr{P}(X)$ is one. The intersection $T[S]$ of all such topologies is the smallest topology on X which contains S, and is called the topology generated by S.

It is easy to figure out what the elements of $T[S]$ must look like. Since $T[S]$ is a topology, then ϕ and X are in $T[S]$, together with all unions of finite intersections of sets in S. Conversely, the set consisting of ϕ, X, and all unions of finite intersections of sets in S is a topology on X containing S, hence (in view of the preceding remarks) must actually be $T[S]$. Put another way, the set consisting of ϕ, X, and all finite intersections of sets in S is a base for $T[S]$; S itself is called a subbase for $T[S]$.

3.2 DEFINITION *S is a subbase for T*
if and only if
$\{\phi, X\} \cup \{\cap A | A \subset S \text{ and } \phi \neq A \text{ and } A \text{ is finite}\}$ is a base for T.

3.3 DEFINITION *T is generated by S*
if and only if
S is a subbase for T.

3.3 EXAMPLE The set of intervals of the form $]-\infty, a[$ and $]a, \infty[$, with a real, forms a subbase for the Euclidean topology on E^1. To see this, note that each basic open set $]a, b[$ (see the discussion preceding Definition 3.1) can be written $]-\infty, b[\cap]a, \infty[$.

3.4 EXAMPLE Let $X = \prod_{i=1}^{\infty} X_i$, where each $X_i = E^1$. Then X may be thought of as the set of all real-valued sequences. Let S consist of all sets $\prod_{i=1}^{\infty} A_i$, where each A_i is open in E^1 and $A_i = E^1$ for all but at most finitely many values of i. Then S generates a topology called the product topology on πX_i. Product topologies will be studied extensively in Chapter III. For now, try to visualize what the open and basic open sets of this topology look like. ∎

Theorem 3.5 summarizes the preceding remarks about subbases.

3.5 Theorem *Let $S \subset \mathscr{P}(X)$. Let $\mathscr{B} = \{\phi, X\} \cup \{\cap A | \phi \neq A \subset S \text{ and } A \text{ is finite}\}$. Let $M = \{\cup B | B \subset \mathscr{B}\}$.*
Then,
(1) M is a topology on X.
(2) $\mathscr{B}$ is a base for M.
(3) S is a subbase for M.
(4) If P is a topology on X and $S \subset P$, then $M \subset P$.

Proof We shall attack (1) and (2) simultaneously. Suppose $x \in a \cap b$, where $a \in \mathscr{B}$ and $b \in \mathscr{B}$. Consider two cases:

i) There are non-empty, finite subsets A and B of S such that $a = \cap A$ and $b = \cap B$.
In this case, note that $x \in \cap(A \cup B) \subset (\cap A) \cap (\cap B)$ and $\cap(A \cup B) \in \mathscr{B}$.

ii) $a = X$ or $b = X$.
If $b = X$, then $x \in a \cap b = a \cap X = a \subset a \cap b$. Similarly if $b = X$.

Hence, by Theorem 3.3, $\mathscr{B}$ is a base for a topology on X. In view of the definition of M, $\mathscr{B}$ is a base for M, and M is a topology on X.

(3) is now immediate by Definition 3.2.

Finally, we prove (4). Suppose P is a topology on X and $S \subset P$. By Definition 1.1 (4), $\mathscr{B} \subset P$. Then, by Definition 1.1, (3), $M \subset P$. ∎

While subbases often conveniently specify a topology, they nevertheless require careful handling. For example, the subbase analogue of Theorem 3.2 is false. This is shown by the following examples.

3.5 Example Let X be any set with at least three elements. Let S consist of all sets $\{x, y\}$ with x and y distinct points in X. Then S is a subbase for the discrete topology $\mathscr{P}(X)$ on X. Now, if $z \in X$, then $\{z\}$ is open, but there is no subbasic open set containing z which is a subset of $\{z\}$. ∎

3.6 Example Let $X = \{1, 2, 3, 4\}$ and let S consist of the two sets $\{1, 2, 3\}$ and $\{2, 3, 4\}$. The topology generated by S consists of the open sets ϕ, X, $\{1, 2, 3\}$, $\{2, 3, 4\}$ and $\{2, 3\}$.

Let $A = \{1, 4\}$. Then, $2 \notin \bar{A}$ by Theorem 2.3 (2), since $\{2, 3\}$ is a neighborhood of 2 which does not intersect A. But, each subbasic open set containing 2 does intersect A. ∎

We shall conclude this section with an example on order topologies, which will be important later. This example also demonstrates the convenience of specifying a topology by exhibiting explicitly just the subbasic open sets (we will see this again in Chapter III when we discuss product spaces).

3.7 EXAMPLE (Order Topologies) An order (linear order) on a nonempty set X is a relation $<$ on X which mirrors the properties of "less than" on the reals. More carefully, we require first that, if $x, y \in X$ and $x \neq y$, then either $x < y$ or $y < x$, but not both. We also require transitivity: if $x < y$ and $y < z$, then $x < z$. However, $x < x$ is always false.

An order $<$ on X induces a topology $T_<$ on X, the subbasic open sets being $\{x | x < \alpha\}$ and $\{x | \alpha < x\}$ for $\alpha \in X$. Order topologies are particularly useful in constructing examples and counterexamples, as we shall see quite often.

A base for $T_<$ is obtained, as usual, by taking finite intersections of the generating subbasic open sets. However, with order topologies, each basic open set can be written as an intersection of just two subbasic sets. A typical basic open set is $\{x | \alpha < x\} \cap \{x | x < \beta\} = \{x | \alpha < x < \beta\} =]\alpha, \beta[$. More intersections with subbasic sets simply yields another basic set of the same form.

Euclidean topology on the reals is the order topology induced by "less than". In this case, subbasic sets are of the form $]-\infty, a[$ and $]a, \infty[$, and a base consists of sets $]-\infty, b[\cap]a, \infty[=]a, b[$.

As a second example, let $X = [0, 1] \times [0, 1]$. If (a, b) and (c, d) are in X, define $(a, b) < (c, d)$ to mean $(a < c)$ or $(a = c$ and $b < d)$. Geometrically, $(a, b) < (c, d)$ if (a, b) is anywhere to the left of (c, d) in X (that is, $a < c$) or if (c, d) is on the vertical line through (a, b), but above (a, b). Typical subbasic sets are shown in Figure 3.1, and two basic neighborhoods are shown in Figure 3.2.

It is interesting to compare the order topology $T_<$ on X with the Euclidean metric topology T_{ϱ_2} on X generated by the "circles" $\{\alpha | \varrho_2(\alpha, \beta) < \varepsilon\} \cap X$, for $\beta \in X$ and $\varepsilon > 0$.

Note that $\{(1, x) | 0 < x \leq 1\} = \{(\alpha, \beta) | (1, 0) < (\alpha, \beta)\}$ is $T_<$-open, but not T_{ϱ_2}-open (Figure 3.3), since $(1, 1)$ is not interior a T_{ϱ_2} point of $\{(1, x) | 0 < x \leq 1\}$.

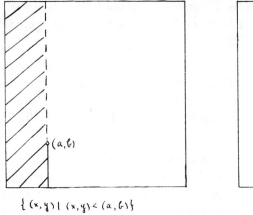
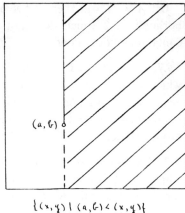

$\{(x,y) \mid (x,y) < (a,b)\}$ $\qquad$ $\{(x,y) \mid (a,b) < (x,y)\}$

FIGURE 3.1

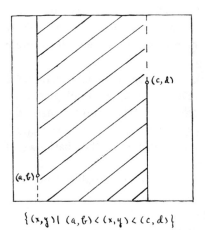
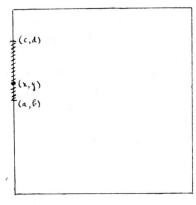

$\{(x,y) \mid (a,b) < (x,y) < (c,d)\}$ $\qquad$ $\{(x,y) \mid (a,b) < (x,y) < (c,d)\}$

FIGURE 3.2

We shall now show that there are also T_{ϱ_2}-open sets which are not $T_<$-open. Let $B = X \cap \{(x,y) \mid \varrho_2((x,y),(0,1)) < 1/4\}$. Then, B is T_{ϱ_2}-open (Figure 3.4a). But B is not $T_<$-open. Look at $(0,1) \in B$. If B were $T_<$-open, then there would be a $T_<$-basic open set M with $(0,1) \in M \subset B$. Now, M must be of the form $\{(x,y) \mid (a,b) < (x,y) < (c,d)\}$ for some (a,b) and (c,d) in X. Since $(0,1) \in M$, then $(a,b) < (0,1) < (c,d)$. Now, $(a,b) < (0,1)$ implies $a = 0$, and $b < 1$, as $a \geqq 0$. And $(0,1) < (c,d)$ implies $0 < c$, as $d \leqq 1$. Thus, $M = \{(x,y) \mid (0,b) < (x,y) < (c,d)\}$, with $c > 0$ (Figure 3.4b). M then contains points inside a strip of width c extending the whole height of X, and so cannot be a subset of B (Figure 3.4b).

Thus the topologies $T_<$ and T_{o_2} are not comparable—neither is a subset of the other.

Ordinal spaces are particularly important order topological spaces. Let ν be a non-zero ordinal. Then ν has a natural order. For $a, b \in \nu$, let $a < b$ if $a \in b$ (this is the same as $a \subsetneq b$). The sets $\{x | x < \alpha\} = [0, \alpha[$ and $\{x | \alpha < x\} =]\alpha, \nu[$ generate the order topology on the set ν of ordinals less than ν.

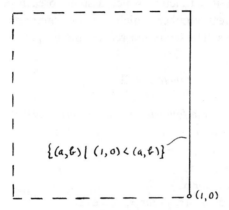

FIGURE 3.3

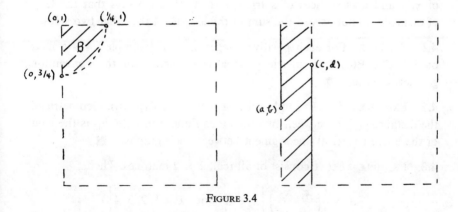

FIGURE 3.4

A base for the order topology on ν consists of two kinds of sets. If $\alpha \in \nu$ and α is a non-limit ordinal, then α has an immediate predecessor, β in ν, and $\alpha = \beta \cup \{\beta\}$. Then, $\{x | x < \alpha \cup \{\alpha\}\} \cap \{x | \beta < x\} = [0, \alpha \cup \{\alpha\}[\cap]\beta, \nu[= \{\alpha\}$ is a basic neighborhood of α. If α is a limit ordinal, then α

2*

has no immediate predecessor, and a typical basic neighborhood of α is $]\gamma, \alpha] = \{x|\gamma < x \leq \alpha\} = \{x|x < \alpha \cup \{\alpha\}\} \cap \{x|\gamma < x\}$, for any $\gamma \in \nu$ with $\gamma < \alpha$. ∎

1.4 Metric spaces

In this section, X is any non-empty set.

A metric on X is a function which assigns to each pair of points in X a non-negative real number which can be interpreted as the distance between the points. The definition reflects the properties we usually expect of distance.

4.1 DEFINITION *ϱ is a metric on X*
if and only if
$\varrho: X \times X \to R$, and the following conditions are satisfied:
(1) $\varrho(x, y) \geq 0$ for each $x, y \in X$.
(2) If x and y are in X, then,
   $\varrho(x, y) = 0 \Leftrightarrow x = y$.
(3) $\varrho(x, y) = \varrho(y, x)$ for $x, y \in X$.
(4) $\varrho(x, y) \leq \varrho(x, z) + \varrho(z, y)$ for $x, y, z \in X$.

Part (4) of the definition is called the triangle inequality. If you think of x, y and z as vertices of a triangle in X, then (4) says that the length of one side cannot exceed the sum of the lengths of the other two sides.

4.1 EXAMPLE For each positive integer n, ϱ_n of Example 1.4 is a metric on E^n. The Euclidean metric ϱ_n measures distance in the conventional geometric sense. ∎

4.2 EXAMPLE d of Example 1.6 is a metric on $C([a, b])$. Geometrically, the distance $d(f, g)$ between two functions f and g in $C([a, b])$ is the length of the longest vertical line segment between their graphs. ∎

4.3 EXAMPLE Let A consist of all real, 2×2 matrices. Then,

$$A = \left\{ \begin{pmatrix} x_1 & x_2 \\ x_3 & x_4 \end{pmatrix} \,\middle|\, x_i \in R \text{ for } i = 1, 2, 3, 4 \right\}.$$

Define

$$\sigma\left(\begin{pmatrix} x_1 & x_2 \\ x_3 & x_4 \end{pmatrix}, \begin{pmatrix} y_1 & y_2 \\ y_3 & y_4 \end{pmatrix}\right) = \max_{1 \leq i \leq 4} |x_i - y_i|.$$

Then, σ is a metric on A. ∎

4.4 EXAMPLE Let $r(f, g) = \int_a^b |f - g|$ whenever $f, g \in C([a, b])$. Then, r is a metric on $C([a, b])$. ∎

For the remainder of this section, let ϱ be a metric on X. Then, ϱ induces a topology on X in the following way. Form the spheres $B_\varrho(x, \varepsilon)$ of radius ε about x, consisting of all points within a ϱ-distance ε of x, for each $\varepsilon > 0$ and x in X. The set of all such spheres is a base for some topology on X. This topology, denoted T_ϱ, is the topology induced on X by ϱ, and the pair (X, T_ϱ), often abbreviated to just (X, ϱ), is called a metric space. Customarily we shall call just X a metric space when ϱ is understood. In this sense, E^n (Example 1.4) and $C([a, b])$ of Example 1.6 are metric spaces.

4.2 DEFINITION

(1) If $x \in X$ and $\varepsilon > 0$, then we let
$B_\varrho(x, \varepsilon) = X \cap \{y | \varrho(x, y) < \varepsilon\}$.
(2) $\mathscr{B}_\varrho = \{B_\varrho(x, \varepsilon) | \varepsilon > 0 \text{ and } x \in X\}$.
(3) $T_\varrho = \{\cup A | A \subset \mathscr{B}_\varrho\}$.

4.1 THEOREM $\mathscr{B}_\varrho$ is a base for the topology T_ϱ on X.

Proof Let $B_\varrho(x, \varepsilon) \in \mathscr{B}_\varrho$ and $B_\varrho(y, \delta) \in \mathscr{B}_\varrho$. Let $z \in B_\varrho(x, \varepsilon) \cap B_\varrho(y, \delta)$. Choose $\alpha = \min(\varepsilon - \varrho(x, z), \delta - \varrho(y, z))$, and observe that $\alpha > 0$, since $\varrho(x, z) < \varepsilon$ and $\varrho(y, z) < \delta$. We now claim that $B_\varrho(z, \alpha) \subset B_\varrho(x, \varepsilon) \cap B_\varrho(y, \delta)$. For, let $q \in B_\varrho(z, \alpha)$. Then, $\varrho(z, q) < \alpha$, so

$$\varrho(x, q) \leq \varrho(x, z) + \varrho(z, q) < \varrho(x, z) + \alpha \leq \varrho(x, z) + (\varepsilon - \varrho(x, z)) = \varepsilon.$$

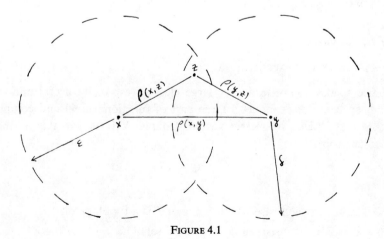

FIGURE 4.1

Then, $B_\rho(z, \alpha) \subset B_\rho(x, \varepsilon)$. Similarly, $B_\rho(z, \alpha) \subset B_\rho(y, \delta)$. By Theorem 3.3, $\mathscr{B}_\rho$ is a base for a topology M on X.

Finally, by Definition 3.1, $T_\rho = M$. ∎

Figure 4.1 illustrates the geometrical motivation behind the proof of Theorem 4.1.

Note that T_ρ could have been defined in exactly the manner of Examples 1.4 and 1.6. Specify, for $A \subset X$, that $A \in T_\rho$ if, given $x \in A$, there is some $\varepsilon > 0$ such that $B_\rho(x, \varepsilon) \subset A$. One would then show that T_ρ is a topology and that $\mathscr{B}_\rho$ is a base for T_ρ.

The following is immediate from Theorem 3.2, but we state it for later reference.

4.2 THEOREM *Let $A \subset X$.*
Then,
(1) $A \in T_\rho$
  if and only if
  if $x \in A$, then, for some $\varepsilon > 0$, $B_\rho(x, \varepsilon) \subset A$.
(2) Let $x \in X$.
  Then,
  $x \in \bar{A}$
  if and only if
  if $\varepsilon > 0$, then $B_\rho(x, \varepsilon) \cap A \neq \phi$.
(3) Let $x \in X$.
  Then,
  x is a T_ρ-cluster point of A
  if and only if
  if $\varepsilon > 0$, then $(B_\rho(x, \varepsilon) - \{x\}) \cap A \neq \phi$.

Proof Left as an exercise. ∎

Metric spaces comprise a very large class of especially useful and important spaces. For example, a large part of functional, real and complex analysis takes place in metric spaces. Chapter VIII is devoted to their particular properties.

Problems

1) Determine in each case whether or not T is a topology on X:
    (a) $X = Z^+$, T consists of ϕ, X, and all sets $\{1, ..., n\}$ for $n \in Z^+$.

(b) X consists of all real-valued functions on $[0, 1]$, T consists of ϕ and all subsets of X which contain a continuous function.

(c) $X = \{(1/n, y) | 0 \leq y \leq 1$ and n is a positive integer$\}$. For each positive integer n, let $M_n = \{(1/m, y) | m > n$ and $1/m \leq y \leq 1\}$. Let T consist of X, ϕ and all the sets M_n, $n \in Z^+$.

(d) Let $\mathscr{M}$ be a topology on A, and $\mathscr{N}$ a topology on B, and $X = A \times B$. Let T consist of all sets $M \times N$, where $M \in \mathscr{M}$ and $N \in \mathscr{N}$.

2) Determine all topologies on the following sets:
   (a) $\{1, 2\}$ (c) $\{1, 2, 3, 4\}$
   (b) $\{1, 2, 3\}$
   * Can you guess a general result for $\{1, 2, ..., n\}$?

3) Let T be a topology on X.
   Then,
   T is the discrete topology
   if and only if
   $\{x\}$ is T-open for each $x X$.

4) Let T_1 and T_2 be topologies on X.
   Then, $T_1 \cap T_2$ is a topology on X.
   But show by example that $T_1 \cup T_2$ need not be a topology on X.

5) Give an example of a space (X, T) such that, for some $x \in X$, $\{x\}$ is neither open nor closed.

6) Show by example that a union of closed sets may be
   (a) open
   (b) closed
   (c) neither open nor closed.

7) Show by example that an intersection of open sets may be
   (a) open
   (b) neither open nor closed.

8) Let $A = \{(1/m, 1/n) | m$ and n are positive integers$\}$.
   Consider A as a subset of Euclidean space E^2.
   Find: Int (A), Bdry (A), Cl (A), Der (A), Is (A). Determine whether A is open, closed, open and closed, or neither open nor closed.

9) Let $A = \{(x, \sin(1/x)) | 0 < x < 1\}$, considered as a subset of E^2.
   Determine Bdry (A), Cl (A), Is (A), Int (A), Der (A), and tell whether A is open, closed, open and closed, or neither open nor closed.

10) Let T be a topology on X. Prove the following:
   (a) $\text{Bdry}_T\left(\text{Bdry}_T\left(\text{Bdry}_T(A)\right)\right) = \text{Bdry}_T\left(\text{Bdry}_T(A)\right)$.
   (b) $\text{Bdry}_T\left(\text{Bdry}_T(A)\right) \subset \text{Bdry}_T(A)$.
   (c) $\text{Bdry}_T(A) \cup \text{Bdry}_T(B) = \text{Bdry}_T(A \cup B) \cup \text{Bdry}_T(A \cap B)$
   $\cup \left(\text{Bdry}_T(A) \cap \text{Bdry}_T(B)\right)$.
   (d) $\text{Bdry}_T(A \cup B) \subset \text{Bdry}_T(A) \cup \text{Bdry}_T(B)$.
   (e) $\text{Bdry}_T(A \cap B) \subset \text{Bdry}_T(A) \cap \text{Bdry}_T(B)$.
   (f) $\text{Bdry}_T\left(\text{Int}_T(A)\right) \subset \text{Bdry}_T(A)$.
   (g) $\text{Int}_T\left(\text{Int}_T(A)\right) = \text{Int}_T(A)$.
   (h) $\text{Int}_T(A \cap B) = \text{Int}_T(A) \cap \text{Int}_T(B)$.
   (i) $\text{Bdry}_T(A) = \phi$ if and only if $A \in T$ and A is T-closed.
   (j) $\text{Der}_T(A \cup B) = \text{Der}_T(A) \cup \text{Der}_T(B)$.
   (k) $\text{Der}_T\left(\text{Der}_T(A)\right) \subset A \cup \text{Der}_T(A)$.
   (l) $\text{Cl}_T\left(\text{Int}_T\left(\text{Bdry}_T(A)\right)\right) = \text{Cl}_T\left(A \cap \text{Int}_T\left(\text{Bdry}_T(A)\right)\right)$
   $= \text{Cl}_T\left(\left(\text{Int}_T\left(\text{Bdry}_T(A)\right)\right) - A\right)$.

11) Show by example that equality need not hold in 10)—b, d, e, f, k.

12) Let X be any set, and $f\colon \mathscr{P}(X) \to \mathscr{P}(X)$.
Suppose that
$f(\phi) = \phi,$
$A \subset f(A),$
$f(f(A)) = f(A),$
$f(A \cup B) = f(A) \cup f(B),$
for each $A, B \subset X$.
   Then, there is exactly one topology T on X such that $\text{Cl}_T(A) = f(A)$ for each $A \subset X$.
   f is called a Kuratowski closure operator on X.

13) Let X be any set, and $f\colon \mathscr{P}(X) \to \mathscr{P}(X)$ such that
$f(X) = X,$
$f(A) \subset A,$
$f(f(A)) = f(A),$
$f(A \cap B) = f(A) \cap f(B),$
for each $A, B \subset X$.
   Then, there is exactly one topology T on X such that $\text{Int}_T(A) = f(A)$ for each $A \subset X$.
   f is called an interior operator on X.

14) Let $X \neq \phi$.
Suppose, if $x \in X$, there is a set $\mathcal{O}_x$ of subsets of X such that:
$\mathcal{O}_x \neq \phi$,
$x \in \cap \mathcal{O}_x$,
$A \in \mathcal{O}_x$ and $B \in \mathcal{O}_x$ implies $A \cap B \in \mathcal{O}_x$,
$A \in \mathcal{O}_x$ and $A \subset B \subset X$ implies $B \in \mathcal{O}_x$,
$A \in \mathcal{O}_x$ implies for some $B \in \mathcal{O}_x$, $B \subset A$ and $B \in \mathcal{O}_y$ for each $y \in B$.
Then, there is a unique topology T on X such that, for each $x \in X$,
$\mathcal{O}_x = \{M | M \subset X$ and $x \in t \subset M$ for some $t \in T\}$.
$\mathcal{O}_x$ is called a neighborhood system of x (in this context, a neighborhood of x is considered to be any set containing an open neighborhood of x).

15) Let X be any set, and $g: \mathcal{P}(X) \to \mathcal{P}(X)$. Suppose:
$g(\phi) = \phi$,
$g(A \cup B) = g(A) \cup g(B)$,
$x \in g(A) \Rightarrow x \in g(A - \{x\})$,
$g(g(A)) \subset A \cup g(A)$,
for each $A, B \subset X$.
Then, there is exactly one topology T on X such that $\text{Der}_T(A) = g(A)$ for each $A \subset X$.

16) Let X be any set, and $h: \mathcal{P}(X) \to \mathcal{P}(X)$. Suppose
$h(X) = \phi$,
$h(A) = h(X - A)$,
$h(A \cup B) \subset h(A) \cup h(B)$,
$A \subset B$ implies $h(A) \subset B \cup h(B)$.

Then, there is exactly one topology T on X such that $\text{Bdry}_T(A) = h(A)$ for each $A \subset X$.

17) Let T be a topology on X and $A \subset X$.
Then,
A is T-open
if and only if
$\text{Bdry}_T(A) \cap A = \phi$.

18) Prove that the set of all intervals $]a, b[$ with a, b rational is a base for the Euclidean space E^1.
Is this base also a subbase for the Euclidean topology?

19) Show that the set of half-open intervals $]a, b]$, with a, b real and $a \leq b$, constitutes a base for a topology T on the reals. How does T compare with the Euclidean topology on the reals? Does the set of half-open intervals $]a, b]$, with a, b rational, also constitute a base for T?

20) Let $X = \{(1/n, y) | n \in Z^+ \text{ and } 0 < y \leq 1\}$.
If n is a positive integer, let M_n consist of all points $(1/t, y)$ in X with $t \leq n$ and $1/n < y \leq 1$. Describe the open sets in the topology T generated by the set of all sets M_n, $n = 1, 2, \ldots$

21) Show that the union of two bases for a topology T on X is also a base for T.

But show by example that the intersection of two bases for T may not be a base for any topology on X, even when the intersection is not empty.

22) Is the union (intersection) of two subbases for a topology on X a subbase for the same topology?

23) Let a topology T have a countable base, and let $\mathscr{C}$ be any base for T. Show that $\mathscr{C}$ has a countable subset $\mathscr{O}$ which is also a base for T.

24) Let X consist of all real sequences (functions from Z^+ to E^1).

Let $\varrho(x, y) = \sum_{i=1}^{\infty} \dfrac{|x_i - y_i|}{2^i(1 + |x_i - y_i|)}$

for $x, y \in X$.
Show that ϱ is a metric on X.

Let $d(x, y) = \sum_{i=1}^{\infty} \dfrac{|x_i - y_i|}{i!(i + |x_i - y_i|)}$

Show that d is also a metric on X, and compare T_ϱ with T_d.

25) Let ϱ and η be metrics on X.
Show that $\varrho + \eta$ is a metric on X, where $(\varrho + \eta)(x, y) = \varrho(x, y) + \eta(x, y)$.
How does $T_{\varrho+\eta}$ compare with T_ϱ and T_η?

26) Let (X, T) be the space of problem $1 - a$. Can a metric ϱ be defined on X such that $T_\varrho = T$? (This problem becomes trivial later on).

27) Give an example of a metric space (X, ϱ) such that, for some $x \in X$ and $\varepsilon > 0$,
$$\overline{B_\varrho(x, \varepsilon)} \neq \{y | \varrho(x, y) \leq \varepsilon\}$$
and
$$\text{Bdry}\,(B_\varrho(x, \varepsilon)) \neq \{y | \varrho(x, y) = \varepsilon\}.$$

This problem shows that the geometric intuition from our experience in E^1, E^2 and E^3 cannot be carried over into metric spaces in general.

28) Let $X = E^1 - \{0\}$, and $d(x, y) = \varrho(1/x, 1/y)$ for $x, y \in X$. Show that d is a metric on X and describe geometrically the open spheres $B_d(x, \varepsilon)$.

29) Let ϱ be a metric on X.
    If A is a nonempty subset of X, and $x \in X$, let
    $$d(x, A) = \inf \{\varrho(x, y) | y \in A\}.$$
    Show that
    $$x \in \text{Cl}_T (A) \quad \text{if and only if} \quad d(x, A) = 0.$$

CHAPTER II

Continuity and homeomorphism

2.1 Continuity

IN CALCULUS, THE ε, δ definition provides a useful, geometrically meaningful concept of continuity at a point. A function f, say from E^n to E^1, is continuous at p if, given an ε-nghd. V of $f(p)$, there is some δ-nghd. of p which f takes into V. That is, points close to p, as measured by ϱ_n, map to points close to $f(p)$, measured by ϱ_1.

This generalizes immediately to metric spaces: simply replace ϱ_n and ϱ_1 with the relevant metrics.

Now suppose that $f: X \to Y$, and T and M are arbitrary topologies on X and Y respectively. In place of ε and δ neighborhoods, we now have open sets in X and Y as determined by T and M. This suggests we replace spherical neighborhoods, determined by metrics, with T-open and M-open sets. The resulting notion of continuity at a point is one of the basic concepts of topology.

For the remainder of this section, let (X, T) and (Y, M) be topological spaces, and let $f: X \to Y$.

1.1 DEFINITION *Let $x \in X$.*
Then,
f is (T, M) continuous at x
if and only if
if $f(x) \in V \in M$, then, for some U, $x \in U \in T$ and $f(U) \subset V$.

When T and M are clear from context, we usually just say that f is continuous at x. It is important in any given instance to understand which topologies are being used, since f may be continuous at x relative to T and M, and not continuous relative to, say, T' and M'.

1.1 EXAMPLE Let M be any topology on Y and T the discrete topology on X. Then any $f: X \to Y$ is continuous at x for any $x \in X$. For, given any M-neighborhood V of $f(x)$, then $\{x\}$ is a T-neighborhood of x which f maps into V. ∎

1.2 EXAMPLE Let T be the discrete, M the indiscrete topology on X. Let $f: X \to X$ be the identity function given by $f(x) = x$ for each $x \in X$. Then f is (T, M) continuous at x, for each $x \in X$, by Example 1.1. But f is not (M, T) continuous at x, for any x in X, if X has at least two elements. To see this, note that $\{f(x)\}$ is a T-nghd. of $f(x)$. The only M-nghd. of x is X, and $f(X) \subset \{f(x)\}$ is not true if X has a point other than x.

This example shows how even a function as uncomplicated as the identity function can fail to be continuous at a point relative to certain topologies. ∎

1.3 EXAMPLE Let $X = \{a, b, c, d\}$. Let T consist of the sets ϕ, X, $\{a, d\}$, $\{b, d\}$, $\{d\}$, and $\{a, b, d\}$.

Let $Y = \{1, 2, 3\}$ and let M consist of the sets ϕ, Y, $\{2\}$, $\{3\}$, and $\{2, 3\}$.
Let $f: a \to 1$
$b \to 1$
$c \to 2$
$d \to 3$.

Then, f is continuous at a. For, $f(a) = 1$, and the only neighborhood of 1 is Y. Choose any neighborhood V of a, and certainly $f(V) \subset Y$.

But, f is not continuous at c. For, $\{2\}$ is a nghd. of $f(c) = 2$, but the only nghd. of c is X, and $f(X) \subset \{2\}$ is false. ∎

1.4 EXAMPLE Let $X = C([a, b])$, with T the sup norm topology.
Let $Y = E^1$, with M the Euclidean topology.

Let $H(g) = \int_a^b g$ for each $g \in C([a, b])$. Then, H is continuous at each point of $C([a, b])$.

To see this, let $g \in C([a, b])$. Let V be any nghd. of $H(g)$ in E^1. We must produce a T-open set U such that $g \in U$ and $H(U) \subset V$.

Note that, for some $\varepsilon > 0$, $H(g) \in \,]H(g) - \varepsilon, H(g) + \varepsilon[\, \subset V$, by definition of the Euclidean topology on E^1 (or by Theorem 4.2 (1). I).

Let $U = B_d(g, \varepsilon/(b - a))$, d the metric of Example 1.6, Chapter I. We shall show that $H(U) \subset \,]H(g) - \varepsilon, H(g) + \varepsilon[$. Let $h \in B_d(g, \varepsilon/(b - a))$. Then, $d(g, h) < \varepsilon/(b - a)$, so $\sup\{|g(x) - h(x)| \mid a \leq x \leq b\} < \varepsilon/(b - a)$. Then,

$$|H(g) - H(h)| = \left|\int_a^b (g - h)\right| \leq \int_a^b |g - h|$$

$$\leq (b - a) \sup\{|g(x) - h(x)| \mid a \leq x \leq b\}$$

$$\leq (b - a)\varepsilon/(b - a) = \varepsilon,$$

implying that
$$H(g) - \varepsilon < H(h) < H(g) + \varepsilon. \blacksquare$$

1.5 EXAMPLE Let $X = (E^2 \cap \{(x, y) | x^2 + y^2 = 1\}) - \{(0, 1)\}$. Then X is the circumference of the unit circle about the origin in E^2, with the point (0, 1) removed.

Let T consist of all sets $X \cap A$, with A open in E^2. It is easy to check that T is a topology on X (note Definition 1.1.III and Theorem 1.1.III next chapter).

Let $Y = E^1$.

Define a map $f: X \to Y$ by putting, for $p \in X$, $f(p)$ equal to the first coordinate of the point of intersection of the horizontal axis with the line from (0, 1) through p (see Figure 1.1).

By simple geometry, $f(x, y) = x/(1 - y)$ whenever $(x, y) \in X$.

Intuitively, f appears continuous at any p in X, since points close to p map to points relatively close to $f(p)$. More carefully, choose any p in X, and let $f(p) \in V$, where V is open in Y. For some $\varepsilon > 0$, $]f(p) - \varepsilon, f(p) + \varepsilon[\subset V$, so it suffices to produce some U open in X such that $f(p) - \varepsilon < f(q) < f(p) + \varepsilon$ for each $q \in U$. Figure 1.2 illustrates one way of doing this. Draw lines from (0, 1) to $(f(p) - \varepsilon, 0)$ and $(f(p) + \varepsilon, 0)$. Then draw a circle about p of radius δ sufficiently small that it lies inside the sector

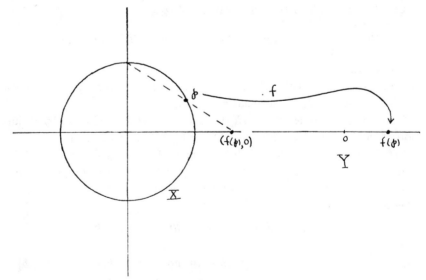

FIGURE 1.1

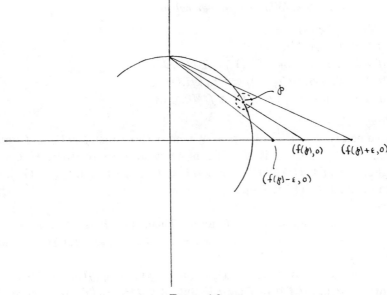

FIGURE 1.2

of these lines. Then, $B_{\varrho_2}(p, \delta) \cap X$ is a T-nghd. of p which f maps into $]f(p) - \varepsilon, f(p) + \varepsilon[$, hence into V. ∎

1.6 EXAMPLE Suppose $f: E^n \to E^1$, and f is continuous at $(x_1, ..., x_n) \in E^n$ in the sense of Definition 1.1. The reader can check that, given $\varepsilon > 0$, there is some $\delta > 0$ such that $|f(x_1, ..., x_n) - f(y_1, ..., y_n)| < \varepsilon$ whenever $\sum_{i=1}^{n}(x_i - y_i)^2 < \delta^2$.

If, conversely, given $\varepsilon > 0$, then there is such a δ, then f is continuous at $(x_1, ..., x_n)$ in the sense of Definition 1.1.

Hence Definition 1.1 reverts to the usual calculus concept of continuity at a point when $X = E^n$ and $Y = E^1$. ∎

1.2 DEFINITION *f is (T, M) continuous*
if and only if
f is (T, M) continuous at x for each $x \in X$.

As usual, if T and M are clear, we shall usually just say that f is continuous.

There are many equivalent formulations of the notion of continuity. Theorem 1.1 lists those which we shall have occasion to use.

1.1 THEOREM *The following are equivalent:*

(1) f *is* (T, M) *continuous.*
(2) If $V \in M$, then $f^{-1}(V) \in T$.
(3) If V is M-closed, then $f^{-1}(V)$ is T-closed.
(4) If $A \subset X$, then $f(\text{Cl}_T(A)) \subset \text{Cl}_M(f(A))$.
(5) If $A \subset Y$, then $\text{Cl}_T(f^{-1}(A)) \subset f^{-1}(\text{Cl}_M(A))$.

Proof $(1) \Rightarrow (2)$:

Assume (1). Let V be M-open. We must show that $f^{-1}(V)$ is T-open. Let $x \in f^{-1}(V)$. Then, f is (T, M) continuous at x, so, for some U, U is a T-nghd. of x and $f(U) \subset V$. Then, $x \in U \subset f^{-1}(V)$, so $x \in \text{Int}_T(f^{-1}(V))$. By Theorem 2.2.I, $f^{-1}(V)$ is T-open.

$(2) \Rightarrow (3)$:

Assume (2). Suppose V is M-closed. Then, $Y - V$ is M-open, so, by (2), $f^{-1}(Y - V) \in T$. But, $f^{-1}(Y - V) = X - f^{-1}(V)$, so $f^{-1}(V)$ is T-closed.

$(3) \Rightarrow (4)$:

Assume (3). Let $A \subset X$. By Theorem 2.3 (4).I, $\text{Cl}_M(f(A))$ is M-closed. By (3), $f^{-1}(\text{Cl}_M(f(A)))$ is T-closed. But, $A \subset f^{-1}(\text{Cl}_M(f(A)))$. Hence, by Theorem 2.3 (8).I, $\text{Cl}_T(A) \subset f^{-1}(\text{Cl}_M(f(A)))$. Then, $f(\text{Cl}_T(A)) \subset \text{Cl}_M(f(A))$.

$(4) \Rightarrow (5)$:

Assume (4). Let $A \subset Y$. Then, $f^{-1}(A) \subset X$. By (4), $f(\text{Cl}_T(f^{-1}(A))) \subset \text{Cl}_M(f(f^{-1}(A))) \subset \text{Cl}_M(A)$. Then, $\text{Cl}_T(f^{-1}(A)) \subset f^{-1}(f(\text{Cl}_T(f^{-1}(A)))) \subset f^{-1}(\text{Cl}_M(A))$.

$(5) \Rightarrow (1)$:

Assume (5). Let $x \in X$ and $f(x) \in V \in M$. We must produce a T-nghd. U of x such that $f(U) \subset V$.

Let $U = X - \text{Cl}_T(f^{-1}(Y - V))$. By Theorem 2.3 (4).I, U is T-open. We shall simplify the expression for U. By Theorem 2.3 (5).I, $\text{Cl}_M(Y - V) = Y - V$. By (5), $\text{Cl}_T(f^{-1}(Y - V)) \subset f^{-1}(\text{Cl}_M(Y - V)) = f^{-1}(Y - V)$. Then, by Theorem 2.3 (1).I, $\text{Cl}_T(f^{-1}(Y - V)) = f^{-1}(Y - V)$. Hence, $U = X - f^{-1}(Y - V)$.

Now, if $x \notin U$, then $x \in f^{-1}(Y - V)$, so $f(x) \notin V$, a contradiction. Then, $x \in U$.

Suppose now that $y \in U$ and $f(y) \notin V$. Then, $f(y) \in Y - V$, so $y \in f^{-1}(Y - V)$, a contradiction.

Hence, $f(U) \subset V$, and f is (T, M) continuous at x. ∎

Theorem 1.1 (2) is used very often as the topological definition of continuity. We chose the approach of Definitions 1.1 and 1.2 simply because of its obvious intuitive appeal.

We have just shown that continuity of f is characterized by the action of f^{-1} on the open sets of Y. If $\mathscr{B}$ is a base for M, then the open sets in Y are determined by $\mathscr{B}$. It is therefore reasonable to expect that continuity of f can be tested by examining $f^{-1}(b)$ for just the basic open sets b. Similarly, one can test continuity using just subbasic open sets.

1.2 Theorem

(1) Let $\mathscr{B}$ be a base for M.
  Then,
  (i) If $x \in X$, then,
    f is (T, M) continuous at x
    if and only if
    if $f(x) \in b \in \mathscr{B}$, then there is some $V \in T$ such that $x \in V$ and $f(V) \subset b$.
  (ii) f is (T, M) continuous
    if and only if
    $f^{-1}(b) \in T$ for each $b \in \mathscr{B}$.

(2) Let S be a subbase for M.
  Then,
  (i) If $x \in X$, then,
    f is (T, M) continuous at x
    if and only if
    if $f(x) \in s \in S$, then there is some $V \in T$ such that $x \in V$ and $f(V) \subset s$.
  (ii) f is (T, M) continuous
    if and only if
    $f^{-1}(s) \in T$ for each $s \in S$.

Proof Left as an exercise. ∎

A word of caution about Theorem 1.2. If f is (T, M) continuous and $\mathscr{B}^1$ is a base for T, we cannot conclude in Theorem 1.2 (1, ii) that $f^{-1}(b) \in \mathscr{B}^1$ if $b \in \mathscr{B}$. In fact, the set of subsets $f^{-1}(b)$, for $b \in \mathscr{B}$, need not be a base for T. Similarly, the sets $f^{-1}(s)$, for $s \in S$, need not constitute a subbase for T.

1.7 Example Let $X = Y = E^1$. Let $f(x) = 1$ for each $x \in E^1$. Then, f is continuous. To see this, note that, if A is open in E^1, then

$$f^{-1}(A) = \begin{cases} \phi & \text{if } 1 \notin A \\ E^1 & \text{if } 1 \in A, \end{cases}$$

so that $f^{-1}(A)$ is always open in E^1.

But, $\{\phi, X\}$ is neither a base nor a subbase for the Euclidean topology on the reals. ∎

Care must also be taken not to read too much into Theorem 1.1 (2). If f is (T, M) continuous, and A is open in X, it does not follow that $f(A)$ is open in Y. An f enjoying this property (continuous or not) is called an open map. The function of Example 1.7 is continuous, but is not an open map. The concept of an open map will arise in later work, so we include the definition here for reference.

1.3 DEFINITION *f is a (T, M) open map
if and only if
if A is T-open, then $f(A)$ is M-open.*

We conclude this section with the intuitively evident fact that a composition of continuous maps is continuous.

1.3 THEOREM *Let f be (T, M) continuous.
Let $g\colon Y \to L$, and let P be a topology on L.
Suppose that g is (M, P) continuous.
Then,
$g \circ f$ is (T, P) continuous.*

Proof Let $A \in P$. By Theorem 1.1, $g^{-1}(A) \in M$. Similarly, by continuity of f, $f^{-1}(g^{-1}(A)) \in T$. But, $f^{-1}(g^{-1}(A)) = (g \circ f)^{-1}(A)$. Hence $g \circ f$ is (T, P) continuous by Theorem 1.1. ∎

2.2 Homeomorphism

It is often useful, and sometimes an end in itself, to compare two spaces in order to see which properties they have in common and perhaps learn something about one space from what is already known about the other.

A continuous map from X to Y does not usually convey much information about Y from X. In Example 1.1, for instance, any $f\colon X \to Y$ is continuous, regardless of the topology on Y. However, if f is a bijection, and if both f and $\text{inv} f$ are continuous, then immediately $f^{-1}(M) = T$ and $f(T) = (\text{inv} f)^{-1}(T) = M$. In this event, all properties defined in terms of open sets (and this includes virtually all properties of topological interest) are shared by both spaces. The topologist regards such spaces as essentially the same, or homeomorphic, and the map f is called an homeomorphism.

For the remainder of this section, let (X, T) and (Y, M) be topological spaces, and let $f\colon X \to Y$.

2.1 DEFINITION *f is a (T, M) homeomorphism if and only if*

(1) f is a bijection.
(2) f is (T, M) continuous.
(3) inv f is (M, T) continuous.

2.2 DEFINITION *(X, T) is homeomorphic to (Y, M) if and only if there is a (T, M) homeomorphism $g: X \to Y$.*

When there is no ambiguity, we simply say that X is homeomorphic to Y, and that f is an homeomorphism.

2.1 EXAMPLE Let X, Y, T, M and f be as in Example 1.5. Then, f is an homeomorphism, and X is homeomorphic to Y.

In this particular example, it is possible to get a geometric feeling for the homeomorphism. Pry X open from $(0, 1)$ and stretch it out to cover Y. This can be done smoothly without tearing, and the map is continuous in the sense that points originally close together in X remain relatively close to each other after they reach Y. (This process is given analytically by f in Example 1.5). The inverse process of shrinking the line Y and rolling it up to form X is likewise a smooth one. Analytically,

$$(\text{inv} f)(t) = \left(\frac{2t}{t^2 + 1}, \frac{t^2 - 1}{t^2 + 1} \right)$$

for each $t \in Y$, and it is easy to show that inv f is continuous.

If we let $X' = X \cup \{(0, 1)\}$ and $T' = \{X' \cap A | A$ is open in $E^1\}$, then the spaces (X', T') and (Y, M) are not homeomorphic. At the present stage of development, this is not so easy to show. However, after Theorem 1.13 in Chapter VII, come back to this example and observe that (X', T') is compact, but (Y, M) is not. ∎

2.2 EXAMPLE Let (A, σ) be the metric space of real, 2×2 matrices of Example 4.3.I. Then, (A, σ) is homeomorphic to Euclidean space E^4. To see this, define $\varphi: A \to E^4$ by:

$$\varphi \left(\begin{pmatrix} x_1 & x_2 \\ x_3 & x_4 \end{pmatrix} \right) = (x_1, x_2, x_3, x_4).$$

Immediately φ is a bijection. To show that φ is continuous, let $\begin{pmatrix} a_1 & a_2 \\ a_3 & a_4 \end{pmatrix} \in A$, and let P be any basic nghd. of (a_1, a_2, a_3, a_4), say

$P = B_{\varrho_4}((a_1, a_2, a_3, a_4), \varepsilon)$. Now, $B_\sigma\left(\begin{pmatrix} a_1 & a_2 \\ a_3 & a_4 \end{pmatrix}, \varepsilon/2\right)$ is a basic nghd. of $\begin{pmatrix} a_1 & a_2 \\ a_3 & a_4 \end{pmatrix}$. Further, if $\begin{pmatrix} x_1 & x_2 \\ x_3 & x_4 \end{pmatrix} \in B_\sigma\left(\begin{pmatrix} a_1 & a_2 \\ a_3 & a_4 \end{pmatrix}, \varepsilon/2\right)$, then $\varrho_4\left(\varphi\begin{pmatrix} x_1 & x_2 \\ x_3 & x_4 \end{pmatrix},\right.$
$\left.(a_1, a_2, a_3, a_4)\right)^2 = \sum_{i=1}^{4} (x_i - a_i)^2 < \dfrac{\varepsilon^2}{4} + \dfrac{\varepsilon^2}{4} + \dfrac{\varepsilon^2}{4} + \dfrac{\varepsilon^2}{4} = \varepsilon^2$, so
$\varphi\left(B_\sigma\left(\begin{pmatrix} a_1 & a_2 \\ a_3 & a_4 \end{pmatrix}, \varepsilon/2\right)\right) \subset B_{\varrho_4}((a_1, a_2, a_3, a_4), \varepsilon)$.

This proves the continuity of φ. Continuity of inv φ follows in similar fashion. ∎

2.3 Example Let $X = \{a, b, c\}$ and let T consist of the sets ϕ, X, $\{a, b\}, \{b, c\}$, and $\{b\}$.

Let $Y = \{1, 2, 3\}$ and let M consist of the sets ϕ, Y, $\{1\}$, $\{2\}$, and $\{1, 2\}$.

Then, (X, T) is not homeomorphic to (Y, M). In general, showing that two spaces are not homeomorphic is very difficult. In this case, the method of exhaustion works. Write down all the bijections of X to Y. They are:

f_1: $a \to 1$ f_2: $a \to 2$ f_3: $a \to 3$
    $b \to 2$ $b \to 3$ $b \to 1$
    $c \to 3$ $c \to 1$ $c \to 2$

f_4: $a \to 1$ f_5: $a \to 2$ f_6: $a \to 3$
    $b \to 3$ $b \to 1$ $b \to 2$
    $c \to 2$ $c \to 3$ $c \to 1$

Now just check each one. f_1 is not continuous, as $\{1\} \in M$, but $f_1^{-1}(\{1\}) = \{a\} \notin T$. f_2 is not continuous, as $\{1\} \in M$ but $f_2^{-1}\{1\}) = \{c\} \notin T$, and so on. In fact, there are not even any continuous bijections of X to Y. ∎

As with continuity, the notion of an homeomorphism has many equivalent formulations. The one we shall use most often is given in terms of the open map concept of Definition 1.3.

2.1 Theorem *f is a (T, M) homeomorphism if and only if f is a (T, R) continuous, open bijection.*

Proof Left as an exercise. ∎

For convenience, let us agree to write $(X, T) \cong (Y, M)$ (or usually just $X \cong Y$) whenever the two spaces are homeomorphic. It is a simple matter to show that $\cong$ is an equivalence relation on the class of all topological spaces.

2.2 Theorem

(1) $(X, T) \cong (X, T)$.
(2) $(X, T) \cong (Y, M) \Leftrightarrow (Y, M) \cong (X, T)$.
(3) $(X, T) \cong (Y, M)$ and $(Y, M) \cong (L, P) \Rightarrow (X, T) \cong (L, P)$.

Proof Left as an exercise. ∎

All of the concepts developed in the first two sections of Chapter I are preserved by homeomorphism. This fact will be used very frequently in the sequel.

2.3 Theorem *Let $f: X \to Y$ be a (T, M) homeomorphism. Then,*

(1) U *is T-open* $\Leftrightarrow f(U)$ *is M-open*.
(2) U *is T-closed* $\Leftrightarrow f(U)$ *is M-closed*.
 Further, if $A \subset X$, then,
(3) $f(\text{Int}_T(A)) = \text{Int}_M(f(A))$.
(4) $f(\bar{A}) = \overline{f(A)}$.
(5) $f(\text{Bdry}_T(A)) = \text{Bdry}_M(f(A))$.
(6) $f(\text{Der}_T(A)) = \text{Der}_M(f(A))$.
(7) $f(\text{Is}_T(A)) = \text{Is}_M(f(A))$.

Proof Left as an exercise. ∎

There is not a great deal more to be said about homeomorphism for now. Very formally, topology is the study of properties invariant under homeomorphism (topological invariants). Of course, we haven't yet developed the major invariants used to distinguish spaces. We will see some in later chapters. It is also impossible to convey at this stage of development the importance of homeomorphism as a tool for studying certain spaces in the light of others whose properties are better understood. A good example of this is the Urysohn Metrization Theorem of Chapter VIII, section 10. Another is Theorem 4.6, Chapter V, which characterizes completely regular spaces.

Problems

1) Let (X, ϱ) be a metric space, and $\phi \neq A \subset X$. Define $f: X \to E^1$ by $f(x) = d(x, A)$ (see problem 29, Ch. I), for each $x \in X$. Then, f is continuous with respect to T_ϱ and the Euclidean topology on the reals.

2) Let (X, ϱ) be a metric space. Let T be the set of all sets $A \times B$, where $A, B \in T_\varrho$. Then, T is a topology on $X \times X$.
Is $\varrho: X \times X \to E^1$ continuous with respect to this topology and the usual topology on E^1?

3) Let $f: X \to Y$ be (T, M) continuous.
Let $x \in \text{Cl}_T(A)$.
Then, $f(x) \in \text{Cl}_M(f(A))$.

4) Give an example of an open map which is not continuous.

5) Give an example of a continuous injection whose inverse is not continuous.

6) Give an example of a map which is both closed and open, but not continuous ($f: X \to Y$ is (T, M) closed if $f(A)$ is M-closed whenever A is T-closed).

7) Let $f: X \to Y$ and $g: Y \to S$. If $g \circ f$ is (T, N) continuous, and g is (M, N) continuous, is f (T, M) continuous?
If $g \circ f$ is (T, N) continuous, and f is (T, M) continuous, is g (M, N) continuous?

8) Let $f: X \to Y$.
Show that
$$f \text{ is a } (T, M) \text{ closed map}$$
if and only if
$$\text{Cl}_M(f(A)) \subset f(\text{Cl}_T(A)) \text{ for each } A \subset X.$$
(See problem 6 for the definition of a closed map).

9) Let (X, T_α) be a space for each $\alpha \in \mathcal{O}$. Let $f_\alpha: S \to X_\alpha$ for each $\alpha \in \mathcal{O}$.
Then,
the smallest topology T on S such that each f_α is (T, T_α) continuous is that generated by $\{f_\alpha^{-1}(G) | \alpha \in \mathcal{O} \text{ and } G \in T_\alpha\}$.
Further, if $g: Y \to S$, then g is (M, T) continuous if and only if $f_\alpha \circ g$ is (M, T_α) continuous for each $\alpha \in \mathcal{O}$.

10) Let (X_α, T_α) be a space for each $\alpha \in \mathcal{O}$. Let $f_\alpha: X_\alpha \to S$ for each $\alpha \in \mathcal{O}$.
Then,
the largest topology M on S such that each f_α is (T_α, M) continuous is $\bigcap_{\alpha \in \mathcal{O}} M_\alpha$, where $M_\alpha = \{A | A \subset S \text{ and } f_\alpha^{-1}(A) \in T_\alpha\}$.

Further, if $g: S \to Y$, then g is (M, N) continuous if and only if $g \circ f_\alpha$ is (T_α, N) continuous for each $\alpha \in \mathcal{O}$.

11) Let $f: X \to Y$ be (T, M) continuous. Let $g: Y \to X$ be (M, T) continuous. Suppose that $f \circ g = 1_Y$ and $g \circ f = 1_X$ (1_Y is the identity map on Y, defined by $1_Y(y) = y$ for each $y \in Y$). Then, f and g are homeomorphisms.

12) If A is a subset of E^1, give A the metric topology $T(A)$ induced by $\varrho_1 | A \times A$. Then,
    (a) $([0, 1], T([0, 1])) \cong ([a, b]), T([a, b]))$ for any a, b real with $a < b$.
    (b) $(]0, 1[, T(]0, 1[))$ is homeomorphic to $(]a, b[, T(]a, b[))$ for any a, b real with $a < b$.
    (c) E^1 is homeomorphic to $(]-1, 1[, T(]-1, 1[))$.

13) If $A \subset E^2$, let $T(A)$ be the metric topology on A induced by $\varrho_2 | A \times A$.
    (a) Let A consist of the set of points making up the sides of any triangle in the plane. Then, $(A, T(A)) \cong (S^1, T(S^1))$, where $S^1 = \{(x, y) | x^2 + y^2 = 1\}$.
    *(b) Let $U = \{(x, \sin(1/x)) | 0 < x \leq 1/\pi\}$. Then, $(U, T(U))$ is homeomorphic to $(]0, 10], T(]0, 10]))$.
    *(c) $(U, T(U))$ is not homeomorphic to $(S^1, T(S^1))$.

14) Rework problem 2, Ch. I, finding all the non-homeomorphic topologies on the given sets.

15) Let $f: X \to Y$ be a (T, M) homeomorphism.
    (a) If $\mathcal{B}$ is a base for T, then $f(\mathcal{B})$ is a base for M. (Here, $f(\mathcal{B}) = \{f(b) | b \in \mathcal{B}\}$).
    (b) If $\mathcal{S}$ is a subbase for T, then $f(\mathcal{S})$ is a subbase for M.

16) Is a (T, M) open bijection necessarily a (T, M) homeomorphism? Is a (T, M) closed bijection a (T, M) homeomorphism?

17) Let $f: X \to Y$ be a (T, M) homeomorphism. Let $g: Y \to S$ and let N be a topology on S.
    Then,
    g is (M, N) continuous if and only if $g \circ f$ is (T, N) continuous.

18) Let f be a continuous injection: $E^1 \to E^1$.
    Show that $f(E^1)$ is open in E^1.

CHAPTER III

Relative, product, identification and quotient topology

THERE ARE SEVERAL constructions of new spaces from one or more given ones which play a vital role in topology. In this chapter we shall consider some of the more important and basic ones.

3.1 Relative topology

Let T be a topology on X, and A a subset of X. There is a topology T_A on A which is in a certain sense compatible with the given topology on X. T_A consists of all intersections of T-open sets with A, and is called the relative topology of T on A. Whenever A is referred to as a subspace of X, it is understood that T_A is the topology on A.

1.1 DEFINITION $T_A = \{t \cap A | t \in T\}$.

1.1 THEOREM *Let T be a topology on X. Let $A \subset X$. Then,*
T_A is a topology on A.

Proof Immediately, $T_A \subset \mathscr{P}(A)$ and $\phi = \phi \cap A \in T_A$ and $A = X \cap A \in T_A$.

Suppose now that $B \subset T_A$. If $b \in B$, then there is some $t_b \in T$ with $b = t_b \cap A$. Then, $\bigcup_{b \in B} t_b \in T$, and we have

$$\cup B = \bigcup_{b \in B} (t_b \cap A) = \left(\bigcup_{b \in B} t_b\right) \cap A \in T_A.$$

Finally, if $B \neq \phi$ and B is finite, then $\bigcap_{b \in B} t_b \in T$, so that

$$\cap B = \bigcap_{b \in B} (t_b \cap A) = \left(\bigcap_{b \in B} t_b\right) \cap A \in T_A. \blacksquare$$

1.1 EXAMPLE The space (X, T) of Example 1.5.II is a subspace of E^2. ∎

1.2 EXAMPLE Let $A =]0, 1]$, considered as a subspace of E^1. Note that $]\frac{1}{2}, 1]$ is open in A (i.e., T_A-open), since $]\frac{1}{2}, 1] = A \cap]\frac{1}{2}, 2[$, but $]\frac{1}{2}, 1]$ is neither open nor closed in E^1. ∎

1.3 EXAMPLE Let X be any set and T the discrete (indiscrete) topology on X. Then T_A is the discrete (indiscrete) topology on A for each $A \subset X$. ∎

1.4 EXAMPLE Let T be the cofinal or Euclidean topology on the reals, and $A = \{1, 2, 3, 4, 5\}$. Then T_A is the discrete (and also the cofinal) topology on A. ∎

For the remainder of this section, let (X, T) be a space and A a subset of X.

It is often convenient to know that closed sets in the subspace are intersections of A with closed sets in the space.

1.2 THEOREM *Let $B \subset A$.*
Then,
B is T_A-closed
if and only if
for some F, F is T-closed and $B = A \cap F$.

Proof Suppose that B is T_A-closed. Then, $A - B \in T_A$, so $A - B = A \cap G$ for some T-open G. Then, $X - G$ is T-closed, and $B = A \cap (X - G)$.

Conversely, suppose that F is T-closed and $B = A \cap F$. Then, $A - B = A - (A \cap F) = A \cap (X - F) \in T_A$, as $X - F \in T$. ∎

As Example 1.2 shows, T_A-open (or closed) sets need not be T-open (or closed). There is, however, an obvious situation in which each set open in the subspace is open in X, and one in which each set closed in A is closed in X.

1.3 THEOREM
(1) If A is T-open, then $T_A \subset T$.
(3) If A is T-closed, and $B \subset A$, then,
 B is T_A-closed $\Leftrightarrow B$ is T-closed.

Proof Left as an exercise. ∎

If $B \subset A$, then also $B \subset X$, and we can consider B both as a subspace (B, T_B) of (X, T) and as a subspace $(B, (T_A)_B)$ of (A, T_A). Fortunately, in a potentially confusing situation, the spaces (B, T_B) and $(B, (T_A)_B)$ turn out to be the same—sets T_B-open in B are also $(T_A)_B$-open in B, and conversely. Put another way, a subspace of a subspace is a subspace.

1.4 THEOREM *Let $B \subset A$.*
Then,
$(T_A)_B = T_B$.

Proof Let $\alpha \in (T_A)_B$. For some $M \in T_A$, $\alpha = M \cap B$. Since $M \in T_A$, there is some $G \in T$ with $M = A \cap G$. Then, $\alpha = (A \cap G) \cap B = G \cap (A \cap B) = G \cap B \in T_B$.

Conversely, suppose that $\beta \in T_B$. For some $P \in T$, $\beta = B \cap P$. Note that $A \cap P \in T_A$. Then, $\beta = (A \cap B) \cap P = B \cap (A \cap P) \in (T_A)_B$. ∎

If $B \subset X$, then we can calculate the set $\text{Der}_T(B)$ of T-cluster points of B. If at the same time $B \subset A$, then we can calculate the set $\text{Der}_{T_A}(B)$ of cluster points of B in the subspace A. One pleasant feature of the relative topology is that the T_A-cluster points of B are just those T-cluster points of B which lie in A. As an immediate consequence, the T_A-closure of B is the intersection of A with the T-closure of B.

1.5 THEOREM *Let $B \subset A$.*
Then,
(1) $\text{Der}_{T_A}(B) = A \cap \text{Der}_T(B)$.
(2) $\text{Cl}_{T_A}(B) = A \cap \text{Cl}_T(B)$.

Proof of (1) Let $x \in \text{Der}_{T_A}(B)$. Immediately, $x \in A$. Let $x \in V \in T$. Then, $x \in V \cap A \in T_A$, so $((V \cap A) - \{x\}) \cap B \neq \phi$ by Theorem 2.3 (2).I. Then also $(V - \{x\}) \cap B \neq \phi$, so $x \in \text{Cl}_T(B)$.

Conversely, let $y \in A \cap \text{Der}_T(B)$. Let $y \in U \in T_A$. Write $U = C \cap A$ for some T-open set C. Since $y \in C \in T$ and $y \in \text{Der}_T(B)$, then $(C - \{y\}) \cap B \neq \phi$. Since $B \subset A$, then $(C - \{y\}) \cap B \subset A$, so $A \cap (C - \{y\}) \cap B = ((A \cap C) - \{y\}) \cap B = (U - \{y\}) \cap B \neq \phi$. Thus $y \in \text{Der}_{T_A}(B)$, proving (1).

Proof of (2) We have, by (1),

$$\text{Cl}_{T_A}(B) = B \cup \text{Der}_{T_A}(B) = B \cup (A \cap \text{Der}_T(B)) = A \cap (B \cup \text{Der}_T(B))$$
$$= A \cap \text{Cl}_T(B). \blacksquare$$

Unfortunately, a similar theorem cannot be proved for Int and Bdry. The reader can show that $A \cap \text{Int}_T(B) \subset \text{Int}_{T_A}(B)$ and $\text{Bdry}_{T_A}(B) \subset A \cap \text{Bdry}_T(B)$, whenever $B \subset A$. However, anyone attempting an honest proof of the reverse inclusions will reach an impasse. The following examples show why.

1.5 EXAMPLE Let $A = B = Q$, the set of rationals, and $X = E^1$, with T the Euclidean topology. Then, $\text{Int}_{T_Q}(Q) = Q$. For, if $x \in Q$, then $x \in]x - 1, x + 1[\cap Q \in T_Q$, and certainly $]x - 1, x + 1[\cap Q \subset Q$.

But, $\text{Int}_T(Q) = \phi$, since any open interval in E^1 has irrational elements. ■

1.6 EXAMPLE Let $X = E^1$, T the Euclidean topology. Let $B = [0, \frac{1}{2}]$ and $A = [0, 1]$. Then, 0 is a T-boundary point of B in A, so $0 \in \text{Bdry}_T(B) \cap A$.

But 0 is not a boundary point of B considered as a subset of the subspace A. For, $0 \in]-1, \frac{1}{2}[\cap A \in T_A$, but $(]-1, \frac{1}{2}[\cap A) \cap (A - B) = \phi$, violating Definition 2.3 (1).I. ■

If A is T-open, then the set inclusions of the preceding discussion become equalities. Note the role of the assumption that A is T-open in proving the set inclusion just one way—the other way is always true.

1.6 THEOREM *Let $B \subset A$ and $A \in T$.*
Then,
(1) $\text{Int}_{T_A}(B) = A \cap \text{Int}_T(B)$.
(2) $\text{Bdry}_{T_A}(B) = A \cap \text{Bdry}_T(B)$.

Proof of (1) Let $x \in A \cap \text{Int}_T(B)$. Since $x \in \text{Int}_T(B)$, then for some V, $x \in V \in T$ and $V \subset B$. Then, $x \in A \cap V \in T_A$, and $A \cap V \subset B$, so $x \in \text{Int}_{T_A}(B)$.

Conversely, let $y \in \text{Int}_{T_A}(B)$. For some M, $y \in M \in T_A$ and $M \subset B$. By Theorem 1.3 (1), $y \in M \in T$, so $y \in \text{Int}_T(B)$.

Proof of (2) Left as an exercise. ■

If $\mathscr{B}$ is a base for T, it is easy to show that all intersections of sets in $\mathscr{B}$ with A form a base for the subspace topology on A. Similarly, a subbase for T_A can readily be manufactured from any subbase for T.

1.7 THEOREM
(1) Let $\mathscr{B}$ be a base for T.
  Then, $\{b \cap A | b \in \mathscr{B}\}$ is a base for T_A.
(2) Let S be a subbase for T.
  Then, $\{s \cap A | s \in S\}$ is a subbase for T_A.

Proof Left as an exercise. ■

We conclude this section with a consideration of continuity relative to subspaces.

Suppose (Y, M) is a space and $f: X \to Y$. There are several questions worth asking.

(1) If $f: X \to Y$ is continuous, if $f|A: A \to Y$ continuous in the subspace topology?

(2) If f is continuous at x in A, is $f|A$ continuous at x? Conversely, if $f|A$ is continuous at x in A relative to the subspace topology, is f continuous at x when viewed as a map of the whole space?

(3) Note that $f: X \to Y$ may be thought of as a surjection $f: X \to f(X) \subset Y$. Considering $f(X)$ as a subspace of Y, does $(T, M_{f(X)})$ continuity of f follow from (T, M) continuity?

Questions (1) and (3) are easy to answer in the affirmative, as is the first part of (2). The converse question in (2) can also be answered affirmatively if A is open in X. If A is not open, then the restricted function may have a continuity at x and the entire function a discontinuity. Example 1.7 shows how this can happen.

1.8 THEOREM *Let $f: X \to Y$. Let f be (T, M) continuous. Then, $f|A$ is (T_A, M) continuous.*

Proof Let V be M-open in Y. Then, $f^{-1}(V)$ is T-open, so $A \cap f^{-1}(V) = (f|A)^{-1}(V) \in T_A$, implying by Theorem 1.1.II that $f|A$ is (T_A, M) continuous. ∎

1.9 THEOREM *Let $f: X \to Y$. Let $x \in A$. Then,*

(1) If f is (T, M) continuous at x, then $f|A$ is (T_A, M) continuous at x.

(2) If $f|A$ is (T_A, M) continuous at x and A is T-open, then f is (T, M) continuous at x.

Proof of (1) Left as an exercise.

Proof of (2) Suppose that $f|A$ is (T_A, M) continuous at x and $A \in T$. Let $f(x) \in V \in M$. For some U, $x \in U \in T_A$ and $f(U) = (f|A)(U) \subset V$. Since A is T-open, then $T_A \subset T$ by Theorem 1.3, so $x \in U \in T$ and $f(U) \subset V$. Then, f is (T, M) continuous at x. ∎

1.7 EXAMPLE Let $X = Y = E^1$ and $A = Q$. Let $T = M =$ Euclidean topology. Let

$$f(x) = \begin{cases} 1 & \text{if } x \text{ is rational} \\ -1 & \text{if } x \text{ is irrational,} \end{cases}$$

for each real x.

Now, $(f|Q)(x) = 1$ for each $x \in Q$, so $f|Q$ is clearly (T_Q, M) continuous at x for each x in Q.

But f is not continuous at any real number.

This example also shows that the converse of Theorem 1.9 (1) is false. ∎

1.10 THEOREM *Let f be (T, M) continuous. Then, f is $(T, M_{f(X)})$ continuous.*

Proof Let $V \in M_{f(X)}$. For some P, $P \in M$ and $V = P \cap f(X)$. Since f is (T, M) continuous, then $f^{-1}(P) \in T$. Then, $f^{-1}(V) = f^{-1}(P \cap f(X)) = f^{-1}(P) \cap f^{-1}(f(X)) = f^{-1}(P) \in T$, so f is $(T, M_{f(X)})$ continuous by Theorem 1.1.II. ∎

3.2 Product topology

Suppose we are given a collection of topological spaces, say (X_α, T_α) for α in A. The spaces need have no relation at all to one another. We would like to topologize the Cartesian product $\prod_{\alpha \in A} X_\alpha$.

There are of course various topologies we can put on $\prod_{\alpha \in A} X_\alpha$, for example, the discrete, indiscrete and cofinal topologies. It turns out that these do not yield interesting theorems. All told, we have the following criteria to guide us in making a final choice:

(1) the topology should be mathematically fruitful.

(2) It should have some relation to the given topologies. That is, we should be able to draw conclusions about the product space from the coordinate spaces, and conversely.

(3) We would like the projections $p_\beta: \prod_{\alpha \in A} X_\alpha \to X_\beta$ to be continuous.

(4) In the case A consists of the integers $1, 2, ..., n$ (or, actually, any finite set), and each $X_i = E^1$, we would like the space $\prod_{i=1}^{n} X_i$ to identify (in the sense of homeomorphism) with Euclidean space E^n.

While a satisfactory topology was easy for finite products, Tietze in 1923 was the first to topologize infinite products with any degree of success. He defined a topology on $\prod_{\alpha \in A} X_\alpha$ by specifying the subbasic sets as those of the form $\prod_{\alpha \in A} G_\alpha$, where each G_α is open in X_α. This was a natural choice, since it was known to work very well for finite products. However, subsequent experience has shown Tychonov's 1930 definition to be the more useful one, and this is the one which today is known as the product topology.

In particular, Tychonov's Theorem (Chapter VII, Theorem 1.12) is considered by many the most important theorem in set topology, and is true for Tychonov's, but not for Tietze's, topology.

For the remainder of this section, A is a non-empty set, and (X_α, T_α) is a space for each α in A.

2.1 DEFINITION *The product topology on $\prod_{\alpha \in A} X_\alpha$ is the topology generated by $\{p_\alpha^{-1}(U_\alpha) | \alpha \in A \text{ and } U_\alpha \text{ is } T_\alpha\text{-open}\}$.*

We shall let $\mathbb{P}$ denote this topology whenever there is no danger of confusion. The basic open sets of $\mathbb{P}$ look like $\bigcap_{i=1}^{n} p_{\alpha_i}^{-1}(U_{\alpha_i})$, where $\alpha_1, \ldots, \alpha_n$ are in A and U_{α_i} is open in X_{α_i} for $i = 1, 2, \ldots, n$. Since $p_\beta^{-1}(U_\beta) \cap p_\beta^{-1}(V_\beta) = p_\beta^{-1}(U_\beta \cap V_\beta)$ whenever $\beta \in A$ and $U_\beta, V_\beta \in T_\beta$, we may always assume for convenience that the α_i's are chosen to be distinct in the above expression.

Note that a set $\prod_{\alpha \in A} G_\alpha$, where each G_α is open in X_α and $G_\alpha \neq X_\alpha$ for infinitely many α is not open in $\prod_{\alpha \in A} X_\alpha$ in the topology $\mathbb{P}$. This is immediate by Theorem 3.2.I, since $\prod_{\alpha \in A} G_\alpha$ can contain no basic open set $\bigcap_{i=1}^{n} p_{\alpha_i}^{-1}(U_{\alpha_i})$ as a subset if each (or at least infinitely many) $G_\alpha \neq X_\alpha$. This makes it easy to see that in general, when A is infinite and infinitely many X_α have at least two points, then

$$\text{Tychonov product topology} \subsetneq \text{Tietze product topology,}$$

since the Tietze product topology is generated by

$$\left\{\prod_{\alpha \in A} G_\alpha \,\middle|\, G_\alpha \text{ is open in } X_\alpha \text{ for each } \alpha \in A\right\}.$$

In fact,

$$\bigcap_{i=1}^{n} p_{\alpha_i}^{-1}(U_{\alpha_i}) = \prod_{\alpha \in A} G_\alpha, \quad \text{where} \quad G_\alpha = \begin{cases} U_\alpha, & \alpha = \alpha_i, i = 1, \ldots, n \\ X_\alpha, & \alpha \neq \alpha_i, \end{cases}$$

so each basic Tychonov-open set is Tietze-open. When A is finite, the two topologies coincide.

2.1 EXAMPLE Let $A = Z^+$ and $X_i = E^1$ for each positive integer i. Then, $\prod_{i \in A} X_i = \prod_{i=1}^{\infty} X_i$ is the set of real sequences mentioned in Example 3.4.I.

A subbasic open set is one of the form $p_j^{-1}(U_j)$, where U_j is Euclidean open in E^1. Since $p_j^{-1}(U_j) = \left\{x \,\middle|\, x \in \prod_{i=1}^{\infty} X_i \text{ and } x_j \in U_j\right\}$, then $p_j^{-1}(U_j)$

consists of all $x: Z^+ \to E^1$ with $x_j \in U_j$. For example, if $U_j =]a, b[$, then $p_j^{-1}(U_j)$ consists of all $x: Z^+ \to E^1$ with $a < x_j < b$; x_i may be any real number for $i \neq j$.

A basic open set is a finite intersection of subbasic open sets, say $s = \bigcap_{i=1}^{n} p_{j_i}(U_{j_i})$, where $j_1, \ldots, j_n$ are positive integers and each U_{j_i} is open in E^1. Then, $x \in s$ exactly when $x: Z^+ \to E^1$ and $x_{j_i} \in U_{j_i}$ for $i = 1, \ldots, n$; for $i \neq j_1, \ldots, j_n$, x_i may be any real number. ∎

2.2 EXAMPLE Let $X_i = \{0, 1\}$ for each positive integer i, and let T_i be the discrete topology on X_i for each i.

Then, $\prod_{i=1}^{\infty} X_i$ consists of all sequences of zeros and ones. The product topology $\mathbb{P}$ is not discrete. To see this, let $x_i = 0$ for each $i \in Z^+$. Then, $x \in \prod_{i=1}^{\infty} X_i$. We shall show that $\{x\}$ is not $\mathbb{P}$-open.

If $\{x\} \in \mathbb{P}$, then $\{x\}$ must be a basic open set (note that $\{x\} = p_j^{-1}(U_j)$ is impossible, so $\{x\}$ is not a subbasic open set). Hence $\{x\} = \bigcap_{i=1}^{n} p_{\alpha_i}^{-1}(U_{\alpha_i})$. Let $t_i = 0$ if $i = \alpha_1, \ldots, \alpha_n$, and $t_j = 1$ if j is a positive integer and $j \neq \alpha_i$, $1 \leq i \leq n$. Then, $t \in \bigcap_{i=1}^{n} p_{\alpha_i}^{-1}(U_{\alpha_i})$, but $t \neq x$, a contradiction. ∎

2.3 EXAMPLE Let $X_1 = \{0, 1\}$, and T_1 the Sierpinski topology. Let $X_2 = \{a, b\}$, and T_2 the indiscrete topology.

We shall list the subbasic open sets of the product topology on $\prod_{i=1}^{2} X_i$. They are:

$p_1^{-1}(\phi) = \phi$,

$p_1^{-1}(\{0\}) = \{f | f: \{1, 2\} \to X_1 \cup X_2$ and $f(1) = 0\}$,

$p_1^{-1}(\{0, 1\}) = \{f | f: \{1, 2\} \to X_1 \cup X_2$ and $f(1)$ is 0 or 1$\}$,

$p_2^{-1}(\phi) = \phi$,

$p_2^{-1}(\{a, b\}) = \{f | f: \{1, 2\} \to X_1 \cup X_2$ and $f(2)$ is a or $b\}$.

To be more explicit, let us label the elements of $\prod_{i=1}^{2} X_i$ as follows:

$f_1: \begin{array}{l} 1 \to 0 \\ 2 \to a \end{array}$ $f_2: \begin{array}{l} 1 \to 0 \\ 2 \to b \end{array}$

$f_3: \begin{array}{l} 1 \to 1 \\ 2 \to a \end{array}$ $f_4: \begin{array}{l} 1 \to 1 \\ 2 \to b \end{array}$

The subbasic open sets are now:

$$\phi, \{f_1, f_2\}, \text{ and } \prod_{i=1}^{2} X_i.$$

These are also the basic open sets, and, in fact, all the open sets in the product topology. ∎

When the index set is finite, say $A = \{1, ..., n\}$ as in Example 2.3, then the function set $\prod_{i=1}^{n} X_i$ can be identified with the n-tuple set $X_1 \times X_2 \times \cdots \times X_n$ by thinking of x in $\prod_{i=1}^{n} X_i$ as $(x_1, ..., x_n)$. The identification is a topological one if we replace subbasic open sets $p_i^{-1}(U_i)$ in the product topology with subsets $X_1 \times \cdots \times X_{i-1} \times U_i \times X_{i+1} \times \cdots \times X_n$ of $X_1 \times \cdots \times X_n$.

2.1 Theorem *Let $A = \{1, ..., n\}$, where n is a positive integer. Let τ be the topology on $X_1 \times \cdots \times X_n$ generated by $\{G_1 \times \cdots \times G_n | G_i \in T_i$ for $i = 1, ..., n\}$.*
Then,

$$\left(\prod_{i=1}^{n} X_i, \mathbb{P}\right) \cong (X_1 \times \cdots \times X_n, \tau).$$

Proof Define a map $\varphi: \prod_{i=1}^{n} X_i \to X_1 \times \cdots \times X_n$ by letting $\varphi(x) = (x_1, ..., x_n)$ for each $x \in \prod_{i=1}^{n} X_i$.

Immediately, φ is a bijection.

To show that φ is continuous, let $G_1 \times \cdots \times G_n \in \tau$. Then, $\varphi^{-1}(G_1 \times \cdots \times G_n)$ $= \bigcap_{i=1}^{n} p_i^{-1}(G_i) \in \mathbb{P}$. Then, φ is continuous by Theorem 1.2 (2–ii).II.

If $\bigcap_{i=1}^{n} p_i^{-1}(V_i) \in \mathbb{P}$, then $\varphi\left(\bigcap_{i=1}^{n} p_i^{-1}(V_i)\right) = V_1 \times \cdots \times V_n \in \tau$. Hence, by Theorem 2.1.II, φ is an homeomorphism. ∎

Henceforth we shall always identify a product space $\left(\prod_{i=1}^{n} X_i, \mathbb{P}\right)$ with the space $(X_1 \times \cdots \times X_n, \tau)$ of Theorem 2.1, which is notationally and conceptually simpler. Note that, in Theorem 2.1, the sets $G_1 \times \cdots \times G_n$ actually form a base for τ, since, for example,

$$(G_1 \times \cdots \times G_n) \cap (H_1 \times \cdots \times H_n) = (G_1 \cap H_1) \times \cdots \times (G_n \cap H_n).$$

2.4 Example Let (X_1, T_1) and (X_2, T_2) be as in Example 2.3. Then $\prod_{i=1}^{2} X_i$ can be identified with $X_1 \times X_2 = \{(0, a), (0, b), (1, a), (1, b)\}$. The open sets in τ are: ϕ, $X_1 \times X_2$, and $\{0\} \times X_2 = \{(0, a), (0, b)\}$. Here,

$(0, a)$ corresponds to f_1 in Example 2.3 under the map φ of Theorem 2.1, and $(0, b)$ corresponds to f_2, so the τ-open set $\{(0, a), (0, b)\}$ corresponds to the $\mathbb{P}$-open set $\{f_1, f_2\}$. ∎

2.5 EXAMPLE Let $X_i = E^1$ for $i = 1, ..., n$. Then $\left(\prod_{i=1}^{n} X_i, \mathbb{P}\right)$ is essentially the same as Euclidean n-space E^n. ∎

Returning now to general product spaces, Theorem 2.2 tells how to calculate the $\mathbb{P}$-closure of a subset $\prod_{\alpha \in A} C_\alpha$ of $\prod_{\alpha \in A} X_\alpha$ in terms of the closure in each coordinate space X_α of the αth coordinate set C_α. A similar theorem holds for interior when the number of coordinate spaces is finite.

2.2 THEOREM *Let $C_\alpha \subset X_\alpha$ for each $\alpha \in A$. Then,*

(1) $\operatorname{Cl}_{\mathbb{P}}\left(\prod_{\alpha \in A} C_\alpha\right) = \prod_{\alpha \in A} \operatorname{Cl}_{T_\alpha}(C_\alpha)$.

(2) $\operatorname{Int}_{\mathbb{P}}\left(\prod_{\alpha \in A} C_\alpha\right) \subset \prod_{\alpha \in A} \operatorname{Int}_{T_\alpha}(C_\alpha)$.

(3) If A is finite, then,

$\operatorname{Int}_{\mathbb{P}}\left(\prod_{\alpha \in A} C_\alpha\right) = \prod_{\alpha \in A} \operatorname{Int}_{T_\alpha}(C_\alpha)$.

Proof of (1) Let $x \in \operatorname{Cl}_{\mathbb{P}}\left(\prod_{\alpha \in A} C_\alpha\right)$. We shall use Theorem 2.3 (2).I. Let $\beta \in A$ and $x_\beta \in V \in T_\beta$. Then, $x \in p_\beta^{-1}(V) \in \mathbb{P}$, and $x \in \operatorname{Cl}_{\mathbb{P}}\left(\prod_{\alpha \in A} C_\alpha\right)$, so $\left(\prod_{\alpha \in A} C_\alpha\right) \cap p_\beta^{-1}(V) \neq \phi$. Let $z \in \left(\prod_{\alpha \in A} C_\alpha\right) \cap p_\beta^{-1}(V)$. Then, $p_\beta(z) = z_\beta \in V \cap C_\beta$, so $V \cap C_\beta \neq \phi$. Hence, $x_\beta \in \operatorname{Cl}_{T_\beta}(C_\beta)$, implying that $x \in \prod_{\alpha \in A} \operatorname{Cl}_{T_\alpha}(C_\alpha)$.

Conversely, note first that, if any $C_\alpha = \phi$, then $\prod_{\alpha \in A} \operatorname{Cl}_{T_\alpha}(C_\alpha) = \phi \subset \operatorname{Cl}_{\mathbb{P}}\left(\prod_{\alpha \in A} C_\alpha\right)$. Thus suppose that $C_\alpha \neq \phi$ for each $\alpha \in A$.

Let $y \in \prod_{\alpha \in A} \operatorname{Cl}_{T_\alpha}(C_\alpha)$. To show that $y \in \operatorname{Cl}_{\mathbb{P}}\left(\prod_{\alpha \in A} C_\alpha\right)$, we use Theorem 3.2 (2).I. Let U be a basic $\mathbb{P}$-nghd. of y, say $U = \bigcap_{i=1}^{n} p_{\alpha_i}^{-1}(V_{\alpha_i})$. Since $y \in \prod_{\alpha \in A} \operatorname{Cl}_{T_\alpha}(C_\alpha)$, then $y_{\alpha_i} \in \operatorname{Cl}_{T_{\alpha_i}}(C_{\alpha_i})$ for $i = 1, ..., n$. But, $y \in U$, so $p_{\alpha_i}(y) = y_{\alpha_i} \in V_{\alpha_i} \in T_{\alpha_i}$. Hence, $V_{\alpha_i} \cap C_{\alpha_i} \neq \phi$. Let $z_{\alpha_i} \in V_{\alpha_i} \cap C_{\alpha_i}$. Define $t \in \prod_{\alpha \in A} C_\alpha$ by:

$$t_{\alpha_i} = z_{\alpha_i} \text{ for } i = 1, ..., n,$$

$t_\alpha = $ any element of C_α for $\alpha \in A$ and $\alpha \neq \alpha_i$, $1 \leq i \leq n$.

Then, $t \in U \cap \prod_{\alpha \in A} C_\alpha$, so $U \cap \prod_{\alpha \in A} C_\alpha \neq \phi$, implying that $y \in \operatorname{Cl}_{\mathbb{P}}\left(\prod_{\alpha \in A} C_\alpha\right)$.

Proof of (2) Left as an exercise.

Proof of (3) Left as an exercise. ∎

The reader can construct an example to show that, in general, $\text{Int}_{\mathbb{P}}\left(\prod_{\alpha \in A} C_\alpha\right) \neq \prod_{\alpha \in A} \text{Int}_{T_\alpha}(C_\alpha)$.

Sets of boundary and cluster points in $\prod_{\alpha \in A} X_\alpha$ are usually difficult to calculate in terms of the individual coordinate spaces. An example will clarify the problem.

2.6 EXAMPLE Let $X_1 = X_2 = E^1$, with $T_1 = T_2 =$ Euclidean topology. Using Theorem 2.1, we identify $\left(\prod_{i=1}^{2} X_i, \mathbb{P}\right)$ with E^2.

Now $]1, 2[\times]3, 4[$ is an open square in E^2. If we calculate the boundary of $]1, 2[\times]3, 4[$, we obtain easily (see Figure 2.1)

$$(\{1\} \times [3, 4]) \cup (\{2\} \times [3, 4]) \cup ([1, 2] \times \{3\}) \cup ([1, 2] \times \{4\}),$$

which amounts geometrically to the set of points on the sides of the square.

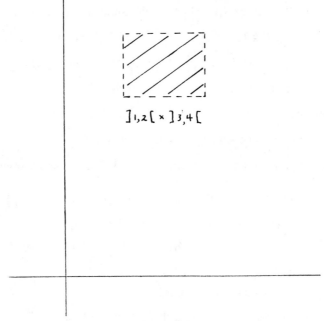

FIGURE 2.1

Observe that this set is quite different from (Bdry (]1, 2[)) × (Bdry (]3, 4[))
= {(1, 3), (1, 4), (2, 3), (2, 4)}, which consists of just the corners of the square.
However, we do have Bdry (]1, 2[×]3, 4[) = (Bdry (]1, 2[) × [3, 4])
∪ ([1, 2] × (Bdry (]3, 4[))) = ((Bdry (]1, 2[)) × $\overline{]3, 4[}$ ∪ ($\overline{]1, 2[}$
× (Bdry (]3, 4[))). ∎

This formula generalizes to arbitrary finite Cartesian products. A similar formula holds for the set of cluster points of a subset of $\prod_{i=1}^{n} X_i$. Be careful in Theorem 2.3 to interpret $\bar{C}_i$ as $\text{Cl}_{T_i}(C_i)$, remembering that the topologies T_i on the sets X_i may be quite different from one another. We shall also denote $\text{Der}_{T_i}(C_i)$ as C'_i in Theorem 2.3, for obvious notational reasons. Note that, without Theorem 2.1, the statement of Theorem 2.3 would be prohibitively complicated.

2.3 THEOREM *Let $A = \{1, ..., n\}$, where n is some positive integer. Let $C_i \subset X_i$ for $i = 1, ..., n$.*
Then,
(1) Bdry $(C_1 × \cdots × C_n) = ((\text{Bdry } C_1) × \bar{C}_2 × \cdots × \bar{C}_n)$
$\cup (\bar{C}_1 × (\text{Bdry } C_2) × \cdots × \bar{C}_n) \cup \cdots \cup (\bar{C}_1 × \cdots × \bar{C}_{n-1} × (\text{Bdry } C_n))$.
(2) Der $(C_1 × \cdots × C_n)$
$= (C'_1 × \bar{C}_2 × \cdots × \bar{C}_n) \cup (\bar{C}_1 × C'_2 × \cdots × \bar{C}_n) \cup \cdots$
$\cup (\bar{C}_1 × \cdots × \bar{C}_{n-1} × C'_n)$.

Proof Left as an exercise. ∎

Suppose $B_\alpha \subset X_\alpha$ for each α in A. Then T_α induces a relative topology T'_α on B_α, and we can consider the product space $\left(\prod_{\alpha \in A} B_\alpha, \mathbb{P}'\right)$ of the spaces (B_α, T'_α). But also $\prod_{\alpha \in A} B_\alpha \subset \prod_{\alpha \in A} X_\alpha$, so $\mathbb{P}$ induces a relative topology $\mathbb{P}_r$ directly on $\prod_{\alpha \in A} B_\alpha$. Fortunately, $\mathbb{P}'$ and $\mathbb{P}_r$ coincide, so that $\prod_{\alpha \in A} B_\alpha$ is the same space, whether viewed as a subspace of $\prod_{\alpha \in A} X_\alpha$ or as a product space in its own right.

2.4 THEOREM *Let $B_\alpha \subset X_\alpha$ for each $\alpha \in A$. Let T'_α be the relative topology of T_α on B_α for each $\alpha \in A$. Let $\mathbb{P}'$ be the product topology on $\prod_{\alpha \in A} B_\alpha$ defined by the spaces (B_α, T'_α).*
Let $\mathbb{P}_r$ be the relative topology of $\mathbb{P}$ on $\prod_{\alpha \in A} B_\alpha$.
Then,
$\mathbb{P}_r = \mathbb{P}'$.

Proof By Theorem 3.4.I, it suffices to show that each basic $\mathbb{P}_r$-open set is a basic $\mathbb{P}'$-open set, and conversely, for some specified bases $\mathscr{B}$ of $\mathbb{P}_r$ and $\mathscr{B}'$ of $\mathbb{P}'$.

By Theorem 1.7 (1), the sets $\left(\prod_{\alpha \in A} B_\alpha\right) \cap \left(\bigcap_{i=1}^{n} p_{\alpha_i}^{-1}(U_{\alpha_i})\right)$, where $U_{\alpha_i} \in T_{\alpha_i}$ for $i = 1, \ldots, n$, form a base $\mathscr{B}$ for $\mathbb{P}_r$.

Let q_β be the projection: $\prod_{\alpha \in A} B_\alpha \to B_\beta$ for each $\beta \in A$.

By Definition 3.2.I, Definition 2.1, and Definition 1.1, the sets $\bigcap_{i=1}^{n} q_{\alpha_i}^{-1}(U_{\alpha_i} \cap B_{\alpha_i})$, where $U_{\alpha_i} \in T_{\alpha_i}$ for $i = 1, \ldots, n$, form a base $\mathscr{B}'$ for $\mathbb{P}'$.

To complete the proof, we have simply to observe that

$$\left(\prod_{\alpha \in B} B_\alpha\right) \cap \left(\bigcap_{i=1}^{n} p_{\alpha_i}^{-1}(U_{\alpha_i})\right) = \bigcap_{i=1}^{n} q_{\alpha_i}^{-1}(U_{\alpha_i} \cap B_{\alpha_i}),$$

whenever n is a positive integer, $\alpha_1, \ldots, \alpha_n \in A$ and $U_{\alpha_i} \in T_{\alpha_i}$ for $i = 1, \ldots, n$. ∎

The projection maps $p_\beta \colon \prod_{\alpha \in A} X_\alpha \to X_\beta$ are all continuous in the Tychonov product topology. Of course, the discrete topology on $\prod_{\alpha \in A} X_\alpha$ also has this property, and is clearly the largest such topology. It turns out that the Tychonov topology is the smallest in which each projection is continuous. A perhaps unexpected feature of the Tychonov topology is that each p_β is an open map as well.

2.5 Theorem

(1) If $\beta \in A$, then $p_\beta \colon \prod_{\alpha \in A} X_\alpha \to X_\beta$ is a $(\mathbb{P}, T_\beta)$ continuous surjection.

(2) If M is a topology on $\prod_{\alpha \in A} X_\alpha$, and if p_β is (M, T_β) continuous for each $\beta \in A$, then $\mathbb{P} \subset M$.

(3) p_β is an open map for each $\beta \in A$.

Proof of (1) Let $\beta \in A$. Immediately, p_β is a surjection. If $G \in T_\beta$, then $p_\beta^{-1}(G) \in \mathbb{P}$ by Definition 2.1, hence p_β is continuous.

Proof of (2) Suppose $p_\beta \colon \prod_{\alpha \in A} X_\alpha \to X_\beta$ is (M, T_β) continuous for each $\beta \in A$. Then, $\{p_\beta^{-1}(V_\beta) | \beta \in A$ and $V_\beta \in T_\beta\} \subset M$ by Theorem 1.1 (2).II. By Definition 2.1 and Theorem 3.5 (4).I, $\mathbb{P} \subset M$.

Proof of (3) Let $\beta \in A$. Note that, if $\bigcap_{i=1}^{n} p_{\alpha_i}^{-1}(U_{\alpha_i})$ is a basic $\mathbb{P}$-open set, then,

$$p_\beta \left(\bigcap_{i=1}^{n} p_{\alpha_i}^{-1}(U_{\alpha_i})\right) = \begin{cases} U_{\alpha_i} \text{ if } \beta = \alpha_i \text{ for some } i,\ 1 \leq i \leq n, \\ X_\beta \text{ if } \beta \neq \alpha_i \text{ for each } i = 1, \ldots, n. \end{cases}$$

Thus, $p_\beta(b) \in T_\beta$ for each basic $\mathbb{P}$-open set b. If now V is any $\mathbb{P}$-open set, then there is some set C of basic open sets with $V = \cup C$. Then, $p_\beta(V) = p_\beta(\cup C) = \bigcup_{b \in C} p_\beta(b) \in T_\beta$. ∎

Sometimes Theorem 2.5 (2) is used as the definition of the product topology. In this approach, $\mathbb{P}$ is by definition the intersection of all topologies on $\prod_{\alpha \in A} X_\alpha$ in which each projection is continuous. One then proves that the sets $p_\beta^{-1}(V_\beta)$ constitute a base for the topology. This approach is motivated by a more general problem in topology: given maps $f_\alpha: X \to Y_\alpha$, and topologies M_α on Y_α, find the smallest topology T on X such that each f_α is continuous (see Problem 9, Ch. II).

It is not in general true that $p_\beta(F)$ is closed in X_β if F is $\mathbb{P}$-closed. The reader can construct an example to show this. (Hint: consider E^2 as a product space and project the graph of a suitable continuous real-valued function onto one of the axes.)

Given a point f in $\prod_{\alpha \in A} X_\alpha$, and some β in A, the subset of $\prod_{\alpha \in A} X_\alpha$ consisting of all g with $g(\alpha) = f(\alpha)$ whenever $\alpha \neq \beta$, may be visualized as a space parallel to the coordinate space X_β. For example, in E^2, the set of points $(x, 3)$ constitutes a space parallel to one copy of E^1. It is not surprising that such a parallel space, or slice, is homeomorphic to X_β. This means that each X_β may be thought of as a subspace of $\prod_{\alpha \in A} X_\alpha$, a fact which is extremely useful when we know something about the product space and wish to study the individual coordinate spaces (see, for example, Theorem 1.7, Ch. V, or Theorem 10.1, Ch. VIII).

2.6 THEOREM Let $f \in \prod_{\alpha \in A} X_\alpha$.
Let $\beta \in A$.
Let $Y = \left(\prod_{\alpha \in A} X_\alpha\right) \cap \{g | g(\alpha) = f(\alpha) \text{ if } \alpha \in A \text{ and } \alpha \neq \beta\}$.
Then,
$(Y, \mathbb{P}_Y) \cong (X_\beta, T_\beta)$.

Proof Let $\varphi = p_\beta | Y$. Immediately $\varphi: Y \to X_\beta$, and φ is a surjection. Further, by Theorem 1.8 and Theorem 2.5 (1), φ is $(\mathbb{P}_Y, T_\beta)$ continuous.

To show that φ is an injection, let $g \in Y$ and $h \in Y$ and suppose $\varphi(g) = \varphi(h)$. Then, $p_\beta(g) = g(\beta) = p_\beta(h) = h(\beta)$. But, $g(\alpha) = h(\alpha)$ whenever $\alpha \in A$ and $\alpha \neq \beta$, so $g = h$.

There remains to show that φ is a $(\mathbb{P}_Y, T_\beta)$ open map. Let $Y \cap \left(\bigcap_{i=1}^{n} p_{\alpha_i}^{-1}(V_{\alpha_i})\right)$ be a basic $\mathbb{P}_Y$-open set, with the indices α_i all distinct (use has been made here of Theorem 1.7 (1)). It is easy to check that:

$$\varphi\left(Y \cap \left(\bigcap_{i=1}^{n} p_{\alpha_i}^{-1}(V_{\alpha_i})\right)\right) = \begin{cases} X_\beta \text{ if } \beta \neq \alpha_j \text{ for each } j \text{ and each } f_{\alpha_j} \in V_{\alpha_j}. \\ \phi \text{ if some } f_{\alpha_i} \notin V_{\alpha_i}. \\ V_{\alpha_k} \text{ if } \beta = \alpha_k \text{ for some } k \text{ and } f_{\alpha_i} \in V_{\alpha_i} \text{ for} \\ \quad i = 1, \ldots, n \text{ and } i \neq k. \end{cases}$$

In any event, $\varphi\left(Y \cap \left(\bigcap_{i=1}^{n} p_{\alpha_i}^{-1}(V_{\alpha_i})\right)\right) \in T_\beta$. It is now immediate that φ is a $(\mathbb{P}_Y, T_\beta)$ open map.

By Theorem 2.1.II, φ is a $(\mathbb{P}_Y, T_\beta)$ homeomorphism. ∎

Suppose $f: Y \to \prod_{\alpha \in A} X_\alpha$, where (Y, M) is any space. Then, $p_\beta \circ f: Y \to X_\beta$ for each $\beta \in A$. If f is continuous, then so is each $p_\beta \circ f$. The converse is less obvious but also true. If each $p_\beta \circ f$ is continuous, so is f. This is the standard way of testing a function into a product space for continuity, and we shall make frequent use of it in the sequel.

$$Y \xrightarrow{f} \prod_{\alpha \in A} X_\alpha$$
$$p_\beta \circ f \downarrow \quad \swarrow p_\beta$$
$$X_\beta$$

2.7 THEOREM Let (Y, M) be a topological space.
Let $f : Y \to \prod_{\alpha \in A} X_\alpha$.
Then,
f is $(M, \mathbb{P})$ continuous
if and only if
$p_\beta \circ f$ is (M, T_β) continuous for each $\beta \in A$.

Proof Suppose that $p_\beta \circ f$ is (M, T_β) continuous for each $\beta \in A$. To show that f is $(M, \mathbb{P})$ continuous, we use Theorem 1.2 (2–ii).II.

Consider any subbasic $\mathbb{P}$-open set $p_\beta^{-1}(V_\beta)$. Now, $f^{-1}(p_\beta^{-1}(V_\beta)) = (p_\beta \circ f)^{-1}(V_\beta) \in M$, by continuity of $p_\beta \circ f$. Hence f is $(M, \mathbb{P})$ continuous.

The converse is immediate by Theorem 1.3.II and Theorem 2.5 (1). ∎

The last two theorems of this section are highly technical, and not as important or widely applicable as Theorem 2.7, but we include them here for later reference.

2.8 THEOREM Let (Y, M) be a topological space.
Let $f: \prod_{\alpha \in A} X_\alpha \to Y$ be $(\mathbb{P}, M)$ continuous.
Let $t \in \prod_{\alpha \in A} X_\alpha$ and $\beta \in A$.

If $x \in X_\beta$, let $t_x \in \prod_{\alpha \in A} X_\alpha$ be defined by:

$$t_x(\alpha) = t(\alpha) \text{ if } \alpha \in A \text{ and } \alpha \neq \beta,$$

$$x \text{ if } \alpha = \beta.$$

Define $g: X_\beta \to Y$ by: $g(x) = f(t_x)$ for each $x \in X$. Then, g is (T_β, M) continuous.

Proof Left as an exercise. ∎

2.9 THEOREM *Let (Y_α, M_α) be a topological space for each $\alpha \in A$. Let $\mathbb{R}$ be the product topology on $\prod_{\alpha \in A} Y_\alpha$.*
Let $f_\alpha: X_\alpha \to Y_\alpha$ be (T_α, M_α) continuous for each $\alpha \in A$.
Define $F: \prod_{\alpha \in A} X_\alpha \to \prod_{\alpha \in A} Y_\alpha$ by:
$(F(x))(\alpha) = f_\alpha(x_\alpha)$ for each $x \in \prod_{\alpha \in A} X_\alpha$ and $\alpha \in A$.
Then,
F is $(\mathbb{P}, \mathbb{R})$ continuous.

Proof Left as an exercise. ∎

We conclude this section with a discussion of the Cantor set in E^1, which is widely used in topology and analysis (particularly measure theory) in constructing examples and counter-examples.

Let $X_n = \{0, 2\}$ for each positive integer n, and let T_n be the discrete topology on X_n. As usual, $\mathbb{P}$ denotes the product topology on $\prod_{n=1}^{\infty} X_n$. Note that $\mathbb{P}$ is not discrete.

Now let K consist of all numbers $\sum_{n=1}^{\infty} \frac{a_n}{3^n}$, where $a \in \prod_{n=1}^{\infty} X_n$. Then, $K \subset [0, 1]$, and is called the Cantor set.

Give K the relative topology k induced by E^1. Then, the space $\left(\prod_{n=1}^{\infty} X_n, \mathbb{P}\right)$ is homeomorphic to (K, k). A suitable homeomorphism is the map φ: $\prod_{n=1}^{\infty} X_n \to K$ given by $\varphi(x) = \sum_{n=1}^{\infty} \frac{x_n}{3^n}$ for each $x \in \prod_{n=1}^{\infty} X_n$.

Thus far we have seen two representations of K, one as a product space and one as a subspace of E^1. The latter representation can also be constructed geometrically. The advantage of doing this is some gain in insight into the topological properties of the Cantor set.

Begin by removing the open middle third $]\frac{1}{3}, \frac{2}{3}[$ from $[0, 1]$. Remove the open middle thirds $]\frac{1}{9}, \frac{2}{9}[$ and $]\frac{7}{9}, \frac{8}{9}[$ from the two remaining segments.

Then remove the open middle thirds from the four remaining segments, and keep going. What is left is the Cantor set. To see this, note that any number x in $[0, 1]$ can be written as a triadic expansion $\sum_{n=1}^{\infty} \frac{a_n}{3^n}$, where each a_n is 0, 1, or 2. In deleting $]\frac{1}{3}, \frac{2}{3}[$, we remove all numbers whose triadic expansion has $a_1 = 1$; in deleting $]\frac{1}{9}, \frac{2}{9}[$ and $]\frac{7}{9}, \frac{8}{9}[$ we remove all $\sum_{n=1}^{\infty} \frac{a_n}{3^n}$ with $a_2 = 1$, and so on. This leaves just the series with each a_i either 0 or 2. More carefully, let

$$A_i = \bigcup_{n=1}^{3^{i-1}} \left]\frac{3^n - 2}{3^i}, \frac{3^n - 1}{3^i}\right[$$

for $n = 1, 2, \ldots$ Then, $K = [0, 1] - \bigcup_{n=1}^{\infty} A_n$.

The reader can verify that K is closed, has no isolated points, each point of K is a cluster point and a boundary point of K, and finally that K has no interior points.

3.3 Identification and quotient topology

Suppose $f: X \to Y$, where X and Y are any sets. Given a topology T on X, is it possible to put a topology M on Y such that f is (T, M) continuous? This is the reverse of the situation described in section 2 as an alternate approach to the definition of the product topology (note also Problem 10, Chapter II).

Immediately, the indiscrete topology $\{\phi, Y\}$ is one possibility (in fact, the smallest one). But this is not a very interesting topology, and has the added disadvantage that f and T play no particular role in the choice. Any $f: X \to Y$ is (T, M) continuous for *any* T if M is indiscrete.

Experience has shown the better choice to be the largest topology on Y in which f is continuous. In view of Theorem 1.1, Ch. II, this is obtained by putting into M all subsets V of Y with $f^{-1}(V) \in T$. This yields the identification topology I_f on Y, with one minor adjustment. It is convenient to assume that f is a surjection. Otherwise, $f(X) \neq Y$, and all subsets of $Y - f(X)$ turn up in I_f, since $f^{-1}(V) = \phi \in T$ if $V \subset Y - f(X)$. This makes I_f discrete as far as the subspace $Y - f(X)$ is concerned, and the discrete topology is not a very useful one. Hence, in discussing the identification topology, the relevant maps will always be surjections.

For the remainder of this section, (X, T) is a space, Y a set, and $f: X \to Y$ a surjection.

3.1 Definition $I_f = \{V | V \subset Y \text{ and } f^{-1}(V) \in T\}$.

3.1 Theorem
(1) I_f is a topology on Y.
(2) f is (T, I_f) continuous.
(3) If S is a topology on Y and f is (T, S) continuous, then $S \subset I_f$.

Proof Left as an exercise. ∎

3.1 Example Let $X = \{0, 1, 2, 3, 4\}$. Let T consist of the sets ϕ, X, $\{0\}$, $\{1, 2\}$, $\{3, 4\}$, $\{4\}$, $\{0, 4\}$, $\{1, 2, 4\}$, $\{0, 1, 2\}$, $\{0, 3, 4\}$, $\{1, 2, 3, 4\}$, $\{0, 1, 2, 4\}$.
Let $Y = \{0, 1, 2\}$.
Define a surjection $f: X \to Y$ by specifying:

f: $0 \to 0$
$1 \to 2$
$2 \to 2$
$3 \to 2$
$4 \to 1$

Then I_f consists of the sets ϕ, Y, $\{0\}$, $\{1\}$, $\{0, 1\}$, and $\{1, 2\}$. Note that the relative topology T_Y consists of the sets ϕ, Y, $\{0\}$, and $\{1, 2\}$, and is different from the identification topology. ∎

3.2 Example Let (X_α, T_α) be a topological space for each $\alpha \in A$, and $\mathbb{P}$ the product topology on $\prod_{\alpha \in A} X_\alpha$.

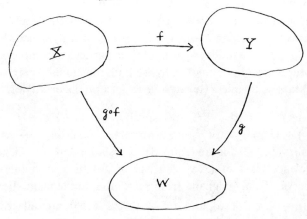

Figure 3.1

Let $\beta \in A$. Then, $p_\beta : \prod_{\alpha \in A} X_\alpha \to X_\beta$ is a surjection. If $V \subset X_\beta$, then $V \in T_\beta \Leftrightarrow p_\beta^{-1}(V) \in \mathbb{P}$. Thus, T_β is exactly the identification topology on X_β induced by $\mathbb{P}$ and p_β. ∎

We are going to prove only one theorem about the identification topology. Suppose we put the identification topology I_f on Y, so that f is (T, I_f) continuous. Suppose $g: Y \to W$, and M is a topology on W. Then g is continuous exactly when $g \circ f$ is continuous. This is in a sense the analogue of Theorem 2.7 for product spaces, which tests continuity by looking at compositions with the projections.

3.2 THEOREM *Let (W, M) be a topological space.*
Let $g: Y \to W$.
Then,
g is (I_f, M) continuous
if and only if
$g \circ f$ is (T, M) continuous.

Proof Suppose that g is (I_f, M) continuous. Let $V \in M$. Then, $g^{-1}(V) \in I_f$, so $f^{-1}(g^{-1}(V)) = (g \circ f)^{-1}(V) \in T$, implying that $g \circ f$ is (T, M) continuous.

Conversely, suppose that $g \circ f$ is (T, M) continuous. Let $V \in M$. Then, $(g \circ f)^{-1}(V) = f^{-1}(g^{-1}(V)) \in T$. But then $g^{-1}(V) \in I_f$, so g is (I_f, M) continuous. ∎

We conclude this section with some important examples, which include the quotient topology and several of the classical topological spaces.

3.3 EXAMPLE Let $\sim$ be an equivalence relation on X. The natural map $\eta: X \to X/\sim$ defined by $\eta(x) = \{y | y \sim x\}$ is a surjection, and so induces an identification topology I_η on $X/\sim$. I_η is customarily called the quotient topology on $X/\sim$. Quotient spaces play a particularly important role in algebraic topology, harmonic analysis, and functional analysis. See, for example, Massey, Spanier, Hewitt and Ross, Loomis or Taylor.

3.4 EXAMPLE Let $X = E^2$, and specify, for (x, y) and (x', y') in E^2 that $(x, y) \sim (x', y') \Leftrightarrow$ for some integers n and m, $x - x' = n$ and $y - y' = m$. Then, $\sim$ is an equivalence relation on E^2. Geometrically, the equivalence class of (x, y) consists of a lattice of points spaced at integer intervals from (x, y) and from each other in the plane. For example, $(\pi, \sqrt{2})$, $(\pi - 4, \sqrt{2} + 6)$ and $(\pi + 10, \sqrt{2} + 55)$ are all in the same equivalence class.

The space $(E^2/\sim, I_\eta)$ is difficult to visualize, but we can give it a more concrete representation by finding a homeomorphic image which is more familiar. Give $S^1 = \{(x, y) | x^2 + y^2 = 1\} \subset E^2$ the relative topology of the plane, and form the product space $(S^1 \times S^1, \mathbb{P})$. Then $(E^2/\sim, I_\eta) \cong (S^1 \times S^1, \mathbb{P})$. The reader can verify that the map $\varphi: E^2/\sim \to S^1 \times S^1$ which takes the equivalence class of (x, y) to $((\cos(2\pi x), \sin(2\pi x)), (\cos(2\pi y), \sin(2\pi y)))$ is an homeomorphism.

The space $(S^1 \times S^1, \mathbb{P})$ is called a torus. ∎

3.5 EXAMPLE Let (X, T) be the product space $([0, 1] \times [0, 1], \mathbb{P})$, where, $[0, 1]$ is considered as a subspace of E^1 (see Fig. 3.2). If $0 < x < 1$, $0 \leq y \leq 1$, $0 < x' < 1$ and $0 \leq y' \leq 1$, define $(x, y) \sim (x', y') \Leftrightarrow x = x'$ and $y = y'$.

If $0 \leq y \leq 1$, define $(0, y) \sim (1, y)$.

Here, $\sim$ leaves points inside the unit square, and on the open horizontal sides, alone (each such point comprises its own equivalence class), and identifies points opposite each other on the vertical sides. This effect can be achieved physically by rolling the square into a cylinder and pasting the vertical sides together. The quotient space $(([0, 1] \times [0, 1]/\sim, I_\eta)$ is called a cylinder.

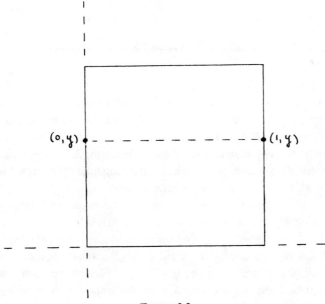

FIGURE 3.2

3.6 EXAMPLE Let (X, T) be the product space $([0, 1] \times\,]0, 1[, \mathbb{P})$, with subbasic open sets $(G_1 \cap [0, 1]) \times (G_2 \cap\,]0, 1[)$, where G_1 and G_2 are open in E^1 (see Figure 3.3).

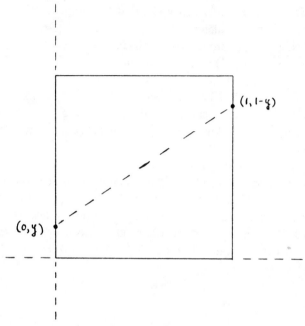

FIGURE 3.3

If $0 < x < 1$, $0 < y < 1$, $0 < x' < 1$ and $0 < y' < 1$, define $(x, y) \sim (x', y') \Leftrightarrow x = x'$ and $y = y'$; and $(0, y) \sim (1, y') \Leftrightarrow y + y' = 1$.

Geometrically, $\sim$ leaves points inside the unit square alone and identifies points $(0, y)$ and $(1, 1-y)$ on opposite vertical sides of the square. This is similar to the cylinder construction of Example 3.5, except that a twist is made before pasting the vertical sides together. The resulting space $(([0, 1] \times\,]0, 1[)/\sim, I_\eta)$ is called a Möbius strip.

A physical model of a Möbius strip may be made by taking a strip of paper (for purely practical reasons, take one much longer in the horizontal than in the vertical length), twisting once and gluing the short edges together.

A Möbius strip is one-sided in the following sense. Before twisting the rectangle of paper, draw a center line down its length on both sides. Mark an x somewhere on the line and a y on the opposite side. Now, in order to go from x to y on the rectangle, you have to cross an edge. The rectangle

RELATIVE, PRODUCT, IDENTIFICATION AND QUOTIENT TOPOLOGY

is two-sided. After twisting and gluing, however, you can travel the center line from x to y without crossing an edge.

Now take scissors and cut along the center line. Then cut the resulting surface again. You will probably be surprised at the result of the second cut. ∎

3.7 EXAMPLE Consider $[0, 1]$ as a subspace of E^1. Define an equivalence relation $\sim$ on $X \times [0, 1]$ as follows:

$$(x, t) \sim (x', t') \Leftrightarrow t = t' = 1.$$

Here, $\sim$ leaves points (x, t) and (x', t'), $0 \leq t < 1$, fixed, and identifies all points $(x, 1)$ and $(x', 1)$ on the top of the "cylinder" $X \times [0, 1]$. This has the effect of pinching the top to one point. The resulting space $((X \times [0, 1])/\sim, I_\eta)$ is called the cone over X. In particular, when $X = S^1$, then $(X \times [0, 1])/\sim$ actually does look like a cone (with a little imagination). ∎

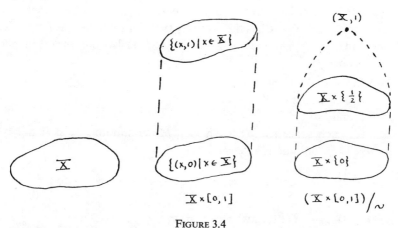

FIGURE 3.4

Problems

1) Let $\mathcal{O}$ be a T-cover of X (i.e., $\mathcal{O} \subset T$ and $\cup \mathcal{O} = X$). Let $B \subset X$. Then $B \in T$ if and only if $B \cap A$ is T_A-open for each $A \in \mathcal{O}$.

2) Let $\mathcal{O}$ be a cover of X by T-closed sets. Suppose $\mathcal{O}$ is neighborhood finite (that is, each point of X has a neighborhood which has a non-empty intersection with at most finitely many sets in $\mathcal{O}$). Let $B \subset X$. Then,
B is T-closed if and only if $B \cap A$ is T_A-closed for each $A \in \mathcal{O}$.

3) Let $\mathcal{O} \subset \mathcal{P}(X)$ such that $\cup \mathcal{O} = X$ and either:
   a) $\mathcal{O} \subset T$
   or
   b) $\mathcal{O}$ is nghd. finite and each set in $\mathcal{O}$ is T-closed.
   Let (Y, M) be a space.
   Let $f_A: A \to Y$ be (T_A, M) continuous for each $A \in \mathcal{O}$.
   Let $f_A|A \cap B = f_B|A \cap B$ whenever $A, B \in \mathcal{O}$.
   Then, there is a unique (T, M) continuous $F: X \to Y$ such that $F|A = f_A$ for each $A \in \mathcal{O}$. (F extends each f_A to X.)

4) Let (X, T) be a space and $A \subset X$. A map $r: X \to A$ is a retraction if r is (T, T_A) continuous and $r|A = 1_A$. (Think of r as collapsing X to A, while leaving A fixed.)
   When such an r exists, then A is said to be a retract of X.
   (a) S^1 is a retract of $E^2 - \{(0,0)\}$.
   (b) If (X, T) has the fixed point property (i.e., any (T, T) continuous function $f: X \to X$ has the property that, for some $x_f \in X$, $f(x_f) = x_f$), and if A is a retract of X, then (A, T_A) also has the fixed point property.
   (c) Let $A \subset X$. Then,
   A is a retract of X
   if and only if
   for each space (S, M), each (T_A, M) continuous $f: A \to S$ has a (T, M) continuous extension.

5) Let $\mathcal{O} \subset \mathcal{P}(X)$, and suppose, if $A \in \mathcal{O}$, $T(A)$ is a topology on A.
   Suppose
   (1) $T(A)_{A \cap B} = T(B)_{A \cap B}$ whenever $A, B \in \mathcal{O}$.
   (2) $A \cap B$ is T_A- and T_B-open (or closed) for each $A, B \in \mathcal{O}$. Then,
   (a) $W = \{U | U \cap A \in T(A)$ for each $A \in \mathcal{O}\} \cap \mathcal{P}(X)$ is a topology on X (this is the weak topology induced on X by $\mathcal{O}$), and $W_A = T(A)$ for each $A \in \mathcal{O}$.
   (b) If $f: X \to Y$, then f is (W, M) continuous if and only if $f|A$ is $(T(A), M)$ continuous for each $A \in \mathcal{O}$.

6) Prove that a finite product of discrete spaces is discrete.

7) Let (X_i, ϱ_i) be a metric space for each $i = 1, \ldots, n$.

Let $\varrho(f, g) = \sup \{\varrho_i(f_i, g_i) | 1 \leq i \leq n\}$ for $f, g \in \prod_{i=1}^{n} X_i$. Then,

(a) ϱ is a metric on $\prod_{i=1}^{n} X_i$.

(b) T_ϱ agrees with the product topology induced by $T_{\varrho_1}, ..., T_{\varrho_n}$.

*8) Let X consist of all sequences $x: Z^+ \to E^1$ with $|x_i| \leq 1/i$ for $i = 1, 2, ...$ Let $\varrho(x, y) = \left[\sum_{i=1}^{\infty} (x_i - y_i)^2\right]^{1/2}$ for $x, y \in X$.
Then, the metric space (X, ϱ) is homeomorphic to the product space $\left(\prod_{j=1}^{\infty} Y_j, \mathbb{P}\right)$, where each $Y_j = [0, 1]$, considered as a subspace of E^1.

9) Let $A_\alpha \subset X_\alpha$ for each $\alpha \in \mathcal{O}$.
Then, $\prod_{\alpha \in \mathcal{O}} A_\alpha$ is $\mathbb{P}$-closed in $\prod_{\alpha \in \mathcal{O}} X_\alpha$ if and only if each A_α is T_α-closed in X_α.

*10) Let (X, T) and (Y, M) be spaces.
Let $f: X \to Y$, and let $G = \{(x, f(x)) | x \in X\}$. Then,
f is (T, M) continuous
if and only if
the map $g: X \to G$ defined by $g(x) = (x, f(x))$ for each $x \in X$, is an homeomorphism of (X, T) onto $(G, \mathbb{P}_G)$, where $\mathbb{P}$ is the product topology induced on $X \times Y$ by T and M.

*11) Let (X_α, T_α) be a space for each $\alpha \in \mathcal{O}$. Let $\mathcal{O} = \bigcup_{\beta \in \mathcal{B}} I_\beta$, where $I_\beta \cap I_\gamma = \phi$ for β, γ in $\mathcal{B}$ and $\beta \neq \gamma$.
Let $Y_\beta = \prod_{\gamma \in I_\beta} X_\gamma$ for each $\beta \in \mathcal{B}$, and $\mathbb{P}_\beta$ the product topology on Y_β.
Let $\mathbb{R}$ denote the product topology on $\prod_{\beta \in \mathcal{B}} Y_\beta$, induced by the topologies $\mathbb{P}_\beta$ on Y_β.
Then, $\left(\prod_{\alpha \in \mathcal{O}} X_\alpha, \mathbb{P}\right) \cong \left(\prod_{\beta \in \mathcal{B}} Y_\beta, \mathbb{R}\right)$.
(This is the theorem on associativity of topological products).

12) Call a (T, M) continuous surjection $f: X \to Y$ an identification map exactly when $M = I_f$.
   (a) If $f: X \to Y$ is a (T, M) continuous surjection, and if there is an (M, T) continuous $g: Y \to X$ such that $f \circ g = 1_Y$, then f is an identification map.

(b) If $f: X \to Y$ is a (T, M) continuous surjection, then
f is an identification map
if and only if
for each space (S, W) and each function $g: Y \to S$, then $g \circ f$ is (T, W) continuous implies g is (M, W) continuous.

13) Let $\approx$ be an equivalence relation on X, T a topology on X.
Let $f: X/\approx \to Y$.
Then,
f is (I_η, W) continuous
if and only if
$f \circ \eta$ is (T, W) continuous.

14) Let (X, T) and (Y, M) be spaces. Let $\approx$ be an equivalence relation on X, and $\sim$ on Y. Let $f: X \to Y$. Suppose, if $x, y \in X$, and $x \approx y$, then $f(x) \sim f(y)$.
Define $f^*: X/\approx \to Y/\sim$ by $f^*([x]_\approx) = [f(x)]_\sim$.
Then,

(a) f^* is continuous with respect to the quotient topologies on $X/\approx$ and $Y/\sim$.

(b) $f^* \circ \eta_\approx = \eta_\sim \circ f$, where $\eta_\approx: X \to X/\approx$ and $\eta_\sim: Y \to Y/\sim$ are the natural maps.

(c) If f is an identification map, then so is f^*.

15) Let $A \subset X$. Then, any retraction of X onto A is an identification map. (See problem 4 for the definition of a retraction.)

*16) Show that the following three spaces are homeomorphic:

(a) The product space $S^1 \times S^1$, the topology on S^1 being the relative topology induced by E^2.

(b) The quotient space $E^2/\sim$ of Example 3.4.

(c) The subspace $\{(x, y, z) | ((x^2 + y^2)^{1/2} - 2)^2 + z^2 = 1\}$ of E^3.

*17) Define an equivalence relation $\approx$ on E^1 by: $x \approx y$ if and only if $x - y \in Z$.
Consider S^1 as a subspace of E^2, and give $E^1/\approx$ the quotient topology. Show that $E^1/\approx \cong S^1$.

*18) Show that $E^{n+1} - \{(0, \ldots, 0)\} \cong S^n \times E^1$ for $n \geq 1$.

*19) Let I^n denote the n-cube $\{(x_1, ..., x_n) | 0 \leq x_i \leq 1 \text{ for } i = 1, ..., n\}$. Let $\dot{I}^n = I^n \cap (x_1, ..., x_n)|x_j = 0 \text{ or } 1 \text{ for some } j, 1 \leq j \leq n\}$.
Define an equivalence relation on I^n by:
$x \approx y$ if and only if $(x, y \in \dot{I}^n)$ or $(x, y \in I^n - \dot{I}^n \text{ and } x = y)$.
Then,
$I^n/\approx \, \cong S^n$.
(Try picturing this geometrically for $n = 1, 2$.)

20) Show that $\{(x, y) | x^2 + y^2 < 1\}$, an open subset of the product space $E^2 = E^1 \times E^1$, cannot be written as $U \times V$ for any open sets U and V in E^1.

21) Let $C_\alpha \subset X_\alpha$ for each $\alpha \in A$. Suppose that $C_\alpha \neq X_\alpha$ for infinitely many α. Then, $\prod_{\alpha \in A} C_\alpha$ has no interior points in the product topology on $\prod_{\alpha \in A} X_\alpha$.

CHAPTER IV

Convergence

4.1 Sequential convergence

THE BASIC PROCESS of analysis is convergence. A generalization of sequential convergence of real sequences to the setting of an arbitrary topological space (X, T) can easily be achieved by replacing E^1 by X and open intervals by T-open sets.

For the remainder of this chapter, T is a topology on X, and $X \neq \phi$. For completeness, we begin with the definition of a sequence to X.

1.1 DEFINITION *S is a sequence to X*
if and only if
$S: Z^+ \to X$.

1.2 DEFINITION *Let S be a sequence to X.*
Then,

(1) Let $x \in X$.
  Then,
  S is T-convergent to x
  if and only if
  if $x \in V \in T$, then, for some $K > 0$, $S_n \in V$ whenever $n \in Z^+$ and $n \geq K$.

(2) S is T-convergent
  if and only if
  for some $x \in X$, S is T-convergent to x.

1.1 EXAMPLE Let $X = E^1$, and let S be a sequence to X. If V is a nghd. of x, then $]x - \varepsilon, x + \varepsilon[\subset V$ for some $\varepsilon > 0$. S converges to x exactly when, corresponding to the ε nghd. $]x - \varepsilon, x + \varepsilon[$ of x, there is some $K_\varepsilon > 0$ such that $S_n \in]x - \varepsilon, x + \varepsilon[$ for $n \in Z^+$ and $n \geq K_\varepsilon$. That is, $|S_n - x| < \varepsilon$ for n sufficiently large. In this case, Definition 1.2 reduces to the usual calculus notion of $\lim_{n \to \infty} S_n = x$. ∎

1.2 EXAMPLE Let T be the discrete topology on X. Then only the eventually constant sequences converge in X. For suppose S is a sequence to X which T-converges to x. Since $\{x\}$ is a nghd. of x in the discrete topology, then, for some $K > 0$, $S_n \in \{x\}$ for $n \geq K$. Then, eventually all $S_n = x$. ∎

1.3 EXAMPLE Let T be the indiscrete topology on X. Then, every sequence to X converges to every point in X. For, if $S: Z^+ \to X$, and $x \in X$, then the only T-nghd. of x is X, and $S_n \in X$ for all $n \in Z^+$. ∎

1.4 EXAMPLE Let $X = C([a, b])$, with T the sup-norm topology. Let S be a sequence to $C([a, b])$ and $g \in X$. We claim: S is T-convergent to g exactly when S converges to g uniformly on $[a, b]$ in the usual calculus sense.

Recall: $S \to g$ uniformly on $[a, b]$ if, given $\varepsilon > 0$, there is some $K_\varepsilon > 0$ such that $|S_n(x) - g(x)| < \varepsilon$ whenever $x \in [a, b]$ and $n \in Z^+$ such that $n \geq K_\varepsilon$.

Now suppose S is T-convergent to g. Let $\varepsilon > 0$. Then, $g \in B_d(g, \varepsilon) \in T$, so for some $K_\varepsilon > 0$, $S_n \in B_d(g, \varepsilon)$ for $n \geq K_\varepsilon$. Then, for $n \geq K_\varepsilon$, $d(S_n, g)$ = $\sup \{|S_n(x) - g(x)| \, | \, a \leq x \leq b\} < \varepsilon$, so that $|S_n(x) - g(x)| < \varepsilon$ for $n \geq K_\varepsilon$ and $a \leq x \leq b$. This means that $S_n \to g$ uniformly on $[a, b]$.

Conversely, if $S_n \to g$ uniformly on $[a, b]$, and $g \in V \in T$, choose $\varepsilon > 0$ such that $B_d(g, \varepsilon) \subset V$. For some $K_\varepsilon > 0$, $|S_n(x) - g(x)| < \varepsilon$ if $n \geq K_\varepsilon$ and $a \leq x \leq b$. Then, $d(S_n, g) < \varepsilon$ for $n \geq K_\varepsilon$, hence $S_n \in B_d(g, \varepsilon) \subset V$ whenever $n \geq K_\varepsilon$. This means that S is T-convergent to g. ∎

1.5 EXAMPLE Let X be the set of real numbers and $T = \{\,]-x, x[\,|\,x \geq 0\} \cup \{X\}$. It is easy to check that T is a topology on X. If S is any sequence to X, then S is T-convergent exactly when the set of numbers S_n ($n \in Z^+$) is bounded. In this event, if $-\beta \leq S_n \leq \beta$ for each n in Z^+, then S T-converges to each x with $|x| \geq \beta$. ∎

As examples 1.3 and 1.5 show, we have lost uniqueness of limits in generalizing the notion of sequential convergence to arbitrary topological spaces. This is a serious problem, which we shall treat in some detail in Chapter V. For the time being, we shall prove that limits of sequences are always unique in metric spaces.

1.1 THEOREM *Let ϱ be a metric on X.*
Let S be a sequence to X.
Let S be T_ϱ-convergent to x and to y in X.
Then,
$x = y$.

Proof If $x \neq y$, let $\varepsilon = \frac{1}{2}\rho(x, y)$. Then, $\varepsilon > 0$, and $B_\rho(x, \varepsilon) \cap B_\rho(y, \varepsilon) = \phi$. Since $x \in B_\rho(x, \varepsilon)$, then for some $K_x > 0$, $S_n \in B_\rho(x, \varepsilon)$ for $n \geq K_x$.
Since $y \in B_\rho(y, \varepsilon) \in T$, then for some $K_y > 0$, $S_n \in B_\rho(y, \varepsilon)$ for $n \geq K_y$.
Choose some positive integer m with $m \geq K_x$ and $m \geq K_y$. Then, $S_m \in B_\rho(x, \varepsilon) \cap B_\rho(y, \varepsilon)$, a contradiction. This implies that $x = y$. ∎

At least in metric spaces, then, we can write $\lim_T S = x$ whenever S is T-convergent to x. However, this symbol is ambiguous if two different elements of X can be put on the right.

There is another, more serious, unsatisfactory aspect to sequential convergence in arbitrary topological spaces. In some spaces (E^n for example) a point x can be characterized as belonging to the closure of a set A exactly when some sequence to A converges to x. We shall show that this is true for any metric space. But Example 1.6 shows that it is not true in general.

1.2 THEOREM *Let ρ be a metric on X.*
Let $x \in X$ and $A \subset X$.
Then,
$x \in \bar{A}$
if and only if
there is a sequence to A which T_ρ-converges to x.

Proof Suppose first that $x \in \bar{A}$. If n is a positive integer, then $B_\rho(x, 1/n) \cap A \neq \phi$, by Theorem 2.3 (2).I. Choose $S_n \in B_\rho(x, 1/n) \cap A$ for each positive integer n. This defines a sequence S to A, and it is easy to check that S is T_ρ-convergent to x.

Conversely, let S be a sequence to A and let S be T_ρ-convergent to x. Let $x \in V \in T$. For some $K > 0$, $S_n \in V$ for each $n \geq K$. Choose any positive integer $m \geq K$. Then, $S_m \in V \cap A$, so $V \cap A \neq \phi$, implying that $x \in \bar{A}$. ∎

The implications of this theorem are quite far reaching. The topology T_ρ is completely determined by the sequences to X. For suppose $A \subset X$ and $A \neq \phi$. By Theorem 1.2, $A = \bar{A}$ exactly when each convergent sequence to A has its limit in A. But A is T_ρ-closed exactly when $A = \bar{A}$. Thus the sequences characterize the closed sets, hence also the open sets, since B is open exactly when $X - B$ is closed. In short: if we were told all the T_ρ-convergent sequences to X and their limits, we could reconstruct the topology T_ρ.

But Theorem 1.2 fails in general, as shown by the following example.

1.6 EXAMPLE Let X be the set of real-valued functions on $[0, 1]$, with the topology of Example 1.7, Chapter I.

Let A be the set of all f in X such that $f(x) = 0$ for all but a finite (but non-zero) number of points in $[0, 1]$, at which $f(x) = 1$.

Let $g(x) = 1$ for each x in $[0, 1]$. Then, $g \notin A$. We claim that $g \in \bar{A}$, but that there is no sequence to A which converges to g in the given topology.

First, $g \in \bar{A}$. For suppose $U(g; x_1, ..., x_n; \varepsilon)$ is a basic nghd. of g. Choose f in A by letting $f(x) = 0$ for $x \neq x_i$ and $f(x_i) = 1$ for $i = 1, ..., n$. Then, $f \in U(g; x_1, ..., x_n; \varepsilon) \cap A$, implying that $g \in \bar{A}$.

Now suppose that S is a sequence to A and S converges to g. If $0 \leq x \leq 1$, then $U(g; x; \frac{1}{2})$ is a nghd. of g, so $S_n \in U(g; x; \frac{1}{2})$ for some positive integer n. Then, $|S_n(x) - g(x)| = |S_n(x) - 1| < \frac{1}{2}$. Since $S_n \in A$, then $S_n(x)$ is either 0 or 1, hence in this case $S_n(x) = 1$. Then, $x \in \{t | S_n(t) = 1\}$. Then, $[0, 1] = \bigcup_{n=1}^{\infty} \{t | S_n(t) = 1\}$. Now, for each positive integer n, $\{t | S_n(t) = 1\}$ is finite, as $S_n \in A$. Then, $\bigcup_{n=1}^{\infty} \{t | S_n(t) = 1\}$ is countable, implying that $[0, 1]$ is countable, a contradiction. Thus, no sequence to A converges to g, even though g is in $\bar{A}$. ∎

The inability of the sequential theory of convergence to distinguish the points of closure of subsets, hence the open and closed sets of the space, is a serious deficiency of the theory. The difficulty is not in our method of generalizing the calculus definition of sequential limit, but is in fact much deeper. Sequential convergence, which was originally conceived in E^n, is adequate for metric spaces, but is not powerful enough to handle the ideosyncracies of non-metric spaces.

The need for a better theory of convergence has led to two major theories, nets and filterbases. In a sense, the two theories are equivalent, although in a given situation one may appear the more convenient. We shall treat them individually, and then discuss the sense in which they are equivalent.

4.2 Nets

The theory of nets given here is a development of ideas originally contained in a paper by E. H. Moore and H. L. Smith. It has since been refined and applied to topology by a number of other mathematicians, but still the theory is often called Moore-Smith convergence.

The idea is to abstract the essential ingredients from sequential convergence. The following chart suggests how this was done (experience was the guide in deciding what was essential).

Sequence	Net
1) A function $S: Z^+ \to X$.	1') A function $S: D \to X$
2) Z^+ is partially ordered by $\leq$.	2') Let $\prec$ be a partial ordering on D.
3) If $n, m \in Z^+$, there is some $k \in Z^+$ with $n \leq k$ and $m \leq k$.	3') If $x, y \in D$, there is some $z \in D$ with $x \prec z$ and $y \prec z$.

The contribution of Moore and Smith lay in making up the list on the right, recognizing those elements of the concept of sequence which make sequential convergence a useful tool in metric spaces.

The following terminology has become fairly standard.

2.1 DEFINITION *D is directed by $\prec$*
if and only if
(1) $D \neq \phi$.
(2) $\prec$ is a partial ordering on D.
(3) If $x \in D$ and $y \in D$, then, for some $z \in D$, $x \prec z$ and $y \prec z$.
If D is directed by $\prec$, we often say that $\prec$ is a direction on D, and call $(D, \prec)$ a directed set.

2.2 DEFINITION *$(S, D, \prec)$ is a net*
if and only if
(1) S is a function on D.
(2) D is directed by $\prec$.

2.3 DEFINITION *$(S, D, \prec)$ is a net to X*
if and only if
$(S, D, \prec)$ is a net and $S: D \to X$.

2.4 DEFINITION *Let $(S, D, \prec)$ be a net to X.*
Then,
$(S, D, \prec)$ is eventually in A
if and only if
for some $n \in D$, $S(k) \in A$ whenever $k \in D$ and $n \prec k$.

It is Definition 2.4 which contains the seeds of the idea of net convergence Thus far nets have no particular connection with topology. Now suppose, for the remainder of this section, that (X, T) is a topological space. Net convergence is defined by direct analogy with sequential convergence, using the terminology of the last definition.

2.5 DEFINITION Let $(S, D, \prec)$ be a net to X and let $x \in X$.
Then,
$(S, D, \prec)$ is T-convergent to x
if and only if
$(S, D, \prec)$ is eventually in each T-nghd of x.

If, for some $x \in X$, a net $(S, D, \prec)$ is T-convergent to x, then, as with sequences, we often say that $(S, D, \prec)$ T-converges in X, without explicitly mentioning x.

2.1 EXAMPLE Every sequence to X is a net, with $D = Z^+$ and $\prec$ as the usual order $\leq$ on Z^+. A sequence converges to x in X exactly when it is eventually in each nghd. of x. ∎

2.2 EXAMPLE Let $f(x, y) = \dfrac{1}{x^2 + y^2}$ for $(x, y) \in E^2$ and $(x, y) \neq (0, 0)$. Define, on $E^2 - \{(0, 0)\}$, $(a, b) \prec (c, d)$ to mean that $a \leq c$ and $b \leq d$. Here, $(a, b) \prec (c, d)$ conveys the idea that (c, d) is at least as far from the origin as (a, d) is. Then, $(f, E^2 - \{(0, 0)\}, \prec)$ is a net to E^1. This net is eventually in each Euclidean nghd. of 0. For consider a basic nghd. $]-\varepsilon, \varepsilon[$ of 0. Let $\left(\dfrac{2}{\sqrt{\varepsilon}}, \dfrac{2}{\sqrt{\varepsilon}}\right) \prec (x, y)$. Then, $\dfrac{4}{\varepsilon} \leq x^2$ and $\dfrac{4}{\varepsilon} \leq y^2$, so
$$\dfrac{1}{x^2 + y^2} \leq \dfrac{\varepsilon}{2} < \varepsilon.$$

In this case, convergence of $(f, E^2 - \{(0, 0)\}, \prec)$ to 0 is equivalent to saying that $\lim\limits_{|(x,y)| \to \infty} f(x, y) = 0$ in the usual calculus sense. ∎

As these two examples suggest, the theory of net convergence includes all the limits ordinarily encountered in calculus.

We shall now show how nets succeed where sequences fail in characterizing the closure points of a set. Note in the proof how the generality of the concepts of net and net convergence make possible the construction of a net converging to the appropriate point, while at the same time there is no way (in the absence of a metric) to produce a sequence converging to the point.

2.1 THEOREM Let $A \subset X$ and $x \in X$.
Then,
$x \in \bar{A}$
if and only if
there is a net to A which T-converges to x.

Proof Suppose $(g, M, \prec)$ is a net to A which T-converges to x. Let V be a T-nghd. of x. Then, $(g, M, \prec)$ is eventually in V. Then, for some $\alpha \in M$, $g_\alpha \in V \cap A$, hence $V \cap A \neq \phi$ and $x \in \bar{A}$.

Conversely, suppose that $x \in \bar{A}$. Let $M = \{V | V$ is a T-nghd. of $x\}$. If $V \in M$ and $U \in M$, define $V \prec U$ to mean $U \subset V$. It is easy to check that $(M, \prec)$ is a directed set. If $x \in V \in T$, then $V \cap A \neq \phi$, as $x \in \bar{A}$. Choose some $f_V \in V \cap A$. This defines a function on M to A, and it is easy to check that the net $(f, M, \prec)$ is T-convergent to x. ∎

In fact, not only closure points, but also cluster points, of A can be distinguished by nets.

2.2 THEOREM *Let $A \subset X$ and $x \in X$. Then,*
x is a T-cluster point of A
if and only if
there is a net to $A - \{x\}$ which T-converges to x.

Proof The proof is similar to that of Theorem 2.1, and is left to the reader. ∎

Of course, a sequence is a net, so in any space (X, T) we can say that x is a cluster point of A if there is a sequence to $A - \{x\}$ converging to x. The converse can be proved for metric spaces (imitate part of the proof of Theorem 1.2), but not for spaces in general, as we saw in Example 1.6.

Unfortunately, nets, like sequences, may fail to have unique limits. For example, if T is the indiscrete topology on X, then every net to X converges to every point in X. As with sequences, we shall defer consideration of this problem to Chapter V, where we shall see that the fault lies not with the theory of convergence, but with the space. In Chapter V, we shall put a very reasonable condition on the space to obtain the notion of a Hausdorff space. A space will turn out to be Hausdorff exactly when convergent nets have unique limits (though this result will not be true for sequences). Since many of the "useful" spaces, at least in analysis, are "at least" Hausdorff, nets will have unique limits in the more commonly encountered spaces.

Net convergence can also be used to characterize continuity. The obvious Lemma which comes first insures that the statements of the next three theorems make sense.

1 LEMMA *Let $f: A \to B$.*
Let $(S, D, \prec)$ be a net to A.
Then,
$(f \circ S, D, \prec)$ is a net to B.

Proof Left as an exercise. ∎

2.3 THEOREM *Let (Y, M) be a topological space.*
Let $f: X \to Y$.
Then,
f is (T, M) continuous
if and only if
if $x \in X$ and $(S, D, \prec)$ is a net to X which T-converges to x, then $(f \circ S, D, \prec)$ is M-convergent to $f(x)$.

Proof Suppose first that f is (T, M) continuous. Let $(S, D, \prec)$ be a net to X which T-converges to $x \in X$. Let $f(x) \in V \in M$.

Since f is (T, M) continuous at x, there is some T-nghd. U of x such that $f(U) \subset V$. For some m, $m \in D$ and $S(n) \in U$ whenever $n \in D$ and $m \prec n$. But then, $f(S(n)) \in V$ whenever $n \in D$ and $m \prec n$, so $(f \circ S, D, \prec)$ is eventually in V.

Conversely, assume that, if $x \in X$ and $(S, D, \prec)$ is a net to X and $(S, D, \prec)$ is T-convergent to x, then $(f \circ S, D, \prec)$ is M-convergent to $f(x)$.

Now suppose that F is M-closed. By Theorem 1.1.II, it suffices to show that $f^{-1}(F)$ is T-closed.

Let $x \in \overline{f^{-1}(F)}$. By Theorem 2.1, there is a net $(S, D, \prec)$ to $f^{-1}(F)$ which T-converges to x.

By assumption, $(f \circ S, D, <)$ is a net to F which M-converges to $f(x)$. Then, again by Theorem 2.1, $f(x) \in \bar{F}$.

Now, F is M-closed, so $\bar{F} = F$, hence $f(x) \in F$. Then, $x \in f^{-1}(F)$, so $f^{-1}(F) = \overline{f^{-1}(F)}$, and $f^{-1}(F)$ is T-closed. ∎

Similarly, net convergence characterizes continuity at a point.

2.4 THEOREM *Let (Y, M) be a topological space.*
Let $f: X \to Y$.
Let $x \in X$.
Then,
f is (T, M) continuous at x
if and only if

if $(S, D, \prec)$ *is a net to* X *which* T-*converges to* x, *then* $(f \circ S, D, \prec)$ *is* M-*convergent to* $f(x)$.

Proof Left as an exercise. ∎

We conclude this section with a consideration of convergence in product spaces. Suppose that $(f, D, \prec)$ is a net to $\prod_{\alpha \in A} X_\alpha$. Then, $(p_\beta \circ f, D, \prec)$ is a net to X_β for each β in A. Theorem 2.5 says that $(f, D, \prec)$ converges to x in $\prod_{\alpha \in A} X_\alpha$ exactly when the projected net $(p_\beta \circ f, D, \prec)$ converges to x_β $(= p_\beta(x))$ for each β in A. Thus, we can determine convergence (and the actual limit) of a product net by examining its projection in each coordinate space.

2.5 Theorem *Let* $A \neq \phi$.
Let (X_α, T_α) *be a space for each* $\alpha \in A$.
Let $\mathbb{P}$ *be the product topology on* $\prod_{\alpha \in A} X_\alpha$.
Let $(S, D, \prec)$ *be a net to* $\prod_{\alpha \in A} X_\alpha$.
Let $x \in \prod_{\alpha \in A} X_\alpha$.
Then,
$(S, D, \prec)$ $\mathbb{P}$-*converges to* x
if and only if
$(p_\alpha \circ S, D, \prec)$ T_α-*converges to* x_α *for each* $\alpha \in A$.

Proof Suppose first that $(S, D, \prec)$ $\mathbb{P}$-converges to x. If $\beta \in A$, then $(p_\beta \circ S, D, \prec)$ T_β-converges to x_β by Theorem 2.4 and by Theorem 2.5 (1).III.

Conversely, suppose that $(p_\beta \circ S, D, \prec)$ T_β-converges to x_β for each β in A. Let $x \in U \in \mathbb{P}$. We must show that $(S, D, \prec)$ is eventually in U.

Produce a basic $\mathbb{P}$-open set $\bigcap_{i=1}^{n} p_{\alpha_i}^{-1}(U_{\alpha_i})$ with $x \in \bigcap_{i=1}^{n} p_{\alpha_i}^{-1}(U_{\alpha_i}) \subset U$. Then, $p_{\alpha_i}(x) = x_{\alpha_i} \in U_{\alpha_i}$ for $i = 1, \ldots, n$. Since $(p_{\alpha_i} \circ S, D, \prec)$ is T_{α_i}-convergent to x_{α_i}, then for some $n_{\alpha_i} \in D$, $(p_{\alpha_i} \circ S)(n) \in U_{\alpha_i}$ whenever $n \in D$ and $n_{\alpha_i} \prec n$, for $i = 1, \ldots, n$.

By obvious induction from Definition 2.1 (3), produce some $K \in D$ such that $n_{\alpha_i} \prec K$ for $i = 1, \ldots, n$.

Now let $j \in D$ and $K \prec j$. Then, $n_{\alpha_i} \prec j$, so $(p_{\alpha_i} \circ S)(j) \in U_{\alpha_i}$, for $i = 1, \ldots, n$, hence $S(j) \in \bigcap_{i=1}^{n} p_{\alpha_i}^{-1}(U_{\alpha_i})$. Then, $(S, D, \prec)$ is eventually in $\bigcap_{i=1}^{n} p_{\alpha_i}^{-1}(U_{\alpha_i})$, hence also in U, proving the theorem. ∎

4.3 Filterbases

The other major theory of convergence is the filterbase theory, which is a refinement of the filter concept originated by H. Cartan and developed extensively by the French Bourbaki school. In some circles filterbases are considered more elegant than nets, and the proof we shall give of Tychonov's Theorem in Chapter VII would seem to support this view, although nets have more intuitive appeal. Nets are clear generalizations of sequences, while filterbases are not clear generalizations of anything familiar to the beginning student. However, filterbase convergence has proved as rich a theory as net convergence, and today both are used extensively in the literature. For an excellent discussion of filterbases (with different terminology) see E. J. McShane, A Theory of Limits, MAA Studies in Mathematics, Vol. I.

In this section, (X, T) is a topological space.

3.1 DEFINITION $\mathscr{F}$ *is a filterbase on X if and only if*
(1) $\mathscr{F}$ is a non-empty set of non-empty subsets of X.
(2) If $a \in \mathscr{F}$ and $b \in \mathscr{F}$, then, for some c, $c \in \mathscr{F}$ and $c \subset a \cap b$.

We make two observations:

(1) the topology on X plays no part in the definition of a filterbase on X; T will be used only to define a notion of convergence of filterbases.

(2) Definition 3.1 is very similar to the condition of Theorem 3.3, Ch. I, for a set of subsets of X to be a topological base.

3.1 EXAMPLE The set of all subsets of X containing a given element x is a filterbase on X. ∎

3.2 EXAMPLE If $x \in X$, the set of all T-neighborhoods of x is a filterbase on X. This is the neighborhood filterbase at x, and is important in our treatment of regular spaces in section 2 of Chapter V. ∎

The following theorem is a trivial consequence of the definition, but will be useful later.

3.1 THEOREM *Let $\mathscr{F}$ be a filterbase on X.*
Let $\alpha_1, ..., \alpha_n \in \mathscr{F}$
Then,
for some $m \in \mathscr{F}$, $m \subset \bigcap_{i=1}^{n} \alpha_i$. Hence also $\bigcap_{i=1}^{n} \alpha_i \neq \phi$.

Proof Left as an exercise. ∎

For the remainder of this section, let $\mathscr{F}$ be a filterbase on X.

3.2 DEFINITION *Let $x \in X$.*
Then,
$\mathscr{F}$ is T-convergent to x
if and only if
if V is a T-nghd. of x, then $a \subset V$ for some $a \in \mathscr{F}$.

3.3 DEFINITION *$\mathscr{F}$ is T-convergent*
if and only if
for some $x \in X$, $\mathscr{F}$ is T-convergent to x.

3.3 EXAMPLE If $x \in X$, then the neighborhood filterbase at x converges to x. This is the key to the filterbase analogue of Theorem 2.1. ∎

3.4 EXAMPLE If T is indiscrete, then $\mathscr{F}$ converges to each point of X. ∎

3.5 EXAMPLE Let $X = \{0, 1, 2\}$ and let T be the discrete topology on X. Let $\mathscr{F}$ consist of the single set $\{0, 1\}$. Then, $\mathscr{F}$ does not converge to any point in X. ∎

We now develop the filterbase analogues of the theorems on net convergence.

3.2 THEOREM *Let $A \subset X$.*
Let $x \in X$.
Then,
$x \in \bar{A}$
if and only if
there is a filterbase on A which T-converges to x.

Proof Suppose first that $x \in \bar{A}$. Let $\mathscr{F} = \{A \cap V | x \in V \in T\}$. It is easy to check that $\mathscr{F}$ is a filterbase on A and that $\mathscr{F}$ T-converges to x.

Conversely, let $\mathscr{F}$ be a filterbase on A which T-converges to x.

Let $x \in V \in T$. For some $a, a \in \mathscr{F}$ and $a \subset V$. Since $a \in \mathscr{F}$, then $\phi \neq a \subset A$, so $A \cap V \neq \phi$, implying that $x \in \bar{A}$. ∎

3.3 THEOREM *Let $A \subset X$.*
Let $x \in X$.
Then,
x is a T-cluster point of A

*if and only if
there is a filterbase on $A - \{x\}$ which T-converges to x.*

Proof Left as an exercise. ∎

Before characterizing continuity in terms of filterbases, we state a technical Lemma which says that any function takes a filterbase to a filterbase.

1 LEMMA *Let $f: A \to B$.
Let $\mathscr{T}$ be a filterbase on A.
Then, $\{f(r)|r \in \mathscr{T}\}$ is a filterbase on B.*

Proof Left as an exercise. ∎

We usually waive notational precision and for convenience denote $\{f(r)|r \in \mathscr{T}\}$ by $f(\mathscr{T})$.

3.4 THEOREM *Let (Y, M) be a topological space.
Let $f: X \to Y$.
Then,
f is (T, M) continuous
if and only if
if $x \in X$ and $\mathscr{F}$ is a filterbase on X which T-converges to x, then $f(\mathscr{F})$ is M-convergent to $f(x)$.*

Proof Suppose first that f is (T, M) continuous and that $x \in X$. Let $\mathscr{F}$ be a filterbase on X which T-converges to x.

If V is an M-nghd. of $f(x)$, then $x \in f^{-1}(V) \in T$ by Theorem 1.1.II.

Since $\mathscr{F}$ T-converges to x, then, for some $a \in \mathscr{F}$, $a \subset f^{-1}(V)$. But then $f(a) \subset V$, so $f(\mathscr{F})$ is M-convergent to $f(x)$.

Conversely, suppose that $f(\mathscr{F})$ is M-convergent to $f(x)$ whenever $x \in X$ and $\mathscr{F}$ is a filterbase on X which T-converges to x.

Let $F \subset Y$ such that F is M-closed. By Theorem 1.1.II, it suffices to show that $f^{-1}(F)$ is T-closed.

Let $y \in \overline{f^{-1}(F)}$. By Theorem 3.2, there is a filterbase $\mathscr{F}$ on $f^{-1}(F)$ which T-converges to y. Then $f(\mathscr{F})$ is a filterbase on F which M-converges to $f(y)$. By Theorem 3.2, $f(y) \in \bar{F}$. But, $F = \bar{F}$, so $f(y) \in F$, hence $y \in f^{-1}(F)$. Then $f^{-1}(F) = \overline{f^{-1}(F)}$ and $f^{-1}(F)$ is T-closed. ∎

3.5 THEOREM *Let (Y, M) be a topological space.
Let $f: X \to Y$.
Let $x \in X$.*

Then,
f is (T, M) continuous at x
if and only if
if $\mathscr{F}$ is a filterbase on X and $\mathscr{F}$ T-converges to x, then $f(\mathscr{F})$ M-converges to $f(x)$.

Proof Left as an exercise. ∎

3.6 Theorem *Let $A \neq \phi$.*
Let T_α be a topology on X_α for each $\alpha \in A$.
Let $\mathbb{P}$ be the product topology on $\prod_{\alpha \in A} X_\alpha$.
Let $\mathscr{F}$ be a filterbase on $\prod_{\alpha \in A} X_\alpha$.
Let $x \in \prod_{\alpha \in A} X_\alpha$.
Then,
$\mathscr{F}$ is $\mathbb{P}$-convergent to x
if and only if
$p_\beta(\mathscr{F})$ is T_β-convergent to x_β for each $\beta \in A$.

Proof Left as an exercise. ∎

Before comparing the net and filterbase theories of convergence, we shall make one observation which will be useful in Chapter V. Our theorems have been phrased in terms of filterbases in general. However, a good theory of convergence can be designed around just neighborhood filterbases (see Example 3.3). For later reference, then, we shall define neighborhood and relative neighborhood filterbases, and restate some of the important theorems in terms of them. Their main use will be in considering an extension problem for continuous functions in the next chapter.

3.4 Definition
(1) If $x \in X$, then we let
$$\mathscr{F}_x = \{V | x \in V \in T\}.$$
(2) If $A \subset X$ and $x \in X$, then we let
$$(\mathscr{F}_x)_A = \{A \cap V | x \in V \in T\}.$$

3.7 Theorem *Let $x \in X$.*
Then,
(1) $\mathscr{F}_x$ is a filterbase on X.
(2) $\mathscr{F}_x$ T-converges to x.
(3) If $x \in \bar{A}$, then $(\mathscr{F}_x)_A$ is a filterbase on A.

Proof Left to the reader. ∎

The assumption that $x \in \bar{A}$ in Theorem 3.7 (3) simply insures that $(\mathscr{F}_x)_A$ does not contain the empty set. In this event, we call $(\mathscr{F}_x)_A$ the relative (to A) nghd. filterbase of x. It is really the neighborhood filterbase of x in the relative topology T_A on A, in the case that $x \in A$.

3.8 Theorem *Let (Y, M) be a topological space.*
Let $f: X \to Y$.
Then,

(1) f is (T, M) continuous
 if and only if
 $f(\mathscr{F}_x)$ is M-convergent to $f(x)$ for each $x \in X$.

(2) If $x \in X$, then
 f is (T, M) continuous at x
 if and only if
 $f(\mathscr{F}_x)$ M-converges to $f(x)$.

Proof Left as an exercise. ∎

4.4 Equivalence of net and filterbase convergence

Nets and filterbases are quite different approaches to the same problem. However, the two theories are equivalent in the following sense:

given a filterbase $\mathscr{F}$, there is a net $(S, D, \prec)$ which has exactly the same limit(s) as $\mathscr{F}$;

and, given a net $(S, D, \prec)$, there is a filterbase $\mathscr{F}$ which has exactly the same limit(s) as $(S, D, \prec)$.

The next two theorems show how to construct the appropriate nets and filterbases to show this equivalence. In their statements, T is a topology on X. For a very complete treatment of net and filterbase convergence, there is an account by R. G. Bartle in the American Mathematical Monthly, Oct. 1955, pp. 551–557.

4.1 Theorem *Let $(S, D, \prec)$ be a net to X.*
Let $\mathscr{F} = \{A | A \subset X \text{ and } (S, D, \prec) \text{ is eventually in } A\}$.
Then,

(1) $\mathscr{F}$ is a filterbase on X.
(2) For each $x \in X$,
 $(S, D, \prec)$ is T-convergent to $x \Leftrightarrow \mathscr{F}$ is T-convergent to x.

Proof of (1) Left as an exercise.

Proof of (2) Suppose first that $(S, D, \prec)$ is T-convergent to x. Let V be a T-nghd. of x. Then, $(S, D, \prec)$ is eventually in V, so $V \in \mathscr{F}$. Since $V \subset V$, then $\mathscr{F}$ T-converges to x.

Conversely, suppose that $\mathscr{F}$ T-converges to y. Let $y \in U \in T$. For some a, $a \in \mathscr{F}$ and $a \subset U$. Since $a \in \mathscr{F}$, then $(S, D, \prec)$ is eventually in a. Hence $(S, D, \prec)$ is eventually in U, so $(S, D, \prec)$ T-converges to y. ∎

4.2 THEOREM *Let $\mathscr{F}$ be a filterbase on X.*
Let $D = \{(x, a) | x \in a \in \mathscr{F}\}$.
Define $(x, a) \prec (y, b)$ for $(x, a) \in D$ and $(y, b) \in D$ to mean that $b \subset a$.
Define $S: D \to X$ by specifying that $S((x, a)) = x$ whenever $(x, a) \in D$.
Then,

(1) $(S, D, \prec)$ is a net to X.
(2) For each $x \in X$,
  $\mathscr{F}$ is T-convergent to $x \Leftrightarrow (S, D, \prec)$ is T-convergent to x.

Proof of (1) This is a routine check and is left to the reader.

Proof of (2) Suppose first that $(S, D, \prec)$ T-converges to x. Let V be a T-nghd. of x. Then, $(S, D, \prec)$ is eventually in V. Then, for some $(y, b) \in D$, $S((t, c)) \in V$ whenever $(t, c) \in D$ and $(y, b) \prec (t, c)$. We claim that $b \subset V$. For suppose that $t \in b$. Then, $(y, b) \prec (t, b)$, so $S((t, b)) = t \in V$. But also $b \in \mathscr{F}$, as $(y, b) \in D$. Hence $\mathscr{F}$ T-converges to x.

Conversely, suppose that $\mathscr{F}$ T-converges to z. Let U be a T-nghd. of z. Then, for some a, $a \in \mathscr{F}$ and $a \subset U$. Let $t \in a$. Then, $(t, a) \in D$.

Suppose now that $(x, g) \in D$ and $(t, a) \prec (x, g)$. Then, $x \in g \subset a \subset U$, so $x \in U$. But then $S((x, g)) = x \in U$, so $(S, D, \prec)$ is eventually in U, hence T-converges to z. ∎

Care must be taken in interpreting the practical sense in which net and filterbase convergence are equivalent. In specific applications, one might be more convenient than the other, depending upon the wording and setting of the problem, and, to some extent, personal tase. Thus it is important for the modern mathematician to be conversant with both theories, as well as the theory of filters developed in the exercises.

4.5 Ultrafilterbases

Nets and filterbases are each sufficient for a good theory of convergence. However, ultrafilterbases provide a delicate instrument for testing certain topological properties. We shall use them in proving the Tychonov and

Ascoli Theorems (Chapters VII and VIII) and the reader may safely skip this section until then without losing any continuity in the subject matter.

5.1 DEFINITION *Let $\mathscr{F}$ and $\mathscr{V}$ be filterbases on X.
Then,
$\mathscr{F} \prec \mathscr{V}$
if and only if
if $a \in \mathscr{F}$, then there is some $b \in \mathscr{V}$ with $b \subset a$.*

Note that $\prec$ is not a partial order on the set of all filterbases on X. In particular, if $\mathscr{F} \prec \mathscr{V}$ and $\mathscr{V} \prec \mathscr{F}$, we cannot conclude that $\mathscr{F} = \mathscr{V}$. However, $\prec$ is a preorder on the class of filterbases on X.

5.1 THEOREM *$\prec$ is a preorder on the class of filterbases on X.*

Proof Left as an exercise. ∎

5.2 DEFINITION *$\mathscr{F}$ is an ultrafilterbase on X*
(1) $\mathscr{F}$ is a filterbase on X.
(2) If $\mathscr{T}$ is a filterbase on X and $\mathscr{F} \prec \mathscr{T}$, then also $\mathscr{T} \prec \mathscr{F}$.

5.2 THEOREM *Let $\mathscr{F}$ be a filterbase on X.
Then,
for some $\mathscr{U}$, $\mathscr{U}$ is an ultrafilterbase on X and $\mathscr{F} \prec \mathscr{U}$.*

Proof Let $P = \{\mathscr{T} | \mathscr{T}$ is a filterbase on X and $\mathscr{F} \prec \mathscr{T}\}$. Then, $P \neq \phi$, as $\mathscr{F} \in P$. We shall apply Zorn's Lemma to produce a maximal element $\mathscr{U}$ of P.

Suppose that C is a chain in P. Let $H = \cup C$. We claim:

(1) H is a filterbase on X.

(2) $\mathscr{F} \prec H$.

(3) If $A \in C$, then $A \prec H$.

The details are routine and are left to the reader. Since each chain in P has an upper bound in P, then P has a maximal element, say $\mathscr{U}$, by Zorn's Lemma. Then, $\mathscr{U} \in P$, so $\mathscr{U}$ is a filterbase on X and $\mathscr{F} \prec \mathscr{U}$. Finally, suppose $\mathscr{T}$ is a filterbase on X and $\mathscr{U} \prec \mathscr{T}$. Then, $\mathscr{F} \prec \mathscr{T}$, as $\mathscr{F} \prec \mathscr{U}$. But then $\mathscr{T} \in P$, so $\mathscr{T} \prec \mathscr{U}$. ∎

The following three theorems are all we shall need on ultrafilterbases.

5.3 THEOREM *Let $\mathscr{M}$ be a filterbase on X.*

Then,
$\mathcal{M}$ *is an ultrafilterbase on* X
if and only if
if $A \subset X$, *then, for some* $m \in \mathcal{M}$, *either* $m \subset A$ *or* $m \subset X - A$.

Proof Suppose first that $\mathcal{M}$ is an ultrafilterbase on X. Suppose, if $\alpha \in \mathcal{M}$, then $\alpha \subset A$ is false. Then, $\alpha \cap (X - A) \neq \phi$ for each $\alpha \in \mathcal{M}$. It is easy to check that $\{\alpha \cap (X - A) | \alpha \in \mathcal{M}\}$ is a filterbase on X, and that $\mathcal{M} \prec \{\alpha \cap (X - A) | \alpha \in \mathcal{M}\}$.

But then $\{\alpha \cap (X - A) | \alpha \in \mathcal{M}\} \prec \mathcal{M}$. Let $\beta \in \mathcal{M}$. Then, for some m, $m \in \mathcal{M}$ and $m \subset \beta \cap (X - A)$, hence $m \subset X - A$.

Conversely, suppose, if $A \subset X$, then, for some m, $m \in \mathcal{M}$ and $m \subset A$ or $m \subset X - A$.

Suppose $\mathcal{F}$ is a filterbase on X and $\mathcal{M} \prec \mathcal{F}$. Let $\alpha \in \mathcal{F}$. Then, $\alpha \subset X$, so, for some $m \in \mathcal{M}$, either $m \subset \alpha$ or $m \subset X - \alpha$. If $m \subset X - \alpha$, then, since $\mathcal{M} \prec \mathcal{F}$, there is some $f \in \mathcal{F}$ such that $f \subset m \subset X - \alpha$. But then $f \cap \alpha = \phi$, which is impossible, as $f, \alpha \in \mathcal{F}$. Hence $m \subset \alpha$, implying that $\mathcal{F} \prec \mathcal{M}$. ∎

5.4 Theorem *Let* $\mathcal{M}$ *be an ultrafilterbase on* X, *and let* $x \in X$. *Then,*
$\mathcal{M}$ *is* T-*convergent to* x
if and only if
$x \in \bigcap_{m \in \mathcal{M}} \bar{m}$.

Proof Suppose first that $x \in \bigcap_{m \in \mathcal{M}} \bar{m}$.

Let $x \in V \in T$. By Theorem 5.3, for some $m \in \mathcal{M}$, either $m \subset V$ or $m \subset X - V$. If $m \subset X - V$, then $m \cap V = \phi$, so $x \notin \bigcap_{\alpha \in \mathcal{M}} \bar{\alpha}$, a contradiction. Hence $m \subset V$, implying that $\mathcal{M}$ T-converges to x.

Conversely, suppose that $\mathcal{M}$ T-converges to x. Let $m \in \mathcal{M}$, and let V be a T-nghd. of x. For some a, $a \in \mathcal{M}$ and $a \subset V$, as $\mathcal{M}$ T-converges to x. Now, $\phi \neq a \cap m \subset V \cap m$, so $V \cap m \neq \phi$. Then, $x \in \bar{m}$, so $x \in \bigcap_{m \in \mathcal{M}} \bar{m}$. ∎

5.5 Theorem *Let* $f: X \to Y$.
Let $\mathcal{M}$ *be an ultrafilterbase on* X.
Then, $f(\mathcal{M})$ *is an ultrafilterbase on* Y.

Proof Left to the reader. ∎

Problems

1) Let (X, ϱ) be a metric space.
   Let $f: X \to X$. Suppose $0 < k < 1$ such that $\varrho(f(x), f(y)) \leq k\varrho(x, y)$ for each $x, y \in X$. Then,
   (a) f is (T_ϱ, T_ϱ) continuous.
   (b) f has a fixed point (i.e., there is some $x \in X$ such that $f(x) = x$). Hint: Let $x \in X$. Show that the sequence denoted by x, $f(x)$, $f(f(x))$, ..., is T_ϱ-convergent in X).
   (c) f has exactly one fixed point.
   Such a function f is called a contracting map.

2) Let R be the set of real numbers, and T the cofinal topology on R. Let $x \in R$.
   Let D_x be the set of all T-neighborhoods of x, R itself excluded.
   (a) If $A, B \in D_x$, define $A \prec B$ if $A \subset B$. Then, $\prec$ is a partial ordering on D_x.
   (b) $(D_x, \prec)$ is a directed set.
   (c) Define $f: D_x \to R$ by putting $f(A) = \inf(R - A)$ if $A \in D_x$. Then, $(f, D_x, \prec)$ is a net to R.
   (d) Investigate the convergence of $(f, D_x, \prec)$.

3) Let $(S, D, \prec)$ be a net to X, T a topology on X, $x \in X$ and W a subbase for T. Then, $(S, D, \prec)$ T-converges to x if and only if $(S, D, \prec)$ is eventually in each subbasic neighborhood of x.

4) Let the real-valued function f be bounded (in the usual calculus sense) on $[a, b]$. Let $\mathscr{P}$ be the set of all partitions of $[a, b]$. (A partition of $[a, b]$ is a set $\{t_0, t_1, ..., t_n\}$, with $n \geq 1$, $a = t_0 < t_1 < \cdots < t_n = b$.)
   If $\{t_0, ..., t_n\} \in \mathscr{P}$ and $\{q_0, ..., q_m\} \in \mathscr{P}$, write $\{t_0, ..., t_n\} \prec \{q_0, ..., q_m\}$ if $n \leq m$ and, for each i, $1 \leq i \leq n$, there is some j, $1 \leq j \leq m$, such that $t_i = q_j$.
   (a) $\prec$ is a direction on $\mathscr{P}$.
   (b) If $\{t_0, ..., t_n\} \in \mathscr{P}$, define
   $$L(t_0, ..., t_n) = \sum_{j=1}^{n} (\inf\{f(x) | t_{j-1} \leq x \leq t_j\})(t_j - t_{j-1}),$$
   and
   $$U(t_0, ..., t_n) = \sum_{j=1}^{n} (\sup\{f(x) | t_{j-1} \leq x \leq t_j\})(t_j - t_{j-1}).$$
   Then, $(L, \mathscr{P}, \prec)$ and $(U, \mathscr{P}, \prec)$ are nets to E^1.

*(c) $(L, \mathscr{P}, \prec)$ converges in E^1 (its limit is the lower Riemann integral $\underline{\int_a^b} f$ of f on $[a, b]$), and $(U, \mathscr{P}, \prec)$ converges in E^1 (its limit is the upper Riemann integral $\overline{\int_a^b} f$ of f on $[a, b]$).

*(d) If f is continuous, then $\underline{\int_a^b} f = \overline{\int_a^b} f = \int_a^b f$.

5) There is a theory of subnets analogous to that of subsequences.
A net $(S^*, D^*, \prec^*)$ is a subnet of the net $(S, D, \prec)$ if there is an $f: D^* \to D$ such that 1) $S^* = S \circ f$, and 2) If $\alpha \in D$, there is some $\alpha^* \in D^*$ such that $\alpha \prec f(\beta^*)$ whenever $\alpha^* \prec^* \beta^* \in D^*$.
Now let (X, T) be any space.
  (a) If $(S, D, \prec)$ is a net to X, and $(S, D, \prec)$ T-converges to x, then so does each subnet of $(S, D, \prec)$.
  (b) However, a subnet of $(S, D, \prec)$ may T-converge to x, and $(S, D, \prec)$ not T-converge to x. Show this by an example.
  (c) Define: x is a T-limit point of the net $(S, D, \prec)$ if $(S, D, \prec)$ is frequently in each T-neighborhood of x, i.e., given a T-nghd. V of x, and any $\alpha \in D$, there is some $\beta \in D$ such that $\alpha \prec \beta$ and $S(\beta) \in V$.
  Show that a net need not converge to a limit point of the net (so that limit points and limits are distinct concepts). However, x is a T-limit point of $(S, D, \prec)$ if and only if there is a subnet of $(S, D, \prec)$ which T-converges to x.

6) A net $(S, D, \prec)$ to X is a universal net if, given $A \subset X$, then $(S, D, \prec)$ is eventually in A or eventually in $X - A$.
  (a) Give examples of nets which are universal and nets which are not universal.
  (b) Each subnet of a universal net is a universal net.
  (c) A universal net is frequently in A if and only if it is eventually in A.
  (d) A universal net converges to each of its limit points.

7) Let $f: X \to Y$ be (T, M) continuous at x. Let $y \in Y$, and suppose that, if V is a T-nghd. of x, then there is some $\xi_V \in V$ such that $f(\xi_V) = y$. Suppose M satisfies Definition 1.1 (1). V.

Then, $f(x) = y$. (That is, a continuous function which takes on a given value in each neighborhood of a point must take on that value at the point also).

8) Let R denote the set of real numbers. Let
$$f_{2n} = \left\{(x, y) \big| x^2 + y^2 < \frac{1}{4n^2}\right\} \cup \left(\left]1, 1 + \frac{1}{2n}\right] \times \{0\}\right) \subset R \times R$$
and
$$f_{2n+1} = \left(\{0\} \times \left]-\frac{1}{4n}, \frac{1}{4n}\right[\right) \subset R \times R \text{ for each } n \in Z^+.$$
Let $\mathscr{F} = \{f_{2n} | n \in Z^+\} \cup \{f_{2n+1} | n \in Z^+\}$.
(a) $\mathscr{F}$ is a filterbase on $R \times R$.
Examine convergence of $\mathscr{F}$ in each of the following topologies.
(b) Euclidean topology on $R \times R$.
(c) Product topology on $E^1 \times R$, with E^1 the Euclidean space and the discrete topology on R.
(d) Topology on $R \times R$ generated by the sets $R \times]-\varepsilon, \varepsilon[$, where $\varepsilon > 0$.

9) Cartan's theory of filters, which preceded the theory of filterbases, may be outlined as follows.
$\mathscr{F}$ is a filter on X if $\mathscr{F}$ is a non-empty set of non-empty subsets of X, and further: 1) $A, B \in \mathscr{F}$ implies $A \cap B \in \mathscr{F}$, and 2) $A \in \mathscr{F}$ and $A \subset B \subset X$ implies $B \in \mathscr{F}$.

(a) Give examples of filterbases which are not filters. Show, however, that each filter is a filterbase.
(b) If $\mathscr{T}$ is a filterbase on X, let
$$\mathscr{F} = \{A | A \subset X \text{ and, for some } G \in \mathscr{T}, G \subset A\}.$$
Then, $\mathscr{F}$ is a filter on X ($\mathscr{F}$ is the filter generated by $\mathscr{T}$).
(c) Let T be a topology on X. A filter $\mathscr{F}$ on X T-converges to x if each T-nghd. of x is in $\mathscr{F}$.
Show that a filterbase $\mathscr{T}$ T-converges to x if and only if the filter $\mathscr{F}$ generated by $\mathscr{T}$ T-converges to x.
(d) Show that filter and filterbase convergence are equivalent (in the sense of section 4 of this chapter).

(e) Let $x \in X$ and $A \subset X$. Then, show directly:
   i) $x \in \bar{A}$ if and only if there is a filter on A T-converging to x.
   ii) $x \in \text{Der}_T(A)$ if and only if there is a filter on $A - \{x\}$ which T-converges to x.

(f) Let $\mathscr{F} = \{A | X - A \text{ is finite}\} \cap \mathscr{P}(X)$. Then, $\mathscr{F}$ is a filter on X (the Fréchet filter). A filter $\mathscr{U}$ is called free if $\cap \mathscr{U} = \phi$. Show that the Fréchet filter is free if X is infinite. Hence, X is infinite if and only if X has a free filter.

(g) If $\mathscr{F}$ and $\mathscr{T}$ are filters on X, define $\mathscr{F} < \mathscr{T}$ to mean: given $f \in \mathscr{F}$, there is some $g \in \mathscr{T}$ with $g \subset f$. This is a partial ordering on the set of filters on X. Maximal elements with respect to this partial ordering are called ultrafilters.
   i) If $\mathscr{F}$ is a filter on X, then there is an ultrafilter $\mathscr{T}$ on X such that $\mathscr{F} < \mathscr{T}$.
   ii) If $\mathscr{T}$ is an ultrafilter on X, and $A, B \subset X$, and $A \cup B \in \mathscr{T}$, then $A \in \mathscr{T}$ or $B \in \mathscr{T}$.
   iii) An ultrafilter $\mathscr{T}$ on X T-converges to x if and only if $x \in \bigcap_{g \in \mathscr{T}} \bar{g}$.

10) Prove that, in the ordinal space $[0, \Omega]$, Ω is a cluster point of $[0, \Omega[$, but that there is no sequence to $[0, \Omega[$ converging to Ω.

CHAPTER V

Separation axioms

5.1 Hausdorff spaces

UP TO THIS POINT we have considered unrestricted topologies. One price paid for complete generality is that neither sequences, nets nor filterbases have uniqueness of limits. The difficulty stems from the fact that a topology may not differentiate points. It may happen that every nghd. of x is also a nghd. of y. In this event, it should not be surprising, in view of the definitions of convergence, that any net or filterbase (or sequence) converging to x will also converge to y. This is why, in mentioning the uniqueness difficulty in the last chapter, we placed the blame on the space, and not on the theory of convergence.

The key to the solution lies in the word "separation". If the topology can separate points, by placing disjoint neighborhoods about distinct points, then we can expect limits to be unique. Such a space is called Hausdorff, or, for historical reasons, a T_2 space.

For the remainder of this chapter, T is a topology on X

1.1 DEFINITION
(1) *T is a Hausdorff (or T_2) topology*
 if and only if
 any two distinct points of X have disjoint T-neighborhoods
(2) *(X, T) is a Hausdorff (or T_2) space*
 if and only if
 T is a Hausdorff topology.
 As usual, when T is clear, we say that X is a Hausdorff space.

1.1 EXAMPLE   If X has more than one point, then the indiscrete topology on X is not Hausdorff. The discrete topology on any set X is Hausdorff. ■

1.2 EXAMPLE   Sierpinski space is not Hausdorff. ■

1.3 EXAMPLE   The cofinal topology on X is Hausdorff exactly when X is finite. ■

1.4 EXAMPLE Let X be the set of real numbers and let T consist of all subsets of X whose complements are countable, together with ϕ. Then, T is a topology on X, and is not Hausdorff. For suppose $x \in A \in T$ and $y \in B \in T$. Note that $(X - A) \cup (X - B) = X - (A \cap B)$. If $A \cap B = \phi$, then $(X - A) \cup (X - B) = X$, impossible if $X - A$ and $X - B$ are both countable.

While T is not Hausdorff, T does have an interesting property which we shall discuss later (Example 1.6). ∎

1.5 EXAMPLE E^n, and, in general, any metric space, is Hausdorff. ∎

1.1 THEOREM *Let ϱ be a metric on X.*
Then,
T_ϱ is Hausdorff

Proof Let x and y be distinct points of X. Then, $\varrho(x, y) > 0$. Let $\varepsilon = \tfrac{1}{2}\varrho(x, y)$. Then, $B_\varrho(x, \varepsilon)$ and $B_\varrho(y, \varepsilon)$ are disjoint T_ϱ-neighborhoods of x and y respectively. ∎

The next theorem lists alternate conditions for a space to be Hausdorff. Note that (1) ⇔ (5) ⇔ (6) gives us not only uniqueness of net and filterbase limits in Hausdorff spaces, but the converse as well: nets and filterbases have guaranteed unique limits only in Hausdorff spaces.

1.2 THEOREM *The following are equivalent:*

(1) T is Hausdorff.
(2) If x and y are distinct points of X, then there is a T-nghd.
  U of x such that $y \notin \overline{U}$.
(3) If $x \in X$, then $\cap \{\overline{U} | x \in U \in T\} = \{x\}$.
(4) $\{(x, x) | x \in X\}$ is closed in the product space $X \times X$.
(5) If $(S, D, \prec)$ is a net to X which T-converges to x and to y, then $x = y$.
(6) If $\mathscr{F}$ is a filterbase on X which T-converges to x and to y, then $x = y$.

Proof For convenience, let $\Delta = \{(x, x) | x \in X\}$.
(1) ⇒ (2):
Assume (1). Let x and y be distinct points of X. Then there are disjoint T-nghds. U of x and V of y. Since $y \in V \in T$ and $U \cap V = \phi$, then $y \notin \overline{U}$.
(2) ⇒ (3):
Assume (2). Immediately, $x \in \cap\{\overline{U} | x \in U \in T\}$.
If $y \in X$ and $y \neq x$, produce by (2) a T-nghd. V of x with $y \notin \overline{V}$. Then, $y \notin \cap\{\overline{U} | x \in U \in T\}$, hence $\cap\{\overline{U} | x \in U \in T\} = \{x\}$.

(3) ⇒ (4):

Assume (3). Let $(a, b) \in (X \times X) - \Delta$. Then, $b \neq a$, so $a \notin \cap\{\bar{U}|b \in U \in T\}$. Then, for some V, $b \in V \in T$ and $a \notin \bar{V}$. Now note that $(a, b) \in (X - \bar{V}) \times V$ and that $(X - \bar{V}) \times V$ is open in $X \times X$. Further, $((X - \bar{V}) \times V) \cap \Delta = \phi$, as $V \subset \bar{V}$. Hence, $(a, b) \in \text{Int}((X \times X) - \Delta)$, implying that Δ is closed in $X \times X$.

(4) ⇒ (5):

Assume (4). Let $(S, D, \prec)$ be a net to X which T-converges to x and to y. If $x \neq y$, then $(x, y) \notin \Delta$. By (4), there are T-open nghds. A of x and B of y with $(x, y) \in A \times B \subset (X \times X) - \Delta$.

Now, $(S, D, \prec)$ is eventually in A and also in B, as $(S, D, \prec)$ converges to x and to y. Hence produce $K \in D$ such that $S(K) \in A \cap B$. But then $(S(K), S(K)) \in (A \times B) \cap \Delta$, a contradiction.

(5) ⇒ (6):

Assume (5). Let $\mathscr{F}$ be a filterbase on X which T-converges to x and to y. By Theorem 4.2.IV, there is a net $(S, D, \prec)$ to X which T-converges to x and to y. Then, by (5), $x = y$.

(6) ⇒ (1):

Assume (6). Let x and y be distinct points of X. Let $\mathscr{F} = \{A \cap B | x \in A \in T \text{ and } y \in B \in T\}$.

If $A \cap B \neq \phi$ whenever $x \in A \in T$ and $y \in B \in T$, then it is easy to check that $\mathscr{F}$ is a filterbase on X which T-converges to x and to y. This is impossible, by (6). Hence, $A \cap B = \phi$ for some T-nghds. A of x and B of y. ∎

Since every sequence is a net, then limits of sequences are also unique in Hausdorff spaces. However, sequences cannot replace nets in the above theorem; a non-Hausdorff space *may* have unique sequential limits. Thus sequential convergence fails to test a space for the Hausdorff separation axiom.

1.6 Example Let (X, T) be the space of Example 1.4. As we saw, T is not Hausdorff. But limits of sequences to X are unique. For suppose $f: Z^+ \to X$, and f converges to x and to y Suppose also that $x \neq y$.

Now, $X - \{y\}$ is a T-nghd. of x, as $X - (X - \{y\}) = \{y\}$ is countable. Then, for some K, $f(n) \in X - \{y\}$ whenever $n \geq K$.

Let $A = \{f(n) | n \in Z^+ \text{ and } n \geq K\}$. Then, $y \in X - A$. Further, $X - (X - A) = A$ is countable, so $X - A$ is a T-nghd. of y. Since f converges to y, there is some $M > 0$ such that $f(n) \in X - A$ whenever $n \geq M$. We now have a contradiction. If $n \in Z^+$ and $n \geq M$ and $n \geq K$, then $f(n) \in A$ and $f(n) \in X - A$, which is impossible. ∎

Hausdorff spaces reflect a number of useful properties of Euclidean space which are not true for arbitrary topological spaces.

First, any finite subset of a Hausdorff space is closed.

Secondly, in a Hausdorff space, each neighborhood of a cluster point of a set actually has infinitely many points of the set in it. This is easy to see geometrically, say, in E^2, by drawing successively smaller circular neighborhoods about the cluster point.

1.3 Theorem *Let T be Hausdorff*
Let A be a finite subset of X.
Then,
A is T-closed.

Proof Let $x \in X$. We shall show that $\{x\}$ is T-closed. If $y \in X - \{x\}$, then there is a T-nghd. V of y with $x \notin V$. Then, $V \subset X - \{x\}$, implying that $X - \{x\}$ is T-open.

If now $A = \{x_1, ..., x_n\} \subset X$, then $X - A = X - \bigcup_{i=1}^{n} \{x_i\} = \bigcap_{i=1}^{n} (X - \{x_i\})$ is T-open, since each $X - \{x_i\}$ is T-open by the above argument. ∎

1.4 Theorem *Let T be Hausdorff.*
Let $A \subset X$ and let x be a T-cluster point of A.
Let V be a T-nghd. of x.
Then,
$V \cap A$ is infinite.

Proof Suppose that $V \cap A$ is finite. Then, $V \cap (A - \{x\})$ is T-closed, by Theorem 1.3.

By Theorem 2.5 (2).I, $V - (V \cap (A - \{x\}))$ is T-open. But, $x \in V - (V \cap (A - \{x\})) = V - (A - \{x\})$. Then, $V - (A - \{x\})$ is a T-nghd. of x. Since x is a T-cluster point of A, then $(((V - (A - \{x\})) - \{x\}) \cap A \neq \phi$, which is absurd. ∎

Of course, Theorem 1.4 does not mean that every Hausdorff space is infinite. It does mean that, in a finite Hausdorff space, no subset has a cluster point. In this event, $A = \bar{A}$ for every $A \subset X$, so every subset is closed, hence also open, and the topology is discrete This characterizes the finite Hausdorff spaces.

The next three theorems deal with the degree to which the property of being Hausdorff is contagious. In brief:

homeomorphic images of Hausdorff spaces are Hausdorff;
subspaces of Hausdorff spaces are Hausdorff;

and Cartesian products are Hausdorff exactly when each coordinate space is Hausdorff. Note how the proof of this makes use of the slice technique mentioned in connection with Theorem 2.6 in Chapter III.

1.5 THEOREM *Let $(X, T) \cong (Y, M)$.*
Then,
T is Hausdorff $\Leftrightarrow$ M is Hausdorff.

Proof Left as an exercise. ∎

1.6 THEOREM *Let $A \subset X$.*
Let T be Hausdorff.
Then,
T_A is Hausdorff.

Proof Left as an exercise. ∎

1.7 THEOREM *Let $A \neq \phi$.*
Let T_α be a topology on X_α for each $\alpha \in A$.
Let $\mathbb{P}$ be the product topology on $\prod_{\alpha \in A} X_\alpha$.
Then,
$\mathbb{P}$ is Hausdorff
if and only if
T_α is Hausdorff for each $\alpha \in A$.

Proof Suppose, for each $\alpha \in A$, T_α is a Hausdorff topology. Let f and g be distinct points of $\prod_{\alpha \in A} X_\alpha$. For some $\beta \in A$, $f(\beta) \neq g(\beta)$. Now, there are disjoint T_β-nghds. U and V of $f(\beta)$ and $g(\beta)$ respectively. Then, $p_\beta^{-1}(U)$ and $p_\beta^{-1}(V)$ are disjoint $\mathbb{P}$-nghds. of f and g, so $\mathbb{P}$ is Hausdorff.

Conversely, suppose that $\mathbb{P}$ is a Hausdorff topology, and let $\beta \in A$. Let $f \in \prod_{\alpha \in A} X_\alpha$. Let $Y = \left(\prod_{\alpha \in A} X_\alpha\right) \cap \{g | g(\alpha) = f(\alpha) \text{ if } \alpha \in A \text{ and } \alpha \neq \beta\}$. By Theorem 1.6, $\mathbb{P}_Y$ is Hausdorff. By Theorem 2.6.III, $(Y, \mathbb{P}_Y) \cong (X_\beta, T_\beta)$. By Theorem 1.5, T_β is a Hausdorff topology. ∎

We conclude this section with two theorems for later reference.

1.8 THEOREM *Let (Y, M) be a topological space.*
Let f and $g: X \to Y$ be (T, M) continuous.
Suppose that M is Hausdorff.
Then,
(1) $\{x | f(x) = g(x)\}$ is T-closed.
(2) $\{(x, f(x)) | x \in X\}$ is closed in $X \times Y$.

Proof of (1) Define a map $\varphi\colon X \to X \times Y$ by $\varphi(x) = (f(x), g(x))$ for each $x \in X$. To show that φ is continuous, let $p_1\colon X \times Y \to X$ and $p_2\colon X \times Y \to Y$ be the projections, and observe that $p_1 \circ \varphi = f$ and $p_2 \circ \varphi = g$ are continuous. Hence, φ is continuous by Theorem 2.7.III.

Now, M is Hausdorff, so $\{(y, y) | y \in Y\}$ is closed in $Y \times Y$ by Theorem 1.2. Then, by Theorem 1.1.II, $\varphi^{-1}(\{(y, y) | y \in Y\})$ is closed in X.

But, $\varphi^{-1}(\{(y, y) | y \in Y\}) = \{x | f(x) = g(x)\}$.

Proof of (2) Define $\xi\colon X \times Y \to Y \times Y$ by letting $\xi(x, y) = (f(x), y)$ for each $(x, y) \in X \times Y$. As in the proof of (1), show that ξ is continuous, hence $\xi^{-1}(\{(y, y) | y \in Y\}) = \{(x, f(x)) | x \in X\}$ is closed in $X \times Y$. ∎

1.9 Theorem *Let $f, g\colon X \to E^1$ be continuous. Then,*
$\{x | f(x) < g(x)\}$ *is T-open.*

Proof Let $h\colon X \to E^2$ be defined by $h(x) = (f(x), g(x))$ for each $x \in X$. Use Theorem 2.7.III, to show that h is continuous.

Since $\{(a, b) | a < b\}$ is open in E^2, then $h^{-1}(\{(a, b) | a < b\}) = \{x | f(x) < g(x)\}$ is open in X. ∎

Actually, this theorem does not require any separation hypotheses, but it is similar in form to Theorem 1.8, and fits as well here as anywhere.

5.2 Regular spaces

The Hausdorff separation axiom was introduced in the first section to insure uniqueness of limits (and, in fact, is equivalent to it). Certain problems, however, require stronger separation axioms if one is to have any hope of success. The problem of existence of continuous extensions is typical of this.

Suppose A is a subspace of X, and f is continuous on A to Y. Can f be extended continuously to some $g\colon X \to Y$? This is among the most difficult (and interesting) problems in topology.

In general, the answer is no, even when the spaces are Hausdorff. The reader can easily supply examples to show this, e.g., $X = E^1, A = \{x | x < 0\}$ and $f(x) = 1/x$ for $x \in A$. However, with a slightly stronger separation axiom on Y, reasonable conditions assuring a positive solution are possible. This result is Theorem 2.6, the final theorem of the section, and may be thought of as the motivation for considering the regularity, or T_3, separation axiom.

For the remainder of this section, T is a topology on X.

2.1 DEFINITION *T is a regular (or T_3) topology if and only if*
(1) T is Hausdorff.
(2) If A is T-closed and $x \in X - A$, then there are disjoint T-open sets U and V with $A \subset U$ and $x \in V$.

2.2 DEFINITION *(X, T) is a regular (or T_3) space if and only if T is a regular topology on X.*

In Definition 2.1 (2), such a set U is called a T-nghd of A.

A regular space is distinguished by its ability to separate closed sets from points This is stronger than the T_2 separation axiom, and on the face of it seems to imply the T_2 axiom by choosing $A = \{y\}$, where $y \neq x$. This reasoning is correct if singletons are always closed in X, but this need not be the case. However, in spaces where $\{y\}$ is closed for each point y, then of course only (2) is needed in the definition of regularity. The reasons for including (1) in the definition are to insure uniqueness of limits, and also to form a logical hierarchy as we proceed through successively stronger separation axioms. In the next sections we shall introduce normal and completely regular spaces, and it is useful and pleasing to have a descending chain: normal $\Rightarrow$ completely regular $\Rightarrow$ regular $\Rightarrow$ Hausdorff.

2.1 EXAMPLE E^n, and, more generally, any metric space is regular. ∎

2.2 EXAMPLE Discrete spaces are regular. ∎

2.3 EXAMPLE A Hausdorff space need not be regular Let R be the set of real numbers, T the usual Euclidean topology on R, Q the set of rationals, and T' the topology on R generated by Q and the sets in T. Then, $T \subset T'$, so (R, T') is a Hausdorff space. But, (R, T) is not regular.

Note that the T'-open sets are $Q \cup (Q$ intersect T-open sets) and (T-open sets) $\cup$ (Q intersect T-open sets). Now, Q is T'-open, so $R - Q$ is T'-closed. Choose any rational number, say 0. Then, $0 \notin R - Q$. If (R, T') were regular, there would be disjoint T'-open U and V with $R - Q \subset U$ and $0 \in V$. Now, for some T-open A and B, $U = A \cup (Q \cap B)$. And, since each T-open set intersects $R - Q$, then for some T-open C, $V = Q \cap C$.

Let $\varepsilon > 0$ such that $Q \cap \,]-\varepsilon, \varepsilon[\,\subset Q \cap C$. Choose an irrational number $\xi \in \,]-\varepsilon, \varepsilon[$. Then, $\xi \in R - Q \subset A \in T$. Since A is T-open, there is some $\delta > 0$ with $]\xi - \delta, \xi + \delta[\,\subset\,]-\varepsilon, \varepsilon[\,\cap A$. Choose a rational number r in $]\xi - \delta, \xi + \delta[$. Then, $r \in A \subset U$, and also $r \in Q \cap \,]-\varepsilon, \varepsilon[\,\subset V$. Then, $U \cap V \neq \phi$, a contradiction. ∎

We precede the proof that metric spaces are regular with a Lemma which will be of use later as well.

1 Lemma *Let ϱ be a metric on X.*
Let F be T_ϱ-closed and $x \in X - F$.
Then,
$\inf \{\varrho(x, y) | y \in F\} > 0$.

Proof Suppose that $\inf \{\varrho(x, y) | y \in F\} = 0$. Given $\varepsilon > 0$, there is some $y_\varepsilon \in F$ with $\varrho(x, y_\varepsilon) < \varepsilon$. Then, $B_\varrho(x, \varepsilon) \cap F \neq \phi$, implying that $x \in \bar{F} = F$, a contradiction. ∎

Often $\inf \{\varrho(x, y) | y \in F\}$ is called the distance from x to F, denoted $d(x, F)$. It is easy to show, but not necessary for our purposes, that $x \in \bar{A} \Leftrightarrow d(x, A) = 0$, where A is any non-empty subset of X.

2. Theorem *Let ϱ be a metric on X.*
Then,
T_ϱ is regular.

Proof Note first that T_ϱ is Hausdorff by Theorem 1.1.

Now let F be T_ϱ-closed and $x \in X - F$. Let $\varepsilon = \inf \{\varrho(x, y) | y \in F\}$. By Lemma 1, $\varepsilon > 0$. To complete the proof, check that:

$$F \subset \bigcup_{y \in F} B_\varrho(y, \varepsilon/4) \in T_\varrho;$$

$$x \in B_\varrho(x, \varepsilon/4) \in T_\varrho;$$

and

$$B_\varrho(x, \varepsilon/4) \cap \left(\bigcup_{y \in F} B_\varrho(y, \varepsilon/4) \right) = \phi. \blacksquare$$

There are several alternate formulations of the notion of regularity. In particular, Theorem 2.2 (2) is important in the proof of the extension theorem (2.6).

2.2 Theorem *Let T be a Hausdorff topology.*
Then, the following are equivalent:
(1) T is regular.
(2) If $x \in X$ and U is a T-nghd. of x, then $\bar{V} \subset U$ for some T-nghd. V of x.
(3) If F is T-closed and $x \in X - F$, then there is a T-nghd. V of x such that $\bar{V} \cap F = \phi$.

Proof (1) ⇒ (2):

Assume (1). Let $x \in X$ and let U be a T-nghd. of x. Since $X - U$ is T-closed, and $x \notin X - U$, there are disjoint, T-open sets V and B such that $x \in V$ and $X - U \subset B$. We claim that $\bar{V} \subset U$.

Note that $V \subset X - B$, as $B \cap V = \phi$. Since $X - B$ is T-closed, then $\bar{V} \subset \overline{X - B} = X - B$. But also $X - U \subset B$, so $X - B \subset X - (X - U) = U$. Hence, $\bar{V} \subset U$.

(2) ⇒ (3):

Assume (2).

Let F be T-closed and $x \in X - F$. Since $X - F$ is a T-nghd. of x, then by (2), there is some T-nghd. V of x with $\bar{V} \subset X - F$. Then, $\bar{V} \cap F = \phi$.

(3) ⇒ (1):

Assume (3).

Let A be T-closed and $x \in X - A$. There is, by (3), a T-nghd. V of x such that $\bar{V} \cap A = \phi$. Then, $A \subset X - \bar{V} \in T$. Further, $V \cap (X - \bar{V}) = \phi$, so T is regular. ∎

The next theorems carry out for regular spaces the program followed for Hausdorff spaces in Theorems 1.5, 1.6 and 1.7.

2.3 Theorem *Let $(X, T) \cong (Y, M)$.*
Then,
T is regular $\Leftrightarrow$ M is regular.
  Proof Left as an exercise. ∎

2.4 Theorem *Let $A \subset X$.*
Let T be regular.
Then,
T_A is regular.
  Proof Left to the reader. ∎

2.5 Theorem *Let $A \neq \phi$.*
Let T_α be a topology on X_α for each $\alpha \in A$.
Let $\mathbb{P}$ be the product topology on $\prod_{\alpha \in A} X_\alpha$.
Then,
$\mathbb{P}$ is regular
if and only if
T_α is regular for each $\alpha \in A$.

  Proof Left as an exercise. ∎

As promised in the introductory remarks, we conclude this section with one solution to the extension problem. Note the role played by the Theorem 2.2 (2) version of the T_3 separation axiom, and also the use of neighborhood filterbases, in the proof.

2.6 Theorem *Let M be a regular topology on Y.*
Let $A \subset X$ and $\bar{A} = X$.
Let $f: A \to Y$ be (T_A, M) continuous.
Then, the following two statements are equivalent:
(1) There is a (T, M) continuous $g: X \to Y$ such that $g|A = f$.
(2) If $x \in X$, then $f((\mathscr{F}_x)_A)$ is M-convergent.

Proof (1) $\Rightarrow$ (2):
Assume (1), and let $x \in X$.

Let $g: X \to Y$ be (T, M) continuous such that $g|A = f$. Note that, since $x \in \bar{A}$, then $(\mathscr{F}_x)_A$ is a filterbase on A by Theorem 3.7 (3).IV. We shall show that $f((\mathscr{F}_x)_A)$ M-converges to $g(x)$.

Let $g(x) \in V \in M$. For some U, $x \in U \in T$ and $g(U) \subset V$, by (T, M) continuity of g at x. Now, $U \cap A \in (\mathscr{F}_x)_A$, and $g(U \cap A) = f(U \cap A) \subset V$. Since $f(U \cap A) \in f(\mathscr{F}_x)_A$, then $f((\mathscr{F}_x)_A)$ M-converges to $g(x)$.

(2) $\Rightarrow$ (1):
Assume (2).

If $y \in X$, then $f((\mathscr{F}_y)_A)$ is M-convergent by hypothesis. Define $g(y)$ to be the M-limit of $f((\mathscr{F}_y)_A)$. Since M is Hausdorff, then $g: X \to Y$.

There are now two things to prove:

(i) $g|A = f$.

(ii) g is (T, M) continuous.

(i): Let $x \in A$. It suffices to show that $f((\mathscr{F}_x)_A)$ is M-convergent to $f(x)$, since M is Hausdorff.

Let $f(x) \in C \in M$. Since f is (T_A, M) continuous, then for some T-open V, we have $x \in V \cap A \in T_A$ and $f(V \cap A) \subset C$. But, $f(V \cap A) \in f((\mathscr{F}_x)_A)$. Then, $f((\mathscr{F}_x)_A)$ M-converges to $f(x)$, hence $f(x) = g(x)$.

(ii):
Let $x \in X$. We shall prove that g is (T, M) continuous at x. Let V be an M-nghd. of $g(x)$. Since M is regular, then by Theorem 2.2 there is an M-open nghd. W of $g(x)$ such that $\bar{W} \subset V$. Now, $f((\mathscr{F}_x)_A)$ is M-convergent to $g(x)$. Hence, for some T-nghd. U of x, we have $f(A \cap U) \subset W$. We claim that $g(U) \subset V$.

Let $t \in U$. Then, $g(t)$ is the M-limit of $f((\mathscr{F}_t)_A)$. But, $t \in \bar{A}$, so $U \cap A \neq \phi$ and it is easy to check that $g(t)$ is also the M-limit of $f((\mathscr{F}_t)_{U \cap A})$.

Now, $f((\mathscr{F}_t)_{U \cap A})$ is a filterbase on $g(U \cap A)$. Hence $g(t) \in \overline{g(U \cap A)}$ by Theorem 3.2.IV. Then, $g(t) \in \overline{g(U \cap A)} = \overline{f(U \cap A)} \subset \overline{W} \subset V$. Hence $g(U) \subset V$, and the proof is complete. ∎

Note that the continuous extension of Theorem 2.6 is unique. For suppose that h is also a continuous extension of f to X. If $x \in X$, then $x \in \bar{A}$. By continuity of h and g, $h((\mathscr{F}_x)_A)$ converges to $h(x)$ and $g((\mathscr{F}_x)_A)$ to $g(x)$. But, if h and g agree with f on A, then $h((\mathscr{F}_x)_A)$ and $g((\mathscr{F}_x)_A)$ are the same filterbase, and in the Hausdorff space Y the limits $h(x)$ and $g(x)$ must be the same.

5.3 Normal spaces

Thus far we have seen how certain problems give rise to separation axioms of varying strength. The T_3 axiom enabled us to prove a reasonable theorem on continuous extensions. In similar fashion, normal spaces were devised to treat questions concerning continuous functions which were found to lie beyond the scope of regular spaces. These questions will be considered shortly.

A space is normal if it is Hausdorff and if disjoint closed sets have disjoint neighborhoods. Immediately, a normal space is regular (this is Theorem 3.3) but not conversely. For convenience, an example of a non-normal, regular space will be deferred until section 4 (Examples 4.1 and 4.2). As with regular spaces, we explicitly assume that a normal space is Hausdorff; when each singleton is closed, this is of course redundant.

3.1 DEFINITION *T is a normal (or T_4) topology*
if and only if
(1) T is Hausdorff.
(2) If A and B are disjoint, T-closed sets, then there are disjoint T-open sets U and V such that $A \subset U$ and $B \subset V$.

3.2 DEFINITION *(X, T) is a normal space*
if and only if
T is a normal topology on X.

3.1 EXAMPLE E^n is normal, as is any metric space. ∎

3.2 EXAMPLE Let ν be a non-empty ordinal. Then ν, with the order topology (Example 3.7.I), is normal.

It is easy to check that v is Hausdorff, and we leave this to the reader. To prove normality, let A and B be disjoint, closed subsets of v. We must produce disjoint, open neighborhoods of A and B.

If $a \in A$, and $[0, a[\cap B = \phi$, let $U_a = [0, a]$. If $[0, a[\cap B \neq \phi$, let b_a be the least upper bound of $[0, a[\cap B$. It is easy to check that $b_a \in B$, using the fact that B is closed. In this case, let $U_a =]b_a, a]$.

Then, U_a is open for each $a \in A$, and $A \subset \bigcup_{a \in A} U_a$.

Now perform a similar construction with B. If $b \in B$, and $[0, b[\cap A = \phi$, let $V_b = [0, b]$. If $[0, b[\cap A \neq \phi$, let $a_b = \text{lub}\,([0, b[\cap A)$. Note that $a_b \in A$, and let $V_b =]a_b, b]$. Then, $\bigcup_{b \in B} V_b$ is a neighborhood of B.

There remains to show that $U \cap V = \phi$.

To prove this, let $\alpha \in A$ and $\beta \in B$, and consider cases.

i) If $[0, \alpha[\cap B = \phi$ and $[0, \beta[\cap A = \phi$, and $\alpha < \beta$, then $\alpha \in [0, \beta[\cap A$, impossible. A similar contradiction arises if $\beta < \alpha$. Thus it is impossible for $[0, \alpha[\cap B = \phi$ and $[0, \beta[\cap A = \phi$.

ii) $[0, \alpha[\cap B = \phi$ and $[0, \beta[\cap A \neq \phi$.

Then, $U_\alpha = [0, \alpha]$ and $V_\beta =]a_\beta, \beta]$. Since $[0, \alpha[\cap B = \phi$, then $\alpha < \beta$. Now, $a_\beta = \text{lub}\,([0, \beta[\cap A)$, and $\alpha \in [0, \beta[\cap A$, so $\alpha \leq a_\beta$. Thus, $U_\alpha \cap V_\beta = \phi$. Similarly if $[0, \alpha[\cap B \neq \phi$ and $[0, \beta[\cap A = \phi$.

iii) $[0, \alpha[\cap B \neq \phi$ and $[0, \beta[\cap A \neq \phi$.

Then, $U_\alpha =]b_\alpha, \alpha]$ and $V\beta =]a_\beta, \beta]$. If $\alpha < \beta$, then $\alpha \in [0, \beta[\cap A$, hence $\alpha \leq a_\beta$, as in ii), and we have $U_\alpha \cap V_\beta = \phi$.

If $\beta < \alpha$, then $\beta \in [0, \alpha[\cap B$, and $\beta \leq b_\alpha$, again as in ii), so $U_\alpha \cap V_\beta = \phi$.

From these cases we now have $U_\alpha \cap V_\beta = \phi$ whenever $\alpha \in A$ and $\beta \in B$, hence $U \cap V = \phi$. ∎

3.1 Theorem *Let ϱ be a metric on X. Then, T_ϱ is normal.*

Proof By Theorem 1.1, T_ϱ is Hausdorff.

Now let A and B be disjoint, T_ϱ-closed sets. If $x \in A$, then $\inf \{\varrho(x, y) | y \in B\} > 0$ by Lemma 1, section 2. Let $\varepsilon_x = \inf \{\varrho(x, y) | y \in B\}$.

Similarly, if $x \in B$, then $\eta_x = \inf \{\varrho(x, y) | y \in A\} > 0$.

Let $U = \bigcup_{x \in A} B_\varrho \left(x, \dfrac{\varepsilon_x}{4} \right).$

Let $V = \bigcup_{x \in B} B_\varrho \left(x, \dfrac{\eta_x}{4} \right).$

Then, U and V are T-open, and $A \subset U$ and $B \subset V$.

To show that $U \cap V = \phi$, suppose that $t \in U \cap V$. For some $x \in A$ and $y \in B$, $t \in B_\varrho\left(x, \dfrac{\varepsilon_x}{4}\right) \cap B_\varrho\left(y, \dfrac{\eta_y}{4}\right)$. Then, $\varrho(x, y) \leq \varrho(x, t) + \varrho(t, y) < \dfrac{\varepsilon_x}{4} + \dfrac{\eta_y}{4}$.

Now, if $\eta_y \leq \varepsilon_x$, then $\varrho(x, y) \leq \dfrac{\varepsilon_x}{2}$, so $\dfrac{\varepsilon_x}{2} \geq \varrho(x, y) \geq \inf\{\varrho(x, z) | z \in B\} = \varepsilon_x > \dfrac{\varepsilon_x}{2}$, a contradiction. A similar contradiction arises if $\varepsilon_x \leq \eta_y$. ∎

As with T_2 and T_3 spaces, there are several equivalent formulations of the T_4 axiom. Theorem 3.2 (2) is of particular importance in proving the major theorems of this section, Urysohn's Lemma and the Urysohn-Tietze Extension Theorem.

3.2 THEOREM *Let T be a Hausdorff topology.*
Then, the following are equivalent:
(1) T is normal.
(2) If A is T-closed, and $A \subset U \in T$, then there is a T-open set V with $A \subset V \subset \bar{V} \subset U$.
(3) If A and B are disjoint, T-closed sets, then there are disjoint T-open sets U and V with $A \subset U$, $B \subset V$, and $\bar{U} \cap \bar{V} = \phi$.

Proof (1) ⇒ (2):

Assume (1). Let A be T-closed and let U be a T-nghd. of A. Note that $X - U$ is T-closed, and $A \cap (X - U) = \phi$. Then, there are disjoint T-open sets V and W such that $A \subset V$ and $X - U \subset W$. We claim that $\bar{V} \subset U$. For, let $x \in \bar{V} \cap (X - U)$. Then, $x \in W \in T$, so $W \cap V \neq \phi$, a contradiction. Hence $\bar{V} \cap (X - U) = \phi$, so $\bar{V} \subset U$.

(2) ⇒ (3):

Assume (2). Let A and B be disjoint, T-closed sets. Then, $A \subset X - B \in T$, so there is by (2) a T-open set U such that $A \subset U \subset \bar{U} \subset X - B$.

Now, $B \subset X - \bar{U} \in T$, so for some T-open V, we have $B \subset V \subset \bar{V} \subset X - \bar{U}$. Finally, it is immediate that $\bar{U} \cap \bar{V} = \phi$.

(3) ⇒ (1):

This is immediate, since $\bar{U} \cap \bar{V} = \phi$ implies $U \cap V = \phi$. ∎

As a matter of record, we state the obvious fact that a normal space is regular.

3.3 THEOREM *Let T be normal.*
Then, T is regular.

Proof Let $x \in X$ and let A be T-closed with $x \notin A$. Since T is normal, then T is Hausdorff. Then, $\{x\}$ is T-closed, and the conclusion is immediate by Definition 3.1 (2). ∎

Of course, normality is a topological invariant, that is, preserved by homeomorphism.

3.4 THEOREM *Let $(X, T) \cong (Y, M)$.*
Then,
T is normal if and only if M is normal.

Proof Left as an exercise. ∎

Thus far things have progressed pretty much as they did with Hausdorff and regular spaces. However, normality is surprisingly pathological in its failure to transmit its separation properties to subspaces and product spaces. The next example shows that a product of two normal spaces need not be normal. A non-normal subspace of a normal space will be exhibited in Chapter VII, when we discuss the Tychonov plank (Example 1.12.VII).

3.3 EXAMPLE Let Ω be the set of countable ordinals. Alternatively, Ω is the first uncountable ordinal. With the order topologies, Ω and $\Omega \cup \{\Omega\}$ are normal, by Example 3.2. We shall now show that the product space $\Omega \times (\Omega \cup \{\Omega\})$ is not normal.

Let $A = \{(x, x) | x \in \Omega\}$, and $B = \{(x, \Omega) | x \in \Omega\}$. The reader can check that A and B are disjoint, closed subsets of $\Omega \times (\Omega \cup \{\Omega\})$ (see Figure 3.1).

Let U be a neighborhood of A, V a neighborhood of B. We shall show that $U \cap V \neq \phi$, hence conclude that $\Omega \times (\Omega \cup \{\Omega\})$ is not normal.

Suppose first, if $x \in \Omega$, then $\Omega \cap \{y | x < y \text{ and } (x, y) \notin U\} \neq \phi$. Then, for each $x \in \Omega$, we can choose $f(x)$ as the least element of $\Omega \cap \{y | x < y \text{ and } (x, y) \notin U\}$.

Choose any $x_0 \in \Omega$, and let $x_{n+1} = f(x_n)$ for each $n \in \omega$. Then, $x_n < f(x_n) \leq x_{n+1} < f(x_{n+1})$ for each $n \in \omega$.

We thus have two non-decreasing sequences to Ω. We shall show that each sequence converges to the same limit. The argument is like that used in calculus for monotone real sequences. Note that $\{x_n | n \in \omega\}$ is bounded above $\left(\text{by} \bigcup_{n \in \omega} \{x_n\}\right)$, hence has a least upper bound, say λ. Note that $x_n < \lambda$ for each $n \in \omega$. We shall show that $x_n \to \lambda$. To see this, consider two cases:

i) λ is a limit ordinal.

SEPARATION AXIOMS

Choose any basic nghd. $]\alpha, \lambda]$ of λ. If $x_n \notin]\alpha, \lambda]$ for each $n \in \omega$, then $x < \lambda \leq \alpha$ by choice of α and λ, and this is impossible. Then, for some k, $k \in \omega$ and $x_k \in]\alpha, \lambda]$. Then, $x_n \in]\alpha, \lambda]$ for each $n \in \omega$ with $n \geq k$.

ii) λ is a non-limit ordinal.

Then, λ has an immediate predecessor, say β. But, then $x_n \leq \beta$ for each $n \in \omega$, so $\lambda \leq \beta$, a contradiction. Thus this case cannot occur, and we are left with case i), in which $\{x_n\}_{n \in \omega}$ converges to λ.

It is now trivial to verify that $\{f(x_n)\}_{n \in \omega}$ converges to λ also, as $x_n < f(x_n) \leq x_{n+1}$ for each $n \in \omega$.

Then, (λ, λ) is a cluster point of $\{(\lambda, f(x)) | x \in \Omega\}$ in $\Omega \times \Omega$. Since $(\lambda, \lambda) \in A \subset U$, it is easy to check that $U \cap \{(x, f(x)) | x \in \Omega\} \neq \phi$, a contradiction.

Thus, for some $x_0 \in \Omega$, $\Omega \cap \{y | x_0 < y \text{ and } (x_0, y) \notin U\} = \phi$. Then, $(x_0, y) \in U$ for each $y \in \Omega$ with $x_0 < y$.

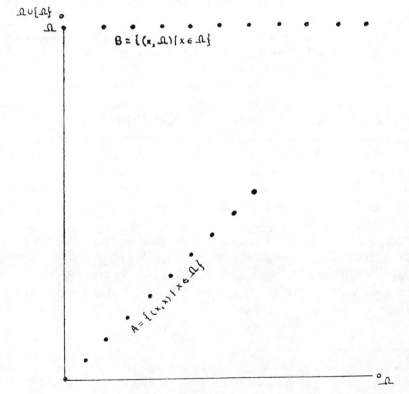

FIGURE 3.1

Now, V is a nghd. of (x_0, Ω), as $(x_0, \Omega) \in B$. Then, there are basic nghds. N_{x_0} of x_0 and N_Ω of Ω such that $(x_0, \Omega) \in N_{x_0} \times N_\Omega \subset V$. We may choose $N_\Omega =]\xi, \Omega]$ for some $\xi \in \Omega$. Choose $y_0 \in]\xi, \Omega[$ with $x_0 < y_0$. Then, $(x_0, y_0) \in N_{x_0} \times N_\Omega \subset V$, so $(x_0, y_0) \in V$. Hence $U \cap V \neq \phi$. ∎

Under special circumstances, it is possible to conclude that a subspace of a normal space is normal. The next theorem is typical of this type of result.

3.5 THEOREM *Let T be normal.*
Let A be T-closed.
Then,
T_A is normal.

Proof Left as an exercise. ∎

Having laid the groundwork, we now turn to the reason for considering the T_4 separation axiom.

Normal spaces came to the attention of Urysohn in the 1920's in connection with the following question: given the space (X, T), are there "enough" real-valued, continuous functions on X?. The word "enough" is purposely vague, but at the very least we would like a guarantee of the existence of some non-constant continuous functions from X to E^1. The T_3 axiom does not provide such a guarantee.* Urysohn was able to show that the T_4 axiom does.

More specifically, with each pair of disjoint, closed subsets of any normal space, Urysohn was able to associate a continuous $f: X \to [0, 1]$ which separates A and B in the sense that f maps A to 0 and B to 1. It also turns out that, conversely, the existence of an Urysohn function for each pair of disjoint closed sets insures normality of the space.

The proof of Urysohn's Lemma is fairly deep, as might be expected of one of the fundamental results of set topology. We precede it by a Lemma which handles some of the more technical details. In proving the Lemma, we make use of the easily established fact that, if $0 \leq x < y \leq 1$, then dyadic numbers (i.e., of the form $m/2^n$) d and d' can be found such that $x \leq d < d' \leq y$.

---

* E. Hewitt, "On Two Problems of Urysohn", *Ann. Math.*, **47**, 1946, pp. 503–509. J. Novak, "A Regular Space on Which Every Continuous Function Is Constant", *Casopsis Pest. Mat. Fys.*, **73**, 1948, pp. 58–68.

1 LEMMA *Let T be a normal topology.
Let A and B be disjoint, T-closed sets.
Then,
there is a set $\{U_t | 0 \leq t \leq 1\}$ of T-open sets such that:*
(1) $A \subset U_0$.
(2) $U_1 \cap B = \phi$.
(3) If $0 \leq x < y \leq 1$, then $\overline{U}_x \subset U_y$.

Proof Define $U_1 = X - B$. Then, $A \subset U_1$ and U_1 is T-open and $U_1 \cap B = \phi$. By Theorem 3.2, there is some T-open set U_0 such that $A \subset U_0 \subset \overline{U}_0 \subset U_1$.

We now proceed to fill in U_t when $0 < t < 1$. We first work with t dyadic, proceeding by induction.

Suppose n is a non-negative integer, and that the sets $U_{k/2^n}$ have been defined for integers k, $0 \leq k \leq 2^n - 1$ (note that U_0 und U_1 have already been defined). Thus, $\overline{U}_{j/2^n} \subset U_{i/2^n}$ for $0 \leq j < i \leq 2^n - 1$. We must define the sets $U_{i/2^{n+1}}$, $0 \leq i \leq 2^{n+1} - 1$. Note first that we need only consider odd values of i. If i is even, say $i = 2j$, then $0 \leq j \leq 2^n - 1$ and $i/2^{n+1} = j/2^n$, and $U_{j/2^n}$ has already been defined by the inductive hypothesis.

Suppose then that i is odd, say $i = 2k + 1 \leq 2^{n+1} - 1$, for some k, $0 \leq k \leq 2^n - 1$. Then, again by Theorem 3.2, since $\overline{U}_{k/2^n} \subset U_{(k+1)/2^n}$, there is some T-open set V such that $\overline{U}_{k/2^n} \subset V \subset \overline{V} \subset U_{(k+1)/2^n}$. Let $U_{(2k+1)/2^{n+1}} = V$. This defines $U_{i/2^{n+1}}$ for all i, $0 \leq i \leq 2^{n+1} - 1$.

By induction, the sets U_d are now defined for all dyadic d.

Now let t be any number in $[0, 1]$.

Define $U_t = \bigcup_{d \leq t} U_d$, the union being over all dyadics d in $[0, 1]$ with $d \leq t$. If t is dyadic, then this definition agrees with that arrived at by induction.

Finally, suppose that $0 \leq x < y \leq 1$. Then there are dyadic numbers $i/2^n$ and $j/2^m$ with $x \leq j/2^m < i/2^n \leq y$. Now, $U_{i/2^n} \subset U_y$ by definition of U_y. Further, for all dyadic d with $0 \leq d \leq x$, we have $d \leq j/2^m$, so $U_d \subset U_{j/2^m}$, hence $U_x \subset U_{j/2^m}$. Then,

$$\overline{U}_x \subset \overline{U}_{j/2^m} \subset U_{i/2^n} \subset U_y,$$

and the Lemma is proved. ∎

We now prove the main theorem, which traditionally is known as Urysohn's Lemma. It is understood in the statement and proof that $[0, 1]$ is considered as a subspace of Euclidean space E^1.

3.6 THEOREM (Urysohn's Lemma) *Let (X, T) be a Hausdorff space. Then, the following are equivalent:*

(1) T is normal.

(2) If A and B are non-empty, disjoint, T-closed sets, then there is a continuous $f: X \to [0, 1]$ such that $f(A) = \{0\}$ and $f(B) = \{1\}$.

Proof $(2) \Rightarrow (1)$:

If A and B are non-empty, disjoint, T-closed subsets of X, then produce by (2) an Urysohn function f and note that $f^{-1}([0, \frac{1}{2}[)$ and $f^{-1}(]\frac{1}{2}, 1])$ are disjoint, T-nghds. of A and B respectively.

$(1) \Rightarrow (2)$:

Assume (1). Let A and B be non-empty, disjoint, T-closed sets. Let the T-open sets U_t, $0 \leq t \leq 1$, be as given by Lemma 1. If $x \in X$, define:

$$f(x) = \begin{cases} 1 \text{ if } x \notin U_1 \\ \inf\{t \mid x \in U_t\} \text{ if } x \in U_1. \end{cases}$$

Immediately, $f: X \to [0, 1]$. If $x \in B$, then $x \notin U_1$, so $f(x) = 1$. If $x \in A$, then $x \in U_0$ and $x \in U_1$, so $f(x) = 0$.

There remains to show that f is continuous.

Let $x \in X$ and let $\varepsilon > 0$. Consider three cases:

i) $f(x) = 0$.

We may assume without loss of generality that $\varepsilon < 1$. Then, $|f(y) - f(x)| = |f(y)| = f(y) \leq \varepsilon/2 < \varepsilon$ for $y \in U_{\varepsilon/2}$.

ii) $f(x) = 1$.

Again, we may assume that $\varepsilon < 1$. If $y \in X - \overline{U}_{1-\varepsilon/2}$, then $y \notin U_j$ for $j \leq 1 - \varepsilon/2$. Then, $f(y) \geq 1 - \varepsilon/2$, so

$$1 - f(y) = |f(x) - f(y)| \leq \varepsilon/2 < \varepsilon.$$

iii) $0 < f(x) < 1$.

We may assume that $\varepsilon \leq f(x)$ and $f(x) + \varepsilon \leq 1$.

Let $y \in U_{f(x)+\varepsilon/2} \cap (X - \overline{U}_{f(x)-\varepsilon/2})$.

Since $y \in U_{f(x)+\varepsilon/2}$, then $y \notin B$, and $f(y) \leq f(x) + \varepsilon/2 < f(x) + \varepsilon$, so $f(y) - f(x) < \varepsilon$.

But also $y \in X - \overline{U}_{f(x)-\varepsilon/2}$, so $y \notin U_j$ for $0 \leq j \leq f(x) - \varepsilon/2$. Then, $(y) \geq f(x) - \varepsilon/2 > f(x) - \varepsilon$, so $f(y) - f(x) > -\varepsilon$.

Then, $-\varepsilon < f(y) - f(x) < \varepsilon$, so $|f(y) - f(x)| < \varepsilon$.

By cases i) through iii), f is continuous at $x \in X$, hence f is continuous. ∎

The idea of the proof of Urysohn's Lemma is summed up in Fig. 3.2. The key is to be able to enclose A in an expanding shell of open sets U_i, with the largest, U_1, still separated from B. If $x \notin U_1$ (and this includes all points of B), $f(x)$ is defined to be 1. If $x \in U_1$, then we look for the smallest index i such that $x \in U_i$, or, more correctly, the infimum of all such indices. Of course, $A \subset U_0$, so each point a of A is in U_0 and $f(a) = 0$. Urysohn's achievement lay in recognizing that the T_4 axiom gave him the open sets U_i which are needed in the proof.

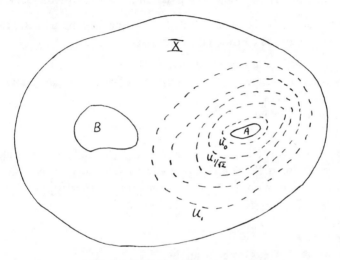

FIGURE 3.2

The very strong connection between the T_4 separation axiom and real continuous functions on X, suggested by Urysohn's Lemma, is very clearly revealed by the next theorem, which should be compared with Theorem 2.6. Theorem 3.7 says that normality is sufficient for the existence of a continuous extension of any real-valued function continuous on any closed subspace.

The converse is also true, so that normality is just the right condition for treating the extension problem on closed subspaces.

Theorem 3.7 is often known as Tietze's Extension Theorem, although there is some confusion about assigning credit here. Historically, Tietze seems to have been the first to define normality, in 1923. Urysohn's Lemma clarified its role in the theory of continuous functions. Tietze proved his extension theorem for metric spaces, which of course are normal, but it was Urysohn who generalized the result to arbitrary normal spaces. Thus

Bourbaki refers to Theorem 3.7 as a theorem of Urysohn, and Dugundji as Tietze's Theorem. Alexandroff calls it the Brouwer-Urysohn Theorem, and Stone the Lebesgue-Urysohn Theorem. It would appear to be fair to credit Tietze with the first proof for a wide class of spaces (metric spaces), and Urysohn with the proof of the general statement.

In the course of the proof, we shall use the Cauchy criterion for the convergence of real sequences, and also the fact that $\lim_{k \to \infty} |S_{n+K} - S_n| = |L - S_n|$ for any real sequence S if $\lim_{n \to \infty} S_n = L$. A preliminary Lemma will absorb some of the technical details. In the Lemma and the theorem, all intervals $[a, b]$ are considered as subspaces of E^1.

2 LEMMA *Let T be a normal topology.*
Let F be a T-closed subset of X and let $f: F \to [-1, 1]$ be continuous. Then,
there is a continuous $g: X \to [-\frac{1}{3}, \frac{1}{3}]$ such that $|g(x) - f(x)| \leq \frac{2}{3}$ for each $x \in F$.

Proof Let $A = f^{-1}([-1, -\frac{1}{3}])$ and $B = f^{-1}([\frac{1}{3}, 1])$. Since f is continuous, A and B are T-closed. Since $A \cap B = \phi$ and T is normal, there is by Urysohn's Lemma some continuous $h: X \to [0, 1]$ such that $h(A) = \{0\}$ and $h(B) = \{1\}$.

Let $t(x) = \frac{2}{3}x - \frac{1}{3}$, for $0 \leq x \leq 1$. This defines an homeomorphism $t: [0, 1] \to [-\frac{1}{3}, \frac{1}{3}]$.

Let $g = t \circ h$.

Then, $g: X \to [-\frac{1}{3}, \frac{1}{3}]$ is continuous.

If $x \in A$, then $g(x) = t(h(x)) = t(0) = -\frac{1}{3}$. Since $f(x) \in [-1, -\frac{1}{3}]$, then $|f(x) - g(x)| \leq \frac{2}{3}$.

If $x \in B$, then $g(x) = t(1) = \frac{1}{3}$. Since $f(x) \in [\frac{1}{3}, 1]$, then $|f(x) - g(x)| \leq \frac{2}{3}$.

Finally, if $x \in F - (A \cup B)$, then $-\frac{1}{3} < f(x) < \frac{1}{3}$, so $|g(x) - f(x)| \leq \frac{2}{3}$. ∎

3.7 THEOREM (Tietze-Urysohn) *Let T be Hausdorff. Then, the following are equivalent:*

(1) T is normal.

(2) If A is T-closed and $f: A \to E^1$ is continuous, then, there is a continuous $F: X \to E^1$ such that $F|A = f$.

Proof (2) ⇒ (1):
Assume (2).
Suppose that A and B are disjoint, T-closed sets. Define $f: A \cup B \to E^1$ by:

$$f(x) = \begin{cases} 0 \text{ if } x \in A \\ 1 \text{ if } x \in B. \end{cases}$$

Then f is easily seen to be continuous. By (2), produce a continuous extension F of f to X.

Then, $F^{-1}(]-\tfrac{1}{2}, \tfrac{1}{3}[)$ and $F^{-1}(]\tfrac{1}{2}, 2[)$ are disjoint, T-open nghds. of A and B respectively. Hence T is normal.

(1) $\Rightarrow$ (2):

Assume (1).

Let A be T-closed.

Suppose $f: A \to E^1$ is continuous.

We consider two cases and proceed in steps.

case 1 f is bounded.

Then, for some $M > 0$, $|f(x)| \leq M$ for each $x \in A$. We may assume without loss of generality that $f: A \to [-1, 1]$. For, $|(1/M)f(x)| \leq 1$ for each $x \in A$, and it is immediate that MF is a continuous extension of f if F is a continuous extension of $(1/M)f$.

Thus, suppose that $f: A \to [-1, 1]$.

i) If n is a non-negative integer, then there is a continuous F_n: $X \to [-1 + (\tfrac{2}{3})^{n+1}, 1 - (\tfrac{2}{3})^{n+1}]$ such that $|F_n(x) - f(x)| \leq (\tfrac{2}{3})^{n+1}$ for each $x \in A$ and, if K is a positive integer, $|F_m(x) - F_n(x)| \leq 2(\tfrac{2}{3})^{K+1}$ for $m, n \geq K$.

We produce the functions F_n by an inductive construction. By Lemma 2, there is a continuous $F_0: X \to [-\tfrac{1}{3}, \tfrac{1}{3}]$ such that $|F_0(x) - f(x)| \leq \tfrac{2}{3}$ for each $x \in A$.

Now suppose n is a non-negative integer, and F_n has been defined. Let $\varphi(x) = (\tfrac{3}{2})^{n+1} (f(x) - F_n(x))$ for each $x \in A$. Then, φ is continuous: $A \to [-1, 1]$. By Lemma 2, there is a continuous $\beta: X \to [-\tfrac{1}{3}, \tfrac{1}{3}]$ such that $|\beta(x) - \varphi(x)| \leq \tfrac{2}{3}$ for each $x \in A$.

Let $F_{n+1}(x) = F_n(x) + (\tfrac{2}{3})^{n+1} \beta(x)$ for each $x \in X$.

The reader can check that F_{n+1} is continuous: $X \to [-1 + (\tfrac{2}{3})^{n+2}, 1 - (\tfrac{2}{3})^{n+2}]$.

Further, if $x \in A$, then

$$|F_{n+1}(x) - f(x)| = |F_n(x) + (\tfrac{2}{3})^{n+1} \beta(x) - f(x)|$$

$$= (\tfrac{2}{3})^{n+1} |\beta(x) - (\tfrac{3}{2})^{n+1} (f(x) - F_n(x))|$$

$$= (\tfrac{2}{3})^{n+1} |\beta(x) - \varphi(x)| \leq (\tfrac{2}{3})^{n+1} (\tfrac{2}{3}) = (\tfrac{2}{3})^{n+2}.$$

Finally, let $x \in X$ and let n be a positive integer. By the definition of F_{n+1}, $|F_{n+1}(x) - F_n(x)| = (\tfrac{2}{3})^{n+1} |\beta(x)| \leq (\tfrac{2}{3})^{n+1} \tfrac{1}{3} = 2^{n+1}/3^{n+2}$. Check by induction that, for $r \geq 1$,

$$|F_{n+r}(x) - F_n(x)| \leq \sum_{j=0}^{r-1} |F_{n+j+1}(x) - F_{n+j}(x)|$$

$$\leq \sum_{j=0}^{r-1} \left(\frac{2^{n+j+1}}{3^{n+j+2}}\right) = \frac{2^{n+1}}{3^{n+2}} \sum_{j=0}^{r-1} \left(\frac{2}{3}\right)^j < \frac{2^{n+1}}{3^{n+2}} \sum_{j=0}^{\infty} \left(\frac{2}{3}\right)^j = \left(\frac{2}{3}\right)^{n+1}.$$

Then, for $m, n \geq K$,

$$|F_m(x) - F_n(x)| \leq |F_m(x) - F_K(x)| + |F_n(x) - F_K(x)| \leq (\tfrac{2}{3})^{K+1} + (\tfrac{2}{3})^{K+1}$$
$$= 2(\tfrac{2}{3})^{K+1}.$$

ii) If $x \in X$, then $\lim_{n \to \infty} F_n(x)$ exists, and $-1 \leq \lim_{n \to \infty} F_n(x) \leq 1$.

Existence follows from i). Since $(\tfrac{2}{3})^{K+1} \to 0$ as $K \to \infty$, then $\{F_n(x)\}_{n=0}^{\infty}$ is a Cauchy sequence to E^1, hence converges. Since $-1 \leq F_n(x) \leq 1$ for each n, then $-1 \leq \lim_{n \to \infty} F_n(x) \leq 1$ also.

iii) Let $F(x) = \lim_{n \to \infty} F_n(x)$ for each $x \in X$.

Then, F is continuous on X to $[-1, 1]$.

All that requires proof is the continuity. Let $x \in X$ and $\varepsilon > 0$. Note that $|F(z) - F_n(z)| = \lim_{k \to \infty} |F_{n+K}(z) - F_n(z)| \leq 2(\tfrac{2}{3})^{n+1}$ for each $n \in Z^+$ and $z \in X$.

Now, if $y \in X$, then for any $n \in Z^+$, we have

$$|F(x) - F(y)| \leq |F(x) - F_n(x)| + |F_n(x) - F_n(y)| + |F_n(y) - F(y)|.$$

Choose n sufficiently large that $2(\tfrac{2}{3})^{n+1} < \varepsilon/3$. Since F_n is continuous, there is a T-nghd. V of x such that $|F_n(x) - F_n(y)| < \varepsilon/3$ if $y \in V$. Then, for each $y \in V$, $|F(x) - F(y)| < \varepsilon/3 + \varepsilon/3 + \varepsilon/3 = \varepsilon$, implying that F is continuous at x, hence continuous.

iv) $F|A = f$.

Let $x \in A$ and $\varepsilon > 0$. Then, for each $n \in Z^+$,

$$|F(x) - f(x)| \leq |F_n(x) - F(x)| + |F_n(x) - f(x)| \leq |F_n(x) - F(x)| + (\tfrac{2}{3})^{n+1}.$$

Choose n sufficiently large that $(\tfrac{2}{3})^{n+1} < \varepsilon/2$ and $|F_n(x) - F(x)| < \varepsilon/2$. Then, $|F(x) - f(x)| < \varepsilon$, implying that $F(x) = f(x)$.

This completes the proof of the theorem in case 1.

case 2 f is not bounded.

Define $g(x) = \dfrac{f(x)}{1 + |f(x)|}$ for each $x \in A$.

The reader can complete the proof by filling in the details of the remaining steps.

v) g is continuous: $A \to [-1, 1]$.

vi) There exists a continuous $G\colon X \to [-1, 1]$ such that $G|A = g$ (apply case 1 to g).

vii) There is a continuous $h\colon X \to [0, 1]$ such that $h(a) = 1$ whenever $a \in A$ and $h(b) = 0$ whenever $G(b) = 1$ or $G(b) = -1$. (Apply Urysohn's Lemma to A and $G^{-1}(\{-1\} \cup \{1\})$.)

viii) Let $F(x) = \dfrac{G(x)\,h(x)}{1 - |G(x)\,h(x)|}$.

Then, F is continuous: $X \to E^1$ and $F|A = f$.
This completes the proof of the theorem. ∎

5.4 Completely regular spaces

We have attempted to motivate the separation axioms somewhat along the following lines:

    Hausdorff ——— uniqueness of limits

    regular ⎫
    ⎬——— existence and extension of continuous functions.
    normal ⎭

Normal spaces saw the strongest separation axiom bring the greatest gain in solving the extension problem. But there was a loss as well, in the hereditary properties. Products and subspaces of normal spaces need not be normal. It was found, however, that subspaces and products of normal spaces, if weaker than normal, are nonetheless stronger than regular—they are completely regular.

4.1 DEFINITION *T is completely regular
if and only if*
(1) T is Hausdorff.
(2) If A is T-closed and $y \in X - A$, then there is a continuous $f\colon X \to [0, 1]$ such that $f(A) = \{0\}$ and $f(y) = 1$.

4.2 DEFINITION *(X, T) is a completely regular space
if and only if
T is a completely regular topology on X.*

Note that we could without any essential difference stipulate in Definition 4.1 (2) that $f(y) = 0$ and $f(A) = \{1\}$. To see this, let $g(x) = 1 - f(x)$ for each $x \in X$, in the definition, and note that $g(y) = 0$ and $g(A) = \{1\}$.

Sometimes completely regular spaces are referred to in the literature as Tychonov spaces. It has also been suggested (though no one wants the credit) that they be called $T_{3.5}$ spaces.

Rather than give examples of completely regular spaces at this time, we shall immediately put this new separation axiom in its proper place with the others.

4.1 Theorem

(1) If T is normal, then T is completely regular.

(2) If T is completely regular, then T is regular.

Proof of (1) In view of Theorem 1.3, (1) is immediate upon taking $B = \{x\}$ in Urysohn's Lemma.

Proof of (2) Suppose that T is completely regular. Let $x \in U \in T$. It suffices by Theorem 2.2 to produce a T-nghd. V of x such that $\bar{V} \subset U$.

To do this, note that $X - U$ is T-closed and $x \notin X - U$. Then, there is a continuous $f: X \to [0, 1]$ such that $f(X - U) = \{0\}$ and $f(x) = 1$. Let $V = f^{-1}(]\tfrac{1}{2}, 1])$. Then, V is a T-nghd. of x, and it is easy to check that $\bar{V} \subset U$. ∎

Thus, for example, each metric space, being normal, is completely regular. There are regular spaces which are not completely regular. A fairly simple one of recent vintage, due to John Thomas, is given in the next example.

4.1 Example If $n \in Z$, let $L_{2n} = \{(2n, y) | 0 \leq y \leq \tfrac{1}{2}\}$, and $S_1 = \bigcup_{n \in Z} L_{2n}$. For

$$k, n \in Z \text{ and } k \geq 2, \text{ let } p_{2n+1,k} = \left(2n + 1, \frac{k-1}{k}\right),$$

and

$$T_{2n+1,k} = \left\{\left(2n + 1 + t, 1 - t - \frac{1}{k}\right) \middle| 0 < t \leq \frac{k-1}{k}\right\}$$

$$\cup \left\{\left(2n + 1 - t, 1 - t - \frac{1}{k}\right) \middle| 0 < t \leq \frac{k-1}{k}\right\}.$$

Let

$$S_2 = \{p_{2n+1,k} | k, n \in Z \text{ and } k \geq 2\}, \text{ and } S_3 = \bigcup_{\substack{k, n \in Z \\ k \geq 2}} T_{2n+1,k}.$$

Finally, choose $p_+ = (0, 10)$ and $p_- = (0, -10)$.

SEPARATION AXIOMS

Figure 4.1 indicates a geometric realization of X, where $X = S_1 \cup S_2 \cup S_3 \cup \{p_+, p_-\}$.

Specify a base for a topology T on X by:

i) If $x \in S_3$, a basic nghd. of x is $\{x\}$.

ii) A basic nghd. of $p_{2n+1,k}$ is any subset of X containing $p_{2n+1,k}$ and all but finitely many points of $T_{2n+1,k}$.

iii) A basic nghd. of $(2n, y) \in L_{2n}$ is any subset of X containing $(2n, y)$ and all but finitely many points (α, y) of X with $|\alpha - 2n| < 1$, i.e., $2n - 1 < \alpha < 2n + 1$.

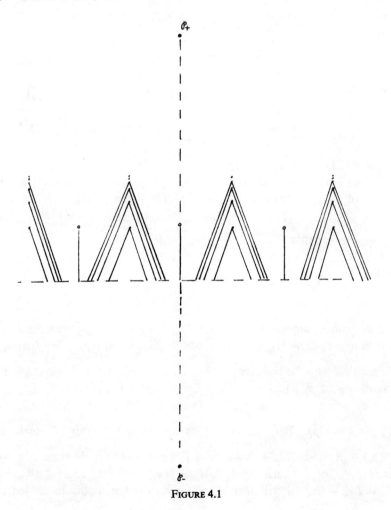

FIGURE 4.1

iv) A basic nghd. of p_+ consists of p_+ and all points (x, y) in X with $x < c$ for c some real number.

v) A basic nghd. of p_- consists of p_- and all points (x, y) in X with $x > c$ for some real number c.

The reader can verify that the resulting set of subsets of X is a base for a topology T on X, and that T is regular. We claim that the space (X, T) is not completely regular.

To prove this, let $f: X \to E^1$ be continuous. We claim that $f(p_+) = f(p_-)$. Note, if $x \in X$, then

$$\{y | f(y) = f(x)\} = \bigcap_{j=1}^{\infty} \{\xi \mid |f(\xi) - f(x)| < 1/j\}.$$

Now,

$$T_{2n+1,k} \cap \{x | f(x) \neq f(p_{2n+1,k})\} = T_{2n+1,k} - \{x | f(x) = f(p_{2n+1,k})\}$$

$$= T_{2n+1,k} - \bigcap_{j=1}^{\infty} \{\xi \mid |f(\xi) - f(p_{2n+1,k})| < 1/j\}$$

$$= \bigcup_{j=1}^{\infty} [T_{2n+1,k} - \{\xi \mid |f(\xi) - f(p_{2n+1,k})| < 1/j\}].$$

Since $\{\xi \mid |f(\xi) - f(p_{2n+1,k})| < 1/j\}$ is a T-nghd. of $p_{2n+1,k}$ in X, then $\{\xi \mid |f(\xi) - f(p_{2n+1,k})| < 1/j\}$ contains $p_{2n+1,k}$ and all but finitely many points of $T_{2n+1,k}$. Thus, $T_{2n+1,k} - \{\xi \mid |f(\xi) - f(p_{2n+1,k})| < 1/j\}$ is finite, so $\bigcup_{j=1}^{\infty} [T_{2n+1,k} - \{\xi \mid |f(\xi) - f(p_{2n+1,k})| < 1/j\}]$ is countable. Then, $T_{2n+1,k} \cap \{x | f(x) \neq f(p_{2n+1,k})\}$ is countable.

Let $H_{2n+1,k} = \{y | \text{for some } x, (x, y) \in T_{2n+1,k} \text{ and } f(x, y) \neq f(p_{2n+1,k})\}$ and

$$M_{2n+1} = \bigcup_{k=1}^{\infty} H_{2n+1,k}.$$

Now, choose any point $p \in L_{2n}$ (or L_{2n+2}), say $p = (\alpha, \beta)$, with $\beta \notin M_{2n+1}$. This can be done as M_{2n+1} is countable. We claim that $f(p) = \lim_{k \to \infty} f(p_{2n+1,k})$.

To see this, let p_k be the point on $T_{2n+1,k}$ with $|p_k - p| < 1$ and the vertical coordinate of p_k equal to β (see Figure 4.2). Now, $f(p_k) = f(p_{2n+1,k})$ by choice of p_k, as $\beta \notin M_{2n+1}$. Thus, $\lim_{k \to \infty} f(p_{2n+1,k}) = \lim_{k \to \infty} f(p_k)$. But, $\lim_{k \to \infty} f(p_k) = f(p)$. For, let $\varepsilon > 0$. We seek a number K such that $|f(p_k) - f(p)| < \varepsilon$ for $k \geq K$. Since f is continuous at p, there is a basic nghd. V_p of p such that $|f(\xi) - f(p)| < \varepsilon$ for each $\xi \in V_p$. Say $V_p = L_{2n} - \{x_1, \ldots, x_r\}$, with $x_1, \ldots, x_r$ on the horizontal line of length 2 centered at p. Choose K sufficiently large that $p_k \neq x_i$ for $k \geq K$. Then, $p_k \in V_p$ for $k \geq K$, so $|f(p_k) - f(p)| < \varepsilon$, hence $\lim_{k \to \infty} f(p_{2n+1,k}) = f(p)$.

Let $f(p) = c$. Then $f(p)$ has the same value c for all except possibly countably many $p \in L_{2n} \cup L_{2n+2}$. It is now easy to check that $f(p) = c$ for all but possibly countably many p in L_{2j} for all j.

Now, given any nghd. V of p_+, there is some $\xi \in V$ such that $f(\xi) = c$. Hence by continuity $f(p_+) = c$. Similarly, $f(p_-) = c$.

It is now immediate that (X, T) is not completely regular. Choose a closed set C containing p_+ and not containing p_-. There can be no continuous $f \colon X \to [0, 1]$ with $f(C) = \{0\}$ and $f(p_-) = 1$. ∎

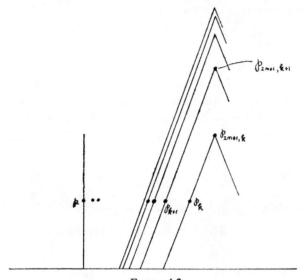

FIGURE 4.2

We shall give an example of a non-normal, completely regular space shortly, after we have examined the notion of complete regularity in more detail.

The following open set alternative to the definition is sometimes useful.

4.2 THEOREM *T is completely regular*
if and only if
if $x \in X$ and V is a T-nghd. of x, then there is a continuous $f \colon X \to [0, 1]$ such that $f(x) = 1$ and $f(X - V) = \{0\}$.

Proof Immediate upon taking $A = X - V$ in Definition 4.1 (2). ∎

The invariance and hereditary properties of completely regular spaces are exactly what we should like them to be. Complete regularity is a topological invariant, and subspaces of both normal and completely regular spaces are completely regular. Finally, a product space is completely regular exactly when each coordinate is completely regular.

4.3 Theorem *Let $(X, T) \cong (Y, M)$.*
Then,
T is completely regular $\Leftrightarrow$ M is completely regular.

Proof Left as an exercise. ∎

4.4 Theorem *Let $A \subset X$.*
Let T be completely regular.
Then,
T_A is completely regular.

Proof Note by Theorem 1.6 that T_A is Hausdorff. Now let B be T_A-closed and $y \in A - B$. For some F, F is T-closed and $B = A \cap F$. Since T is completely regular, there exists a continuous $f: X \to [0, 1]$ such that $f(F) = \{0\}$ and $f(y) = 1$. By Theorem 1.8, Chapter III, $f|A: A \to [0, 1]$ is continuous. Further, $(f|A)(B) = \{0\}$ and $(f|A)(y) = 1$. ∎

1 Corollary *Let $A \subset X$.*
Let T be normal.
Then,
T_A is completely regular.

Proof Immediate by Theorem 4.1 (1) and Theorem 4.4. ∎

4.5 Theorem *Let $A \neq \phi$.*
Let T_α be a topology on X_α for each $\alpha \in A$.
Let $\mathbb{P}$ be the product topology on $\prod_{\alpha \in A} X_\alpha$.
Then,
$\mathbb{P}$ is completely regular
if and only if
T_α is completely regular for each $\alpha \in A$.

Proof Suppose first that $\mathbb{P}$ is completely regular. Let $f \in \prod_{\alpha \in A} X_\alpha$ and let $\beta \in A$. Let

$$Y = \left(\prod_{\alpha \in A} X_\alpha\right) \cap \{g \mid g(\alpha) = f(\alpha) \text{ if } \alpha \in A \text{ and } \alpha \neq \beta\}.$$

By Theorem 4.4, $(Y, \mathbb{P}_Y)$ is completely regular. By Theorem 2.6.III, $(Y, \mathbb{P}_Y) \cong (X_\beta, T_\beta)$. By Theorem 4.3, T_β is completely regular.

Conversely, suppose that T_α is completely regular for each $\alpha \in A$. We shall use Theorem 4.2 to show that $\mathbb{P}$ is completely regular.

Let $f \in \prod_{\alpha \in A} X_\alpha$ and let $\bigcap_{i=1}^{n} p_{\alpha_i}^{-1}(V_{\alpha_i})$ be any basic $\mathbb{P}$-nghd. of f. Since T_{α_i} is completely regular for $i = 1, \ldots, n$, then there is a continuous $g_{\alpha_i}: X_{\alpha_i} \to [0, 1]$ such that $g_{\alpha_i}(f_{\alpha_i}) = 1$ and $g_{\alpha_i}(X_{\alpha_i} - V_{\alpha_i}) = \{0\}$. Let

$$F(t) = \inf \{g_{\alpha_i}(t) | 1 \leq i \leq n\} \text{ for each } t \in \prod_{\alpha \in A} X_\alpha.$$

It is easy to check that $F: \prod_{\alpha \in A} X_\alpha \to [0, 1]$ is continuous, and that $F(f) = 1$ and $F\left(\prod_{\alpha \in A} X_\alpha - \bigcap_{i=1}^{n} p_{\alpha_i}^{-1}(V_{\alpha_i})\right) = \{0\}$.

To complete the proof, extend this result to arbitrary $\mathbb{P}$-open sets and use Theorem 4.2. ∎

Note that, in view of Theorem 4.1 (1) and the last theorem, a product of normal spaces is completely regular. We shall exploit this fact to give an example of a completely regular (hence also regular) but non-normal space.

4.2 EXAMPLE The ordinal spaces Ω and $\Omega \cup \{\Omega\}$ are normal (Example 3.2), hence completely regular. Thus their product $\Omega \times (\Omega \cup \{\Omega\})$ is also completely regular. But $\Omega \times (\Omega \cup \{\Omega\})$ is not normal. ∎

Our chart of separation axioms now reads:

normal
⇓
completely regular
⇓
regular
⇓
Hausdorff

and none of the implications is reversible.

In view of the last two theorems, each subspace of a Cartesian product of closed intervals is completely regular. We shall now show the converse: each completely regular space is homeomorphic to a subspace of an interval product. In fact, since each closed interval in E^1 is homeomorphic to $[0, 1]$, we can characterize completely regular spaces as precisely the subspaces of "cubes" built from $[0, 1]$.

4.6 Theorem *Let (X, T) be completely regular. Then, for some set A, (X, T) is homeomorphic to a subspace of $\left(\prod_{\alpha \in A} I_\alpha, \mathbb{P}\right)$, where $I_\alpha = [0, 1] \subset E^1$ for each $\alpha \in A$.*

Proof Let $A = \{f | f \text{ is continuous}: X \to [0, 1]\}$.
Define a map $\varphi: X \to \prod_{\alpha \in A} I_\alpha$ by specifying that $\varphi(x)(t) = t(x)$ for each $x \in X$ and $t \in A$. We leave it for the reader to show that φ is an homeomorphism of X onto the subspace $\varphi(X)$ of $\prod_{\alpha \in A} I_\alpha$. ∎

If the details of the proof become too oppressive, take a look at the proof of Urysohn's Metrization Theorem in Chapter VIII.

A map φ as defined in the proof is sometimes called an evaluation map. Use of evaluation maps is also made in proving Ascoli's Theorem, section 9 of Chapter VIII, and they are discussed briefly in Problem 33, Chapter VII.

Problems

1) Give an example of a non-Hausdorff space in which $\cap \{F | x \in F \text{ and } F \text{ is closed}\} = \{x\}$ for each $x \in X$. Why does this not contradict Theorem 1.2?

2) Suppose, if $x \in X$, there is a T-nghd. V of x such that the space $(\bar{V}, T_{\bar{V}})$ is Hausdorff. Then, (X, T) is Hausdorff.

3) Let T be a topology on X. Then,
   T is Hausdorff
   if and only if
   if x and y are distinct points of X, then there are T-closed sets F_x and F_y such that $F_x \cup F_y = X$, $x \in F_x$, $y \notin F_x$, $y \in F_y$ and $x \notin F_y$.

4) Let $f: X \to Y$ be a continuous surjection. Then Y is Hausdorff if X is.

5) Let $f: X \to Y$ be a continuous, closed surjection. If X is Hausdorff, so is Y.

6) Let $f: X \to Y$ be continuous, and let Y be Hausdorff. If $x, y \in X$, write $x \sim y$ if and only if $f(x) = f(y)$. Then, the quotient space $X/\sim$ is Hausdorff.

7) Let (X, T) be any space, and $\sim$ an equivalence relation on X.
   (a) Suppose (1) $\sim$ is a closed subset of the product space $X \times X$,
       (2) η is an open map.
   Then, $(X/\sim, I_\eta)$ is Hausdorff.
   (b) If $(X/\sim, I_\eta)$ is Hausdorff, then $\sim$ is closed in $X \times X$.

8) Each retract (see problem 4, Ch. III) of a Hausdorff space is closed.

9) Let $f, g: X \to Y$ be continuous, and Y Hausdorff. Suppose that $A \subset X$ such that $\bar{A} = X$ and $f|A = g|A$. Then, $f = g$.

10) Let (X, T) be regular. Let $\sim$ be an equivalence relation on X.
    (a) If $\eta: X \to X/\sim$ is a closed map, then $\sim$ is closed in $X \times X$.
    (b) If η is both open and closed, then $X/\sim$ is Hausdorff in the quotient topology.

11) Suppose, if $x \in X$, there is a T-nghd. V of x such that $(\bar{V}, T_{\bar{V}})$ is regular.
    Then, (X, T) is regular.

*12) Let (X, T) be regular. Let A be an infinite subset of X. Then, there is a countable set $\mathcal{O}$ of T-open sets such that $\bar{U} \cap \bar{V} = \phi$ whenever U, V are distinct elements of $\mathcal{O}$ and, for some $W \in \mathcal{O}$, $A \cap V \neq \phi$ for each $V \in \mathcal{O} - \{W\}$. (Hint: construct $\mathcal{O} = \{V_i | i \in Z^+ \cup \{0\}\}$ inductively, letting $V_0 = \phi$).

13) X is completely normal if each subspace of X is normal. Show that each metric space is completely normal.

14) If $\prod_{\alpha \in \mathcal{O}} X_\alpha$ is normal, then so is each X_α.

*15) Let R be the set of real numbers, with topology T generated by the half-open intervals $]a, b]$. Then, R is normal, but $R \times R$ is not.

16) Show that E^n can replace E^1 in the Urysohn-Tietze Extension Theorem.

*17) Let $\varphi: F \to E^n$ be continuous, where F is a closed subset of a normal space X.
    Suppose that $|\varphi(x)| \leq M$ for each $x \in F$.
    Then, there exists a continuous $\Phi: X \to E^n$ such that $\Phi|F = \varphi$ and $|\Phi(x)| \leq M\sqrt{n}$ for each $x \in X$.

18) A cover $\mathcal{O}$ of X by subsets of X is point-finite if each point of X is an element of at most finitely many sets in $\mathcal{O}$.
Let T be a topology on X. Suppose that T is Hausdorff. Then, (X, T) is normal
if and only if
if $\mathcal{O} = \{G_\alpha | \alpha \in \mathcal{M}\}$ is a point-finite, T-open cover of X, then there is a T-open cover $\mathcal{B} = \{H_\alpha | \alpha \in \mathcal{M}\}$ such that $\overline{H}_\alpha \subset G_\alpha$ for each $\alpha \in \mathcal{M}$ and $H_\alpha \neq \phi$ whenever $G_\alpha \neq \phi$.

19) Let (X, T) be normal. Let F be T-closed. Let $F \subset \bigcup_{i=1}^{n} V_i$, where each $V_i \in T$. Then, there are T-open sets $U_1, ..., U_n$ with $\overline{U}_i \subset V_i$ and $F \subset \bigcup_{i=1}^{n} U_i$.

*20) Let $\{V_1, ..., V_n\}$ be an open cover of X, and let X be a normal space. Then, there are continuous $f_i : X \to [0, 1]$ such that $\sum_{i=1}^{n} f_i(x) = 1$ for each $x \in X$, and $f_i(X - V_i) = \{0\}$, for $i = 1, ..., n$.

*21) Let X be normal. Let A and B disjoint, closed subsets of X. Then,
(a) There exists a continuous $f: X \to [0, 1]$ such that $f(B) = \{1\}$ and $A = f^{-1}(\{0\})$ if and only if A is a G_δ set (i.e. A is a countable intersection of open sets).
Contrast "$A = f^{-1}(\{0\})$" with "$f(A) = \{0\}$" in Urysohn's Lemma.
(b) There exists continuous $f: X \to [0, 1]$ such that $f^{-1}(\{1\}) = B$ and $f^{-1}(\{0\}) = A$ if and only if A and B are both G_δ-sets.

CHAPTER VI

Connectivity

6.1 Connectedness

AS WE MENTIONED in the second chapter, one object of topology is to characterize spaces by distinguishing their topological invariants. One of the more important invariants is connectedness.

A space is connected if it cannot be partitioned into non-empty, disjoint open pieces. For example, on a purely intuitive basis, one has the feeling that E^1 consists of just one piece, and is connected, while E^1 with a point removed consists of two separated pieces, and so is disconnected.

For the remainder of this chapter, T is a topology on X.

1.1 DEFINITION *A and B T-separate X
if and only if*
(1) A and B are T-open
(2) $A \neq \phi$ and $B \neq \phi$.
(3) $A \cap B = \phi$.
(4) $A \cup B = X$.

1.2 DEFINITION *X is T-connected
if and only if
there do not exist A and B which T-separate X.*

1.1 EXAMPLE E^1 is connected in the Euclidean topology. This will follow easily from a later theorem.

But the set of reals is not connected in the discrete topology, since, for example, the rationals and irrationals form a separation in this topology.

The set of reals is also connected in the cofinal topology, but not in the topology generated by the half-open intervals $]a, b]$. ∎

1.2 EXAMPLE Sierpinski space is connected. ∎

1.3 EXAMPLE Any indiscrete space is connected. ∎

If X is not T-connected, we generally say that X is T-disconnected, or, if T is clear, simply that X is disconnected.

There are many alternate formulations of the notion of connectedness. The ones we shall use are given in the next theorem. In (3) of the theorem, and throughout the remainder of the chapter, we always consider $\{0, 1\}$ as a discrete space in speaking of continuous functions on X to $\{0, 1\}$. Note throughout the chapter how handy condition (3) is in devising neat, short proofs.

1.1 Theorem *The following are equivalent:*
(1) X is T-connected.
(2) The only subsets of X which are both T-open and T-closed are ϕ and X.
(3) There is no continuous surjection $f: X \to \{0, 1\}$.

Proof (1) $\Rightarrow$ (2):
Assume (1). If $A \neq \phi$ and $A \neq X$ and A is both T-open and T-closed, then A and $X - A$ separate X, a contradiction.

(2) $\Rightarrow$ (3):
Assume (2). Suppose that $f: X \to \{0, 1\}$ is a continuous surjection. By continuity, $f^{-1}(\{0\})$ is T-open and T-closed, as $\{0\}$ is open and closed in $\{0, 1\}$. Since f is surjective, $f^{-1}(\{0\}) \neq \phi$, and $f^{-1}(\{0\}) \neq X$, a contradiction to (2).

(3) $\Rightarrow$ (1):
Assume (3). If A and B separate X, then the function $f: X \to \{0, 1\}$ defined by:

$$f(x) = \begin{cases} 0 \text{ if } x \in A \\ 1 \text{ if } x \in B \end{cases}$$

is a continuous surjection, contradicting (3). ∎

It is often convenient to be able to speak of a subset of a space as being connected or disconnected. By this we always understand that the subset is being considered as a subspace, that is, connectedness is measured by the relative topology.

1.3 Definition *Let $A \subset X$.*
Then,
A is T-connected
if and only if
A is T_A-connected.

1.4 Example E^2 is connected, but the subset $\{(x, y) | x^2 + y^2 < 1\} \cup \{(x, y) | 3 \leq x^2 + y^2 \leq 5\}$ is not. ∎

The next theorem is often useful in determining whether or not a given set B is connected. If B can be fit between A and $\bar{A}$, and if A is connected, then so is B. In particular, $\bar{A}$ is connected whenever A is, although not conversely (let $A = E^1 - \{0\}$).

1.2 THEOREM *Let A and B be subsets of X.*
Let $A \subset B \subset \bar{A}$.
Suppose that A is T-connected.
Then,
B is T-connected.

Proof Suppose that B is not T-connected. Then, there are disjoint, non-empty, T_P-open sets P and M such that $B = P \cup M$.

Choose $t_P \in T$ and $t_M \in T$ such that $P = B \cap t_P$ and $M = B \cap t_M$. Now,
$$A = B \cap A = (P \cup M) \cap A = (P \cap A) \cup (M \cap A)$$
$$= ((t_P \cap B) \cap A) \cup ((t_M \cap B) \cap A) = (t_P \cap A) \cup (t_M \cap A).$$

Thus, we have disjoint, T_A-open sets $t_P \cap A$ and $t_M \cap A$ whose union is A. We claim that $t_P \cap A \neq \phi$ and $t_M \cap A \neq \phi$.

To show this, note that $t_P \cap B \neq \phi$, so there is some $b \in t_P \cap B$. Since $B \subset \bar{A}$, then $b \in \bar{A}$. Since $b \in t_P \in T$, then $t_P \cap A \neq \phi$.

Similarly, $t_M \cap A \neq \phi$.

But then $t_P \cap A$ and $t_M \cap A$ form a T_A-separation of A, a contradiction. Hence B is T-connected. ∎

1 COROLLARY *Let $A \subset X$ and suppose that A is T-connected.*
Then,
$\bar{A}$ is T-connected.

Proof Choose $B = \bar{A}$ in Theorem 1.2. ∎

We shall give applications of the last theorem and corollary after we have accumulated some more theory.

It is not true in general that a non-empty intersection of connected sets is connected. For example, let A be the annulus $\{(x,y) | 40 < x^2 + y^2 < 300\}$ in E^2, and B the strip $\{(x, y) | -1 \leq x \leq 1$ and y is real$\}$. A look at Figure 1.1 should convince anyone that $A \cap B$ is disconnected.

Of course, a union of connected sets may also be disconnected—but not if they have an element in common. The next theorem is somewhat more general than this.

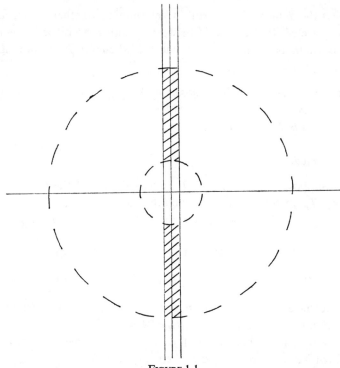

FIGURE 1.1

1.3 THEOREM *Let $B \neq \phi$.*
Let A_α be T-connected for each $\alpha \in B$.
Suppose that $A_\alpha \cap A_\beta \neq \phi$ whenever $\alpha, \beta \in B$.
Then,
$\bigcup_{\alpha \in B} A_\alpha$ is T-connected.

Proof Let $f: \bigcup_{\alpha \in B} A_\alpha \to \{0, 1\}$ be continuous in the relative topology on $\bigcup_{\alpha \in B} A_\alpha$.

Then, $f|A_\alpha: A_\alpha \to \{0, 1\}$ is also continuous for each $\alpha \in B$. Since A_α is T-connected, then $f|A_\alpha$ is not a surjection. Then, $f|A_\alpha: A_\alpha \to \{x\}$, where x is either 0 or 1.

Similarly, if $\beta \in B$, then $f|A_\beta: A_\beta \to \{y\}$, where y is either 0 or 1.

But, $A_\alpha \cap A_\beta \neq \phi$, so, for some t, $t \in A_\alpha \cap A_\beta$.

Then, $(f|A_\alpha)(t) = x = (f|A_\beta)(t) = y$.

But then $f|A_\alpha: X_\alpha \to \{x\}$ for each $\alpha \in B$, so that $f: \bigcup_{\alpha \in B} A_\alpha \to \{x\}$. Then, f is not a surjection, implying that $\bigcup_{\alpha \in B} A$ is T-connected. ∎

Connectedness is preserved by continuous surjections, hence also by homeomorphism. Thus connectedness is a topological invariant, and can be used to distinguish spaces.

1.4 THEOREM *Let (Y, M) be a topological space.*
Let $f: X \to Y$ be a (T, M) continuous surjection.
Suppose that X is T-connected.
Then,
Y is M-connected.

Proof Let $g: Y \to \{0, 1\}$ be continuous. Then, $g \circ f: X \to \{0, 1\}$ is also continuous.

Now, $g \circ f$ is not a surjection, as X is T-connected. Since f is a surjection, then g cannot be surjective, implying that Y is M-connected. ∎

Alternatively, we could prove Theorem 1.4 by noting that any M-separation A and B of Y yields a T-separation $f^{-1}(A)$ and $f^{-1}(B)$ of X.

2 COROLLARY *Let $(X, T) \cong (Y, M)$.*
Then,
X is T-connected $\Leftrightarrow Y$ is M-connected.

Proof Immediate by Theorem 1.4. ∎

Before showing how these theorems are used, we shall characterize the connected sets in E^1.

If $A \subset E^1$, we call A an interval if $c \in A$ whenever $a, b \in A$ and $a < c < b$. For example, ϕ, $\{2\}$, $[1, 3]$, $]0, 1[$ and E^1 are intervals, while the set of irrationals is not an interval.

1.5 THEOREM A is connected in $E^1 \Leftrightarrow A$ is an interval.

Proof Suppose first that $A \subset E^1$ and that A is connected. If $a, b \in A$, and $a < c < b$ such that $c \notin A$, then $A \cap \{x | x < c\}$ and $A \cap \{x | x > c\}$ separate A, a contradiction. Hence A must be an interval.

Conversely, suppose that A is an interval. If A is not connected, then there is a continuous surjection $f: A \to \{0, 1\}$. Then, for some $a \in A$, $f(a) = 0$ and for some $b \in A$, $f(b) = 1$.

Suppose $a < b$. The proof is essentially the same if $b < a$.

Observe that, since $f^{-1}(\{0\})$ is open in E^1, and A is an interval, there is some $x \in]a, b[$ such that $f(x) = 0$.

Similarly, there is some $y \in]a, b[$ with $f(y) = 1$. Let $z = \inf \{y | y > a$ and $f(y) = 1\}$. Then, $a < z < b$. Now, either $f(z) = 0$ or $f(z) = 1$.

If $f(z) = 0$, then $z \in f^{-1}(\{0\}) \in T_A$, so, for some $\varepsilon > 0$, $]z - \varepsilon, z + \varepsilon[\subset f^{-1}(\{0\})$. But then $z < z + \varepsilon$ and $f(x) = 0$ for $a < x < z + \varepsilon$, contradicting the choice of z.

If $f(z) = 1$, then $z \in f^{-1}(\{1\}) \in T_A$. Then, for some $\delta > 0$, $]z - \delta, z + \delta[\subset f^{-1}(\{1\})$. Then, $f(x) = 1$ for $z - \delta < x < z$, again contradicting the choice of z.

Thus there is no continuous surjection of A onto $\{0, 1\}$, and A is connected. ∎

1.5 EXAMPLE $]0, 1[$ and $]0, 1]$, as subspaces of E^1, are not homeomorphic. To prove this, suppose that $\varphi:]0, 1] \to]0, 1[$ is an homeomorphism. Then $\xi:]0, 1[\to]0, 1[- \{\varphi(1)\}$ defined by $\xi(x) = \varphi(x)$ if $x \in]0, 1[$ is also an homeomorphism.

But, $]0, 1[$ is connected, and $]0, 1[- \{x\}$ is not for any $x \in]0, 1[$, by Theorem 1.5. This is a contradiction to Corollary 2.

This example is typical of the use of a topological invariant to determine whether or not two spaces are topologically the same. ∎

1.6 EXAMPLE Let $B = \{(x, \sin(1/x)) | 0 < x \leq 1/\pi\} \cup \{(0, t) | -1 \leq t \leq 1\}$ (see Figure 1.2). Then, B is a connected subset of E^2.

To prove this, note that $f:]0, 1/\pi] \to \{(x, \sin(1/x)) | 0 < x \leq 1/\pi\}$ defined by $f(x) = (x, \sin(1/x))$ for $0 < x \leq 1/\pi$ is a continuous surjection (continuity is most easily proved by Theorem 2.7.III). Now, $]0, 1/\pi]$, being an interval, is connected in E^1, so that $\{(x, \sin(1/x)) | 0 < x \leq 1/\pi\}$ is

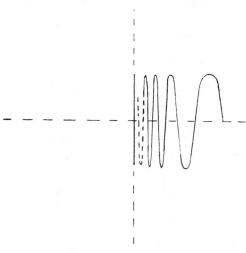

FIGURE 1.2

connected in E^2 by Theorem 1.4. Then, B is connected by Corollary 1, since $\{(x, \sin(1/x)) | 0 < x \leq 1/\pi\} = B$. ∎

We shall now show that a product space is connected exactly when each coordinate space is connected. This type of theorem is quite typical—one deduces a topological invariant of a product space by examining each of the (usually simpler) coordinate spaces. The existence of such theorems (more will be seen in this and the next chapter) is the real justification for the choice of a product topology made in Chapter III. As an illustration, connectedness of E^n becomes a trivial consequence of Theorem 1.5, since E^1 is an interval.

The proof is fairly lengthy, and will be preceded by a technical Lemma.

1 LEMMA *Let $A \neq \phi$.*
Let T_α be a topology on X_α for each $\alpha \in A$.
Suppose that X_α is T_α-connected for each $\alpha \in A$.
Let $\mathbb{P}$ be the product topology on $\prod_{\alpha \in A} X_\alpha$.
Let $f \in \prod_{\alpha \in A} X_\alpha$.
Let
$$Y = \left(\prod_{\alpha \in A} X_\alpha\right) \cap \{g | A \cap \{\alpha | g(\alpha) \neq f(\alpha)\} \text{ is finite}\}.$$
Then,

(1) $\bar{Y} = \prod_{\alpha \in A} X_\alpha$.

(2) If $\beta \in A$, let $\varphi_\beta^f : X_\beta \to \prod_{\alpha \in A} X_\alpha$ be defined by:
$$(\varphi_\beta^f(x))(\alpha) = \begin{cases} f(\alpha) \text{ if } \alpha \in A \text{ and } \alpha \neq \beta \\ x \text{ if } \alpha = \beta, \end{cases}$$
for each $x \in X_\beta$.
Then, φ_β^f is $(T_\beta, \mathbb{P})$ continuous.

(3) Let $S: \prod_{\alpha \in A} X_\alpha \to \{0, 1\}$ be continuous.
Then, $S(g) = S(f)$ whenever $g \in Y$.

(4) If $S(Y) = \{0\}$ (or $S(Y) = \{1\}$), then $S\left(\prod_{\alpha \in A} X_\alpha\right) = \{0\}$ $\left(\text{or } S\left(\prod_{\alpha \in A} X_\alpha\right) = \{1\}\right)$.

Proof of (1) Let $h \in \prod_{\alpha \in A} X_\alpha$, and let $\bigcap_{i=1}^{n} p_{\alpha_i}^{-1}(U_{\alpha_i})$ be a basic $\mathbb{P}$-nghd. of h. Define $q \in \prod_{\alpha \in A} X_\alpha$ by:
$$q(\alpha) = \begin{cases} f(\alpha) \text{ if } \alpha \in A \text{ and } \alpha \neq \alpha_i \quad \text{for } i = 1, \ldots, n, \\ h(\alpha_i) \text{ if } \alpha = \alpha_i, \quad i = 1, \ldots, n. \end{cases}$$

Then, $q \in Y \cap \left(\bigcap_{i=1}^{n} p_{\alpha_i}^{-1}(U_{\alpha_i})\right)$, so $Y \cap \left(\bigcap_{i=1}^{n} p_{\alpha_i}^{-1}(U_{\alpha_i})\right) \neq \phi$, implying that $h \in \bar{Y}$.

Proof of (2) Let $p_\alpha^{-1}(U_\alpha)$ be a subbasic $\mathbb{P}$-open set. Note that, if $x \in X_\beta$, then
$$x \in (\varphi_\beta^f)^{-1}(p_\alpha^{-1}(U_\alpha)) \Leftrightarrow (\varphi_\beta^f)(x) \in p_\alpha^{-1}(U_\alpha)$$
$$\Leftrightarrow p_\alpha((\varphi_\beta^f)(x)) \in U_\alpha \Leftrightarrow ((\varphi_\beta^f)(x))(\alpha) \in U_\alpha.$$

There are three possibilities:

i) $\beta \neq \alpha$ and $f(\alpha) \in U_\alpha$.
Then, $((\varphi_\beta^f)(x))(\alpha) = f(\alpha) \in U_\alpha$ for each $x \in X_\beta$, so $(\varphi_\beta^f)^{-1}(p_\alpha^{-1}(U_\alpha)) = X_\beta \in T_\beta$.

ii) $\beta \neq \alpha$ and $f(\alpha) \notin U_\alpha$.
Then, $((\varphi_\beta^f)(x))(\alpha) = f(\alpha) \notin U_\alpha$ for each $x \in X_\beta$, so $(\varphi_\beta^f)^{-1}(p_\alpha^{-1}(U_\alpha)) = \phi \in T_\beta$.

iii) $\beta = \alpha$.
Then, $((\varphi_\beta^f)(x))(\alpha) = ((\varphi_\beta^f)(x))(\beta) = x$, so $(\varphi_\beta^f)^{-1}(p_\alpha^{-1}(U_\alpha)) = (\varphi_\beta^f)^{-1}(p_\beta^{-1}(U_\beta))) = U_\beta \in T_\beta$.

By Theorem 1.2 (2–ii).II, φ_β^f is $(T_\beta, \mathbb{P})$ continuous.

Proof of (3) We proceed by induction.
If n is a positive integer, let $Q(n)$ be the proposition:
if $g \in Y$ and $A \cap \{\alpha | f(\alpha) \neq g(\alpha)\}$ has n elements, then $S(f) = S(g)$.
To show that $Q(1)$ is true, suppose that $g \in Y$ and $g(\alpha) = f(\alpha)$ for each $\alpha \in A$ except $\alpha = \beta$, where $g(\beta) \neq f(\beta)$. For convenience, denote $\varphi_\beta^f = \varphi$. Note that $\varphi: X_\beta \to \prod_{\alpha \in A} X_\alpha$ is $(T_\beta, \mathbb{P})$ continuous by (2). Now, X_β is T-connected. Then, $S \circ \varphi$ is not a surjection, so $S \circ \varphi: X_\beta \to \{t\}$, where t is either 0 or 1.

Now,
$$(\varphi(f(\beta)))(\alpha) = \begin{cases} f(\alpha) \text{ if } \alpha \neq \beta \\ f(\beta) \text{ if } \alpha = \beta. \end{cases}$$

Thus, $\varphi(f(\beta)) = f$.
Similarly,
$$(\varphi(g(\beta)))(\alpha) = \begin{cases} f(\alpha) = g(\alpha) \text{ if } \alpha \neq \beta \\ g(\beta) \text{ if } \alpha = \beta. \end{cases}$$

Then,
$$(S \circ \varphi)(f(\beta)) = S(f) = t = (S \circ \varphi)(g(\beta)) = S(g),$$

proving $Q(1)$.

Now let n be a positive integer and assume that $Q(n)$ is true. To prove $Q(n + 1)$, let $g \in Y$ and suppose that $A \cap \{\alpha | g(\alpha) \neq f(\alpha)\}$ has $n + 1$ elements, say $\alpha_1, ..., \alpha_{n+1}$.

Define $h \in Y$ by:
$$h(\alpha) = \begin{cases} f(\alpha) \text{ if } \alpha \in A \text{ and } \alpha \neq \alpha_1, ..., \alpha_n, \\ g(\alpha) \text{ if } \alpha = \alpha_1, ..., \alpha_n. \end{cases}$$

Then, $A \cap \{\alpha | h(\alpha) \neq f(\alpha)\} = \{\alpha_1, ..., \alpha_n\}$ has n elements. By the inductive hypothesis, $S(h) = S(f)$.

Now note that $h(\alpha) = g(\alpha)$ if $\alpha \in A$ and $\alpha \neq \alpha_{n+1}$. Also, by (2), $\varphi_{\alpha_{n+1}}^h$: $X_{\alpha_{n+1}} \to \prod_{\alpha \in A} X_\alpha$ is $(T_{\alpha_{n+1}}, \mathbb{P})$ continuous. Since $X_{\alpha_{n+1}}$ is $T_{\alpha_{n+1}}$-connected, then $S \circ \varphi_{\alpha_{n+1}}^h : X_{\alpha_{n+1}} \to \{0, 1\}$ is not a surjection. Then, $S \circ \varphi_{\alpha_{n+1}}^h : X_{\alpha_{n+1}} \to \{z\}$, where z is 0 or 1. Now,
$$(\varphi_{\alpha_{n+1}}^h(h(\alpha_{n+1})))(\alpha) = \begin{cases} h(\alpha) \text{ if } \alpha \neq \alpha_{n+1} \\ h(\alpha_{n+1}) \text{ if } \alpha = \alpha_{n+1}, \end{cases}$$
so
$$(\varphi_{\alpha_{n+1}}^h)(h(\alpha_{n+1})) = h.$$

Similarly,
$$(\varphi_{\alpha_{n+1}}^h(g(\alpha_{n+1})))(\alpha) = \begin{cases} h(\alpha) = g(\alpha) \text{ if } \alpha \neq \alpha_{n+1} \\ g(\alpha_{n+1}) \text{ if } \alpha = \alpha_{n+1}. \end{cases}$$

Then,
$$(\varphi_{\alpha_{n+1}}^h)(g(\alpha_{n+1})) = g.$$

Then,
$$(S \circ \varphi_{\alpha_{n+1}}^h)(h) = S(h) = z = (S \circ \varphi_{\alpha_{n+1}}^h)(g) = S(g).$$

Then,
$$S(f) = S(h) = S(g), \text{ and } Q(n + 1)$$
is true.

Proof of (4) Suppose that $S(Y) = \{0\}$. We wish to show that $S\left(\prod_{\alpha \in A} X_\alpha\right) = \{0\}$.

Suppose instead that, for some g, $g \in \prod_{\alpha \in A} X_\alpha$ and $S(g) = 1$. By continuity of S, $S^{-1}(\{1\})$ is $\mathbb{P}$-open. Since $g \in S^{-1}(\{1\})$, and, by (1), $g \in \bar{Y}$, then $S^{-1}(\{1\}) \cap Y \neq \phi$.

Let $h \in S^{-1}(\{1\}) \cap Y$. Then, $h \in Y$, so $S(h) = 0$. But, $h \in S^{-1}(\{1\})$, so $S(h) = 1$, a contradiction, proving (4) when $S(Y) = \{0\}$.

If $S(Y) = \{1\}$, then $S\left(\prod_{\alpha \in A} X_\alpha\right) = \{1\}$ by interchanging 0 and 1 in the above argument. ∎

The proof of the theorem is now fairly straightforward.

1.6 THEOREM *Let $A \neq \phi$.*
Let T_α be a topology on X_α for each $\alpha \in A$.
Let $\mathbb{P}$ be the product topology on $\prod_{\alpha \in A} X_\alpha$.
Then,
$\prod_{\alpha \in A} X_\alpha$ is $\mathbb{P}$-connected
if and only if
X_α is T_α-connected for each $\alpha \in A$.

Proof Suppose first that $\prod_{\alpha \in A} X_\alpha$ is $\mathbb{P}$-connected. If $\beta \in A$, then $p_\beta : \prod_{\alpha \in A} X_\alpha \to X_\beta$ is a continuous surjection, hence X_β is T_β-connected.

Conversely, suppose that X_α is T_α-connected for each $\alpha \in A$. If $\prod_{\alpha \in A} X_\alpha$ is not $\mathbb{P}$-connected, then there is a continuous surjection $S: \prod_{\alpha \in A} X_\alpha \to \{0, 1\}$. Let $f \in \prod_{\alpha \in A} X_\alpha$. Let

$$Y = \left(\prod_{\alpha \in A} X_\alpha\right) \cap \{g | A \cap \{\alpha | g(\alpha) \neq g(\alpha)\} \text{ is finite}\}.$$

By Lemma 1 (3), $S(Y) = \{S(f)\}$. By Lemma 1 (4), $S\left(\prod_{\alpha \in A} X_\alpha\right) = \{S(f)\}$, a contradiction.

Hence $\prod_{\alpha \in A} X_\alpha$ is $\mathbb{P}$-connected. ∎

Actually, only parts (3) and (4) of the Lemma are needed in the proof of the theorem. But (1) is used in proving (4), and (2) in proving (3).

Unlike product spaces, identification (and hence also quotient spaces) are handled very easily.

1.7 THEOREM *Let $f: X \to Y$ be a surjection.*
Suppose that X is T-connected.
Then,
Y is I_f-connected.

Proof Immediate by Theorem 1.4 and Theorem 3.1 (2).III. ∎

Suppose $x \in X$. Now, X itself may or may not be connected, but $\{x\}$ certainly is, regardless of T. Thus there is at least one connected subspace of X containing x as a point. The largest such subspace is called the component of X (relative to the given topology T).

1.4 DEFINITION *Let $x \in X$.*
Then,
$\text{Comp}_T(x) = \cup\{A | x \in A \subset X \text{ and } A \text{ is } T\text{-connected}\}$.

1.8 Theorem

(1) $\text{Comp}_T(x)$ *is T-connected for each* $x \in X$.
(2) If $x \in A \subset X$ *and* A *is T-connected, then* $A \subset \text{Comp}_T(x)$.
(3) $\bigcup_{x \in X} \text{Comp}_T(x) = X$.
(4) If $x, y \in X$, *then* $\text{Comp}_T(x) \cap \text{Comp}_T(y) = \phi$ *or* $\text{Comp}_T(x) = \text{Comp}_T(y)$.
(5) $\text{Comp}_T(x)$ *is T-closed for each* $x \in X$.
(6) X *is T-connected* $\Leftrightarrow \text{Comp}_T(x) = X$ *for each* $x \in X$.

Proof of (1) Immediate by Theorem 1.3.

Proof of (2) Immediate by Definition 1.4.

Proof of (3) Immediate, as $x \in \text{Comp}_T(x)$ for each $x \in X$.

Proof of (4) Suppose that $\text{Comp}_T(x) \cap \text{Comp}_T(y) \ne \phi$. Then, $\text{Comp}_T(x) \cup \text{Comp}_T(y)$ is T-connected by Theorem 1.3. But, $x \in \text{Comp}_T(x) \cup \text{Comp}_T(y)$ by the definition, so $\text{Comp}_T(x) \cup \text{Comp}_T(y) \subset \text{Comp}_T(x)$ by (2). Then, $\text{Comp}_T(y) \subset \text{Comp}_T(x)$.
Similarly, $\text{Comp}_T(x) \subset \text{Comp}_T(y)$.

Proof of (5) If $x \in X$, then $\text{Comp}_T(x)$ is T-connected by (1).
By Corollary 1, $\overline{\text{Comp}_T(x)}$ is T-connected. By (2), $\overline{\text{Comp}_T(x)} \subset \text{Comp}_T(x)$, implying that $\text{Comp}_T(x)$ is T-closed.

Proof of (6) Left as an exercise. ∎

1.7 Example In a discrete space, $\text{Comp}_T(x) = \{x\}$. ∎

1.8 Example If $f \in \prod_{\alpha \in A} X_\alpha$, then $\text{Comp}_{\mathbb{P}}(f) = \prod_{\alpha \in A} \text{Comp}_{T_\alpha}(f(\alpha))$. ∎

The number of distinct components gives some measure of how disconnected a space is. Intuitively, a space with five components is "more" disconnected than a space with three components. The number of components is a topological invariant, and provides another means of distinguishing spaces.

1.9 Theorem *Let φ be a (T, M) homeomorphism:* $X \to Y$. *Then,*

(1) $\varphi(\text{Comp}_T(x)) = \text{Comp}_M(\varphi(x))$ *for each* $x \in X$.
(2) There is a bijection on the set of T-components of X onto the set of M-components of Y.

Proof Left as an exercise. ∎

130   FUNDAMENTAL CONCEPTS OF TOPOLOGY

1.9 EXAMPLE The subspaces of E^2 shown in Figure 1.3 are not homeomorphic, since one has three components, the other two. ∎

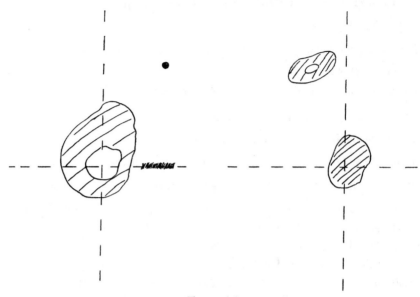

FIGURE 1.3

6.2 Path connectedness

The notion of path connectedness arises very naturally in a number of ways. For example, in complex analysis one often wishes to integrate along a path between two points in a given region. A space having the property that any two points can be joined by a path is called path connected.

In elementary analysis it is usually sufficient to think of a path as a curve or locus (usually in the plane). For example, given functions f, g: $[0, 1] \to E^1$, we would think of $\{(f(t), g(t)) | 0 \leq t \leq 1\}$ as a path from $(f(0), g(0))$ to $(f(1), g(1))$. In topology, however, it is more convenient (for reasons which will become apparent) to think of a path as a function instead of as a locus or image. Experience has also shown it useful to consider only continuous functions, and for uniformity we shall always take the domain as $[0, 1]$, though any closed interval would do just as well.

In this section, T is a topology on X. Continuity of functions on $[0, 1]$ to X is always measured in the relative topology of E^1 on $[0, 1]$, and T.

2.1 Definition *f is a path in X*
if and only if
f is continuous: $[0, 1] \to X$.

2.2 Definition *Let $x \in X$ and $y \in X$.*
Then,
f is a path from x to y in X
if and only if
f is a path in X and $f(0) = x$ and $f(1) = y$.

In Definition 2.2, x is called the initial, and y the terminal point of f. Note that a path f from x to y gives rise to a path g from y to x by letting $g(x) = f(1 - x)$, $0 \leq x \leq 1$.

2.1 Example Let $X = \{0, 1\}$, with the Sierpinski topology. Let $f(x) = 0$ for each $x \in [0, 1]$. Then f is a path from 0 to 0 in X. Here, f is a closed path, having the same initial and terminal points. ∎

2.2 Example Any $f: [0, 1] \to X$ is a path if T is the indiscrete topology on X. ∎

2.3 Definition *X is T-path connected*
if and only if
if $x, y \in X$, then there is a path from x to y in X.

2.3 Example E^n is path connected. ∎

2.4 Example A connected space need not be path connected. A classic example is the following.

Let $A = \{(x, \sin(1/x)) | 0 < x \leq 1/\pi\} \cup \{(0, 0)\}$, considered as a subspace of E^2. As in Example 1.6, A is connected. But A is not path connected: there is no path in A from $(0, 0)$ to any point $(x, \sin(1/x))$ with $0 < x \leq 1/\pi$. In particular, we shall prove that there is no path in A from $(0, 0)$ to $(1/\pi, 0)$.

Suppose that g is such a path. Then, $g: [0, 1] \to A$ and g is continuous, $g(0) = (0, 0)$ and $g(1) = (1/\pi, 0)$. Let p_1 and p_2 be the projections to the x and y axes respectively. Note that $p_1 \circ g$ and $p_2 \circ g$ are continuous.

Let $M = \sup([0, 1] \cap \{t | g(t) = (0, 0)\})$. Then, $0 \leq M < 1$. We claim first that $(p_1 \circ g)(M) = 0$. This is immediate if $M = 0$, so suppose that $0 < M < 1$. If $(p_1 \circ g)(M) > 0$, then there is a nghd. $]M - \delta, M + \delta[$ of M in $[0, 1]$ with $|(p_1 \circ g)(M) - (p_1 \circ g)(t)| < \dfrac{(p_1 \circ g)(M)}{2}$ for $M - \delta$

$< t < M + \delta$. Then, $|(p_1 \circ g)(M)| - |(p_1 \circ g)(t)| < \dfrac{(p_1 \circ g)(M)}{2}$, so $0 < \dfrac{(p_1 \circ g)(M)}{2} < (p_1 \circ g)(t)$ for $M - \delta < t < M$, a contradiction to the choice of M.

Since $(p_1 \circ g)(M) = 0$, then $g(M) = (0, 0)$, so $(p_2 \circ g)(M) = 0$ also. Let $\varepsilon > 0$ with $M + \varepsilon < 1$. By continuity of $p_2 \circ g$, there is a $\delta > 0$ with $M + \delta < 1$ and $|(p_2 \circ g)(t)| < \varepsilon$ if $M < t \leq M + \delta$. Now, $]M, M + \delta]$ is an interval, so $(p_1 \circ g)(]M, M + \delta])$ is also an interval. Further, by choice of M, $g(t) \neq (0, 0)$ if $M < t \leq M + \delta$. Choose N sufficiently large that $2/N\pi \in (p_1 \circ g)(]M, M + \delta])$. Then, $2/n\pi \in (p_1 \circ g)(]M, M + \delta])$ for all integers $n \geq N$. Choose some integer $K \geq N$ with $\sin(K\pi/2) = 1$. Then, $2/K\pi \in (p_1 \circ g)(]M, M + \delta])$, so there is some $x_K \in]M, M + \delta]$ with $(p_1 \circ g)(x_K) = 2/K\pi$. Then, $(p_2 \circ g)(x_K) = \sin(K\pi/2) = 1 > \varepsilon$, a contradiction. ∎

As this example shows, in general connected $\not\Rightarrow$ path connected. However, the implication can be made in the other direction.

2.1 THEOREM *Let X be T-path connected.*
Then,
X is T-connected.

Proof Let $x \in X$. We shall show that $\text{Comp}_T(x) = X$.

If $y \in X$, then there is a continuous $g_y: [0, 1] \to X$ such that $g_y(0) = x$ and $g_y(1) = y$. Since $[0, 1]$ is connected, so is $g_y([0, 1])$ in X. Then, $y \in g_y([0, 1]) \subset \text{Comp}_T(x)$ by Theorem 1.8 (2). ∎

As with connectedness, path connectedness is preserved by continuous surjections, hence is also a topological invariant.

2.2 THEOREM *Let (Y, M) be a topological space.*
Let $f: X \to Y$ be a (T, M) continuous surjection.
Suppose that X is T-path connected.
Then,
Y is M-path connected.

Proof Let $x, y \in Y$.

Produce $a, b \in X$ such that $f(a) = x$ and $f(b) = y$. There is a path h from a to b in X. It is now easy to check that $f \circ h$ is a path from x to y n Y. ∎

1 COROLLARY *Let $(X, T) \cong (Y, M)$.*
Then,
X is T-path connected $\Leftrightarrow$ Y is M-path connected.

Proof Immediate by Theorem 2.2. ∎

2.5 EXAMPLE $E^n \not\cong E^1$ if $n \geq 2$.

To prove this, suppose that $E^1 \cong E^n$, say φ is an homeomorphism of E^1 with E^n. Let $x \in E^1$. Then, $E^1 - \{x\}$ is not path connected, implying that $E^n - \{\varphi(x)\}$ is not path connected, as φ determines an homeomorphism $E^1 - \{x\} \cong E^n - \{\varphi(x)\}$, in the obvious way. This is a contradiction, since it is easy to show that $E^n - \{t\}$ is path connected for any $t \in E^n$. ∎

2.6 EXAMPLE $E^2 \cap \{(x, y) | x^2 + y^2 < 1\}$ and $\{(x, \sin(1/x)) | 0 < x \leq 1/\pi\} \cup \{(0,0)\}$ considered as subspaces of E^2, are not homeomorphic, since one is path connected and the other is not. Both, however, are connected. ∎

A product space is path connected exactly when each coordinate space is path connected. Fortunately, this is not as tedious to prove as the analogous theorem for connectedness.

2.3 THEOREM *Let $A \neq \phi$.*
Let T_α be a topology on X_α for each $\alpha \in A$.
Let $\mathbb{P}$ be the product topology on $\prod_{\alpha \in A} X_\alpha$.
Then,
$\prod_{\alpha \in A} X_\alpha$ is $\mathbb{P}$-path connected
if and only if
X_α is T_α-path connected for each $\alpha \in A$.

Proof Suppose first that X_α is T_α-path connected for each $\alpha \in A$. Let $f, g \in \prod_{\alpha \in A} X_\alpha$. If $\alpha \in A$, then $f_\alpha, g_\alpha \in X_\alpha$, so there is a path φ_α from f_α to g_α in X_α. Define $\Phi: [0, 1] \to \prod_{\alpha \in A} X_\alpha$ by:

$$(\Phi(t))(\alpha) = \varphi_\alpha(t) \text{ for each } \alpha \in A \text{ and } t \in [0, 1].$$

It is now easy to check that Φ is a path from f to g in $\prod_{\alpha \in A} X_\alpha$.

The converse is immediate by Theorem 2.2, since the projections are continuous surjections. ∎

The following is also an immediate consequence of Theorem 2.2.

2.4 Theorem *Let $f: X \to Y$ be a surjection.*
Suppose that X is T-path connected.
Then,
Y is I_f-path connected.

Proof Use Theorem 2.2. ■

As with connected subsets, it is often convenient to be able to talk about a subset of a space being path connected. This is done in the obvious way.

2.4 Definition *Let $A \subset X$.*
Then,
A is T-path connected
if and only if
A is T_A-path connected.

It should be obvious that a subspace of a path connected space need not be path connected. Easy examples abound in, say, E^2. We shall prove a watered-down of Theorem 1.3. A more general theorem (proved by transfinite induction), is possible, but is not necessary for our purposes. A simple induction argument will extend the next theorem to arbitrary finite unions.

2.5 Theorem *Let A and B be T-path connected subsets of X.*
Let $A \cap B \neq \phi$.
Then,
$A \cup B$ is T-path connected.

Proof Let $x, y \in A \cup B$. Let $z \in A \cap B$. There is a path g from x to z in A. There is a path h from z to y in B. Define $f: [0, 1] \to A \cup B$ by:

$$f(t) = \begin{cases} g(t/2) & \text{if } 0 \leq t \leq \tfrac{1}{2} \\ h(2t - 1) & \text{if } \tfrac{1}{2} \leq t \leq 1. \end{cases}$$

We leave it for the reader to check that f is a path from x to y in $A \cup B$. ■

The path component of a point x of X is the largest path connected subset of X having x as an element. The definition and basic properties are entirely analogous to those of components in section 1.

2.5 Definition *Let $x \in X$.*
Then,
$P \operatorname{Comp}_T (x) = \cup \{A | x \in A \subset X \text{ and } A \text{ is } T\text{-path connected}\}$.

2.6 THEOREM
(1) $P\,\mathrm{Comp}_T(x)$ is T-path connected for each $x \in X$.
(2) If $x \in A \subset X$, and A is T-path connected, then $A \subset P\,\mathrm{Comp}_T(x)$.
(3) $\bigcup_{x \in X} P\,\mathrm{Comp}_T(x) = X$.
(4) If $x \in X$ and $y \in X$, then $P\,\mathrm{Comp}_T(x) \cap P\,\mathrm{Comp}_T(y) = \phi$ or $P\,\mathrm{Comp}_T(x) = P\,\mathrm{Comp}_T(y)$.
(5) X is T-path connected $\Leftrightarrow P\,\mathrm{Comp}_T(x) = X$ for each $x \in X$.

Proof Left as an exercise. ∎

Theorem 1.8 (5) does not extend to path components. $P\,\mathrm{Comp}_T(x)$ need not be a closed subset of X.

2.7 EXAMPLE Let $X = \{(x, \sin(1/x)) | 0 < x \leq 1/\pi\} \cup \{(0, 0)\}$, considered as a subspace of E^2. Then, $P\,\mathrm{Comp}\,((1/\pi, 0)) = \{(x, \sin(1/x)) | 0 < x \leq 1/\pi\}$, but this set is not closed in X, since $(0, 0)$ is a cluster point. ∎

We have seen that in general connectedness does not imply path connectedness. It is natural to ask whether the two notions are ever equivalent. We conclude this section with one such instance.

2.7 THEOREM *Let A be open in E^n.*
Then,
A is connected $\Leftrightarrow A$ is path connected.

Proof Left as an exercise. ∎

Problems

1) S^n is connected in E^{n+1} for each $n \geq 1$.

2) Let R be the set of real numbers, and T the topology of finite complements (consisting of ϕ and all sets $R - S$, S-finite). Examine the space (R, T) with regard to connectedness and path-connectedness.

3) Let R be the set of real numbers, and T the topology of countable complements (consisting of ϕ and all sets $R - S$, S countable). Examine the space (R, T) with regard to connectedness and path-connectedness.

4) Let $X = [0, 1] \times [0, 1]$, and $T_<$ the order topology induced on X by the order: $(a, b) < (c, d)$ if $a < c$ or $(a = c$ and $b < d)$. Examine the space $(X, T_<)$ with regard to connectedness and path-connectedness.

5) Give a topological proof of the Intermediate Value Theorem of calculus:

   Let $f: [a, b] \to E^1$ be continuous. Let x and y be distinct points of $[a, b]$ and let c lie between $f(x)$ and $f(y)$ (i.e., $f(x) < c < f(y)$ or $f(y) < c < f(x)$). Then, for some t, $t \in [a, b]$ and $f(t) = c$.

6) Show by example that $A \cup B$, $A \cap B$, $\text{Bdry}_T(A)$, $\text{Der}_T(A)$, and $\text{Int}_T(A)$ may be disconnected, although A and B are connected.

7) Prove that $E^2 - A$ is connected and path-connected for any countable subset A of E^2.

8) Let (A, B) T-separate X. Let C be a T-connected subset of X. Then, $C \subset A$ or $C \subset B$.

9) Let $A = \{(x, y) | x \text{ and } y \text{ are rational}\}$. Is A connected as a subset of E^2?

10) (a) If X is T-connected and $\sim$ is an equivalence relation on X, then $X/\sim$ is I_η-connected.

    (b) If $X/\sim$ is connected, and $[x]_\sim$ is T-connected for each $x \in X$, then X is T-connected.

11) X is T-connected

    if and only if

    if $\mathcal{O}$ is a T-open cover of X, and $U, V \in \mathcal{O}$, then there are finitely many sets $W_1, \ldots, W_n$ in $\mathcal{O}$ such that $U \cap W_1 \neq \phi$, $W_i \cap W_{i+1} \neq \phi$ for $i = 1, \ldots, n - 1$, and $W_n \cap V \neq \phi$.

12) Prove or disprove the following:

    Let $A \subset X$ such that A is T-path connected. Then, $\bar{A}$ is T-path connected.

13) Define: X is T-locally connected if T has a base consisting of T-connected sets.

    (a) Connected does not imply locally connected, and locally connected does not imply connected.

    (b) $\prod_{\alpha \in \mathcal{O}} X_\alpha$ is locally connected if and only if each X_α is locally connected and all but at most finitely many of the spaces (X_α, T_α) are connected.

    (c) Local connectedness is not preserved by continuous surjections, but is preserved by continuous open surjections (hence also by homeomorphism).

(d) X is T-locally connected if and only if each component of an open set is open.

*(e) The Cantor set, considered as a subspace of E^1, is not locally connected.

(f) If X is T-locally connected, and $\sim$ is an equivalence relation on X, then $X/\sim$ is locally connected in the quotient topology.

14) Define: X is T-locally path connected if and only if T has a base consisting of path connected sets.

(a) Path connectedness does not imply local path connectedness, and local path connectedness does not imply path connectedness.

(b) (Connected and locally path connected) implies path connected.

(c) Derive necessary and sufficient conditions on the spaces (X_α, T_α) for the product space $\prod_{\alpha \in \mathcal{O}} X_\alpha$ to be locally path connected.

15) If $x \in X$, call x a slice point if $X - \{x\} = A_x \cup B_x$, where A_x and B_x are non-empty, disjoint, connected, open subsets of X. Prove the following theorem of P. M. Rice:

Let X be a connected, locally connected, separable metric space, containing at least two points. (Separable means that there is a countable subset A with $\bar{A} = X$). Suppose each point of X is a slice point. Then, $X \cong E^1$.

(See P. M. Rice, "A Topological Characterization of the Real Numbers", American Mathematical Monthly, Vol. 76, No. 2, Feb. 1969, p. 184–185).

CHAPTER VII

Compactness and covering theorems

7.1 Compactness

COMPACTNESS IS A topological concept which was originally inspired by properties of point sets in E^n. It was recognized quite early that certain kinds of sets had avantages over others in the analysis of functions. For example, a function continuous on a closed and bounded set achieves a maximum and a minimum.

Unfortunately, when topology was in its formative stages, there was no obvious way to generalize "closed and bounded". However, the Heine-Borel Theorem of classical analysis suggested a solution. A set A in E^n is closed and bounded exactly when every cover of A by open sets can be reduced to a finite cover. That is, if $A \subset \bigcup_{\alpha \in B} t_\alpha$, where the sets t_α are open, then it is possible to select finitely many of the t_α's such that $A \subset t_{\alpha_1} \cup t_{\alpha_2} \cup \cdots \cup t_{\alpha_n}$. This notion of reducibility of open covers generalizes immediately to any space. A set with the property that each open cover has a finite reduction will be called compact.

The word "compact" has had an interesting history. Around 1906, Fréchet used compact to mean that every infinite subset of A has a cluster point in A. This was probably motivated by the Bolzano-Weierstrass Theorem. Alexandroff and Urysohn, about 1924, used the notion of reduction of open covers, but called it bicompactness. In 1940, Bourbaki dropped the prefix "bi", but restricted "himself" to Hausdorff spaces. When examples appeared in differential geometry of interesting non-Hausdorff spaces having the reducibility property for open covers, Bourbaki labelled such spaces quasi-compact. The term compact as we have used it above is now widely accepted.

A modification of Fréchet's original notion is today called Bolzano-Weierstrass (or B-W) compactness (see section 4). As we shall see in Chapter VIII, the two kinds of compactness are equivalent in metric spaces, as are the notions of countable compactness (section 3) and sequential compactness (section 4).

Throughout this chapter let T be a topology on X.

1.1 DEFINITION *A is a T-cover of X
if and only if
(1) $A \subset T$.
(2) $\cup A = X$.*

1.2 DEFINITION *X is T-compact
if and only if
if A is a T-cover of X, then there is a finite subset B of A such that $\cup B = X$.*

Definition 1.2 is often rephrased: X is compact exactly when each open cover of X can be reduced to a finite cover.

1.1 EXAMPLE E^n is not compact. This is easiest to see when $n = 1$, where the open intervals $]n - 1, n + 1[$, as n takes on all integer values, form an open cover which cannot be reduced by as much as one set. A similar construction works in E^n. Take, for example, the spheres $B_{\varrho_n}((x_1, ..., x_n), \sqrt{n})$, where each coordinate x_i of $(x_1, ..., x_n)$ is an integer. ∎

1.2 EXAMPLE Any indiscrete space is compact. ∎

1.3 EXAMPLE A discrete space is compact exactly when it is finite. ∎

1.4 EXAMPLE Let $X = Z^+$, and let T consist of X, ϕ and all sets $[1, n] \cap Z^+$, where $n \in Z^+$. Then, X is not T-compact. ∎

1.5 EXAMPLE Any interval $[a, b]$ in E^1, considered as a subspace of E^1, is compact. This will follow easily from Theorem 1.2. ∎

We can speak of a subset A of X as being T-compact or non T-compact by referring A to the subspace topology induced by T.

1.3 DEFINITION *Let $A \subset X$.
Then,
A is T-compact
if and only if
A is T_A-compact.*

In practice, compactness of A can always be decided without actually going into the relative topology. We call a collection of T-open sets a cover of A if A is a subset of their union (but not necessarily equal to their union). Compactness of A can then be tested by looking at T-covers of A, and T_A-covers never have to be considered. This saves a great deal of bookkeeping in proofs involving compactness of subsets, and will be used often without explicit reference (see, for example, the proofs of Theorem 1.2 and Theorem 1.8).

1.4 DEFINITION Let $A \subset X$.
Then,
$\mathcal{O}$ is a T-cover of A
if and only if
(1) $\mathcal{O} \subset T$.
(2) $A \subset \cup \mathcal{O}$.

1.1 THEOREM Let $A \subset X$.
Then,
A is T-compact
if and only if
each T-cover of A has a finite subset which T-covers A.

Proof Suppose first that A is T-compact. Let $\mathcal{O}$ be a T-cover of A. Then, $\{\alpha \cap A | \alpha \in \mathcal{O}\}$ is a T_A-cover of A. Since A is T_A-compact, there is a finite subset, say $\{\alpha_i \cap A | i = 1, ..., n\}$ of $\{\alpha \cap A | \alpha \in \mathcal{O}\}$ which covers A. Then, $\{\alpha_i | i = 1, ..., n\}$ is a finite subset of $\mathcal{O}$ which covers A.

Conversely, suppose, if $\mathcal{O}$ is a T-cover of A, then there is a finite subset of $\mathcal{O}$ which covers A.

Let $\mathscr{C}$ be a T_A-cover of A. If $c \in \mathscr{C}$, there is some $t_c \in T$ such that $c = A \cap t_c$. Then, $\{t_c \cap A | c \in \mathscr{C}\} = \mathscr{C}$, and $\{t_c | c \in \mathscr{C}\}$ is a T-cover of A. For some $c_1, ..., c_n \in \mathscr{C}$, $A \subset \bigcup_{i=1}^{n} t_{c_i}$. Then, $A \subset \bigcup_{i=1}^{n} (t_{c_i} \cap A)$ so that $\{t_{c_i} \cap A | i = 1, ..., n\}$ is a finite subset of $\mathscr{C}$ which T_A-covers A, implying that A is T_A-compact. ∎

1.6 EXAMPLE In Example 1.4, a subset A of Z^+ is compact exactly when it is bounded above. That is, the only compact sets are the finite subsets of Z^+. Conversely, each finite subset is compact. ∎

1.7 EXAMPLE The empty set is a compact subset of any space, as is $\{x_1, ..., x_n\}$ for any finite number of points $x_1, ..., x_n$ in X. ∎

1.8 EXAMPLE If α is a non-limit ordinal, then the ordinal space $[0, \alpha[$ (with the order topology) is compact. We shall prove this by transfinite induction. Let N be the set of all non-limit ordinals α. Note that 0 is the first element of N, and, if $\alpha \in N$, then $\alpha \cup \{\alpha\} \in N$. If $\alpha \in N$, let P_α be the proposition: $[0, \alpha[$ is compact.

Now, P_0 is true, since $[0, 0[= \phi$ is trivially compact.

Assume then that $\alpha \in N$ and that P_β is true for each $\beta \in N$ with $\beta < \alpha$. It suffices to show that $[0, \alpha[$ is compact.

Let $\mathcal{T}$ be an open cover of $[0, \alpha[$. Now, as α is a non-limit ordinal, then α has an immediate predecessor, say α_p. We now consider two cases:

i) α_p is a non-limit ordinal.

Since $\alpha_p < \alpha$, then $[0, \alpha_p[$ is compact by the inductive assumption. Then there are sets $G_1, ..., G_r \in \mathcal{T}$ with $[0, \alpha_p[\subset \bigcup_{j=1}^{r} G_j$. Now, $\alpha = \alpha_p \cup \{\alpha_p\}$, so $[0, \alpha[= [0, \alpha_p[\cup \{\alpha_p\}$. For some $G \in \mathcal{T}$, we have $\alpha_p \in G$. Then $[0, \alpha[\subset G \cup \left(\bigcup_{j=1}^{r} G_j \right)$.

ii) α_p is a limit ordinal.

Choose some $G_{\alpha_p} \in \mathcal{T}$ with $\alpha_p \in G_{\alpha_p}$. For some $\beta < \alpha_p$, there is a basic neighborhood $]\beta, \alpha_p]$ of α_p with $]\beta, \alpha_p] \subset G_{\alpha_p}$. Now, $\beta \cup \{\beta\}$ is a non-limit ordinal. Further, $\beta \cup \{\beta\} < \alpha_p < \alpha$, so $[0, \beta \cup \{\beta\}[$ is compact. Choose $G_1, ..., G_m \in \mathcal{T}$ with $[0, \beta \cup \{\beta\}[\subset \bigcup_{j=1}^{m} G_j$. But, $[\beta \cup \{\beta\}, \alpha_p]$ $=]\beta, \alpha_p] \subset G_{\alpha_p}$, so $[0, \alpha[= [0, \alpha_p] \subset G_{\alpha_p} \cup \left(\bigcup_{j=1}^{m} G_j \right)$, and the proof is complete. ∎

1.9 EXAMPLE $]0, 1]$ is not compact in E^1. For, $]0, 1] \subset \bigcup_{n=2}^{\infty}]1/n, 2[$, and this infinite union of open intervals cannot be reduced to a finite union and still cover $]0, 1]$.

Note what a difference a point makes—$[0, 1]$ is compact (see Theorem 1.2). The above union does not contain $[0, 1]$, since 0 is not in any of the intervals of the union. If we adjoin to the union one more open set, say P, containing 0, then $0 \in]-\varepsilon, \varepsilon[\subset P$ for some $\varepsilon > 0$. Eventually, for some integer N, $1/N < \varepsilon$. We can then omit from the union all but one of the intervals $]1/n, 2]$ and write $[0, 1] \subset P \cup]1/N, 2[$.

This of course does not prove that $[0, 1]$ is compact, as we have not considered all possible open covers. However, it should provide some feeling for the tremendous difference between $]0, 1]$ and $[0, 1]$ as far as compactness is concerned. ∎

Before considering compactness in general, we shall characterize the compact subsets of E^1. All topological concepts in the theorem are of course referred to the Euclidean topology. Later, with the help of the Tychonov Theorem, we shall be able to extend this characterization to E^n.

1.2 Theorem *Let $A \subset E^1$.*
Then,
A is compact
if and only if
A is closed and bounded.

Proof Suppose first that A is compact. Let $\mathcal{O} = \{]x - 1, x + 1[\, | \, x \in A\}$. Then, $\mathcal{O}$ is an open cover of A. Since A is compact, there are finitely many points $x_1, \ldots, x_r$ in A such that $A \subset \bigcup_{i=1}^{r}]x_i - 1, x_i + 1[$. Then, $A \subset [-\eta, \eta]$, where $\eta = \sup\{|x_i| + 1 \, | \, 1 \leq i \leq r\}$, hence is bounded.

To show that A is closed, suppose that $y \in \bar{A} - A$. We leave it to the reader to derive a contradiction by showing that $\{E^1 - [y - 1/j, y + 1/j] \, | \, j \in Z^+\}$ has no finite subset which covers A.

Conversely, suppose that A is closed and bounded. We consider two cases.

Case 1: $A = [a, b]$ for some real numbers a and b with $a < b$.

Let $\mathcal{O}$ be an open cover of $[a, b]$. Let $M = \{x \, | \, a \leq x \leq b$ and a finite subset of $\mathcal{O}$ covers $[a, x]\}$. Then, $M \neq \phi$, as $a \in M$. Further, M is bounded above by b. Hence, M has a least upper bound, say L, and $a \leq L \leq b$. We shall show that $L = b$.

Note first that $a < L$. To see this, note that, for some v, $v \in \mathcal{O}$ and $a \in v$. Since v is open, then $a \in]a - \varepsilon, a + \varepsilon[\subset v$ for some $\varepsilon > 0$, so that by choosing ε sufficiently small that $a + \varepsilon \leq b$ we have $a + \varepsilon \in M$. Then, $a < a + \varepsilon \leq L$.

Now, since $L \in [a, b]$, there is some $U \in \mathcal{O}$ such that $L \in U$. Since U is open, there is some $\delta > 0$ such that $]L - \delta, L + \delta[\subset U$. If $L < b$, we can choose δ sufficiently small that $L + \delta < b$ and $a < L - \delta$.

Now, by definition of L, there is some $y \in M$ with $L - \delta < y \leq L$. Since $y \in M$, there are open sets $P_1, \ldots, P_t$ in $\mathcal{O}$ such that $[a, y] \subset \bigcup_{i=1}^{t} P_i$. Then, $[a, L + \delta] \subset \left(\bigcup_{i=1}^{t} P_i\right) \cup U$, implying that $L + \delta \in M$, which is impossible.

Hence, $L = b$.

Case 2: A is any closed, bounded subset of E^1.

Since A is bounded, then $A \subset [-\delta, \delta]$ for some real δ. Let $\mathcal{O}$ be any open cover of A. Then, $\mathcal{O} \cup \{E^1 - A\}$ is an open cover of $[-\delta, \delta]$. By case 1, $[-\delta, \delta]$ is compact. Then, there are finitely many sets, say $V_1, \ldots, V_s$, of $\mathcal{O}$ such that $[-\delta, \delta] \subset (E^1 - A) \cup \bigcup_{i=1}^{s} V_i$. Then, $A \subset \bigcup_{i=1}^{s} V_i$, so A is compact. ∎

We now begin the real study of compactness in general. It is a simple matter to show that compactness is preserved by finite unions. Examples can easily be constructed in E^1, with the help of the last theorem, of arbitrary unions of compact sets which are not compact. Intersections of compact sets need not be compact (see Problem 3).

1.3 Theorem *Let A be a finite set of T-compact subsets of X. Then,*
$\cup A$ *is T-compact.*

Proof Left as an exercise. ∎

The next three theorems provide alternate formulations of the notion of compactness, and will be extremely useful later.

The first is a somewhat strange looking condition called the finite intersection property. Historically, this was used by Riesz in 1908 as his definition of compactness.

1.5 Definition *Let F be a set of subsets of X. Then,*
F has the finite intersection property
if and only if
if $\cap F = \phi$, then, for some finite, non-empty subset B of F, $\cap B = \phi$.

1.4 Theorem *The following are equivalent:*
(1) X is T-compact.
(2) If F is a non-empty set of T-closed subsets of X, then F has the finite intersection property.

Proof (1) ⇒ (2):
Assume (1). Let F be a non-empty set of T-closed subsets of X. Suppose that $\cap F = \phi$. We must produce a finite subset of F whose intersection is empty.

Now, $X = X - \cap F = \cup \{X - f | f \in F\}$. If $f \in F$, then $X - f \in T$, so $\{X - f | f \in F\}$ is a T-cover of X. Then, there are finitely many elements $f_1, ..., f_n$ of F such that $X = \bigcup_{i=1}^{n} (X - f_i)$. Choose $B = \{f_1, ..., f_n\}$, and we then have $B \subset F$ and $\cap B = \phi$.

$(2) \Rightarrow (1)$:

Assume (2). Let $\mathcal{O}$ be a T-cover of X. Then, $\mathcal{O} \subset T$ and $X = \cup \mathcal{O}$. Then, $\phi = X - \cup \mathcal{O} = \cap \{X - a | a \in \mathcal{O}\}$. If $a \in \mathcal{O}$, then $X - a$ is T-closed. By (2), there is a finite set of elements $a_1, ..., a_n$ of $\mathcal{O}$ such that $\bigcap_{i=1}^{n}(X - a_i) = \phi$. Then, $X = \bigcup_{i=1}^{n} a_i$, and so X is T-compact. ∎

Note that, upon careful examination of the proof, the last theorem is just a restatement of DeMorgan's Laws as they relate to compactness.

1.5 Theorem *X is T-compact*
if and only if
if $\mathcal{F}$ is a filterbase on X, then $\bigcap_{f \in \mathcal{F}} \bar{f} \neq \phi$.

Proof Suppose first that X is T-compact. Let $\mathcal{F}$ be a filterbase on X. By Theorem 1.4, $\{\bar{f} | f \in \mathcal{F}\}$ has the finite intersection property. If $\bigcap_{f \in \mathcal{F}} \bar{f} = \phi$, then there is a finite subset $\{f_1, ..., f_n\}$ of $\mathcal{F}$ such that $\bigcap_{i=1}^{n} \bar{f}_i = \phi$. This contradicts Theorem 3.1.IV.

Conversely, suppose that $\bigcap_{f \in \mathcal{F}} \bar{f} \neq \phi$ if $\mathcal{F}$ is a filterbase on X. Let $\mathcal{O}$ be a T-open cover of X. If C is a non-empty, finite subset of $\mathcal{O}$, let $B_C = X - \bigcup_{\alpha \in C} \alpha$.

Suppose that $B_C \neq \phi$ for each finite subset C of $\mathcal{O}$. Let $\mathcal{F} = \{B_C | \phi \neq C \subset \mathcal{O}$ and C is finite$\}$. Now, $B_{C \cup D} \subset B_C \cap B_D$ for any non-empty, finite subsets C and D of $\mathcal{O}$. Then, $\mathcal{F}$ is a filterbase on X. Then, $\bigcap_{B_C \in \mathcal{F}} \bar{B}_C \neq \phi$. In particular, if $\alpha \in \mathcal{O}$, then $\phi \neq \{\alpha\} \subset \mathcal{O}$, and $\{\alpha\}$ is finite, so $\phi \neq \bigcap_{B_C \in \mathcal{F}} \bar{B}_C \subset \bigcap_{\alpha \in \mathcal{O}} \bar{B}_{\{\alpha\}} = \bigcap_{\alpha \in \mathcal{O}}(X - \alpha) = X - \bigcup_{\alpha \in \mathcal{O}} \alpha = X - \cup \mathcal{O}$. This is a contradiction, as $X = \cup \mathcal{O}$. Hence, $\phi \in \mathcal{F}$, and there is a finite subset C of $\mathcal{O}$ such that $B_C = \phi$, and $X = \cup C$. ∎

Our final "equivalence" theorem characterizes compactness in terms of convergence of ultrafilterbases. This will very shortly be the key to the proof of the Tychonov Theorem. See also the proof of Theorem 7.6.VIII.

1.6 Theorem *X is T-compact*
if and only if
each ultrafilterbase on X is T-convergent.

Proof Suppose first that X is T-compact. Let $\mathcal{M}$ be an ultrafilterbase on X. By Theorem 1.5, $\bigcap_{m \in \mathcal{M}} \bar{m} \neq \phi$. Let $x \in \bigcap_{m \in \mathcal{M}} \bar{m}$. By Theorem 5.4.IV, $\mathcal{M}$ is T-convergent to x.

Conversely, suppose that each ultrafilterbase on X is T-convergent. We shall use Theorem 1.5 to show that X is T-compact.

Let $\mathscr{F}$ be a filterbase on X. By Theorem 5.2.IV, there is an ultrafilterbase $\mathscr{M}$ on X such that $\mathscr{F} < \mathscr{M}$. By Theorem 5.4.IV, $\bigcap_{m \in \mathscr{M}} \bar{m} \neq \phi$. Now, if $f \in \mathscr{F}$, there is some $m \in \mathscr{M}$ with $m \subset f$, hence $\bar{m} \subset \bar{f}$. Then,

$$\phi \neq \bigcap_{m \in \mathscr{M}} \bar{m} \subset \bigcap_{f \in \mathscr{F}} \bar{f}$$

so X is T-compact. ∎

We now consider the invariance properties of compactness. As with connectedness, compactness is preserved by continuous surjections, hence also by homeomorphism. This provides us with another topological invariant (to go with connectedness, path connectedness, number of components and number of path components).

1.7 THEOREM *Let (Y, M) be a topological space.*
Let $f: X \to Y$ be a (T, M) continuous surjection.
Let X be T-compact.
Then,
Y is M-compact.

Proof Let $\mathcal{O}$ be an M-cover of Y. Then, $\{f^{-1}(b) | b \in \mathcal{O}\}$ is a T-cover of X, by continuity of f.

Since X is T-compact, there is a finite set of elements $b_1, \ldots, b_n$ of $\mathcal{O}$ such that $\bigcup_{i=1}^{n} f^{-1}(b_i) = X$. Then, $\bigcup_{i=1}^{n} b_i = Y$, as f is a surjection. ∎

1 COROLLARY *Let $(X, T) \cong (Y, M)$.*
Then,
X is T-compact $\Leftrightarrow$ Y is M-compact.

Proof Immediate by Theorem 1.7. ∎

1.10 EXAMPLE $[0, 1] \cup]6, 8]$ is not homeomorphic to $[0, 1] \cup [6, 8]$ in E^1, since the latter set is compact by Theorem 1.2, but the former is not. Note that both spaces are disconnected and not path connected, so that testing homeomorphism by the topological invariants of the last chapter fails. ∎

Of course, not every subspace of a compact space is compact. For example, $]\frac{1}{2}, \frac{3}{4}[$ is a non-compact subspace of compact $[0, 1]$. However, a closed subset of a compact space is compact. This will follow easily from a slightly more general statement.

1.8 THEOREM *Let A be T-compact.*
Let B be T-closed.
Then,
$A \cap B$ is T-compact.

Proof Let $\mathcal{O}$ be a T-cover of $A \cap B$. Then, $\mathcal{O} \cup \{X - B\}$ is a T-cover of A, hence can be reduced to a finite T-cover, say $m \cup \{X - B\}$, where $m \subset \mathcal{O}$. Then, m is a T-cover of $A \cap B$, implying that $A \cap B$ is T-compact. ∎

2 COROLLARY *Let X be T-compact.*
Let B be T-closed.
Then,
B is T-compact.

Proof Immediate upon taking $A = X$ in Theorem 1.8. ∎

The converse of Theorem 1.8 (or of Corollary 2) is of course not in general true—a compact subset of a space need not be closed.

1.11 EXAMPLE In the topology T on Z^+ of Example 1.4, $\{0, 1\}$ is compact (as it is finite, see Example 1.6), but is not closed, as $Z^+ - \{0, 1\} \notin T$. ∎

However, when T is a Hausdorff topology, then every compact subset of X is closed, whether X is compact or not. The proof is preceded by a Lemma which will also be of use in proving Theorem 1.10.

1 LEMMA *Let (X, T) be a Hausdorff space.*
Let A be T-compact and $y \in X - A$.
Then,
there are disjoint T-open sets U and V such that $y \in U$ and $A \subset V$.

Proof If $x \in A$, then $x \neq y$, so there are disjoint, T-open sets U_x and V_x such that $x \in V_x$ and $y \in U_x$. Now, $A \subset \bigcup_{x \in A} V_x$, so that $\{V_x | x \in A\}$ is a T-cover of A. This can be reduced to a finite T-cover of A, say $A \subset \bigcup_{i=1}^{n} V_{x_i}$, where $x_1, \ldots, x_n \in A$.

Then,
$$y \in \bigcap_{i=1}^{n} U_{x_i} \in T,$$
$$A \subset \bigcup_{i=1}^{n} V_{x_i} \in T,$$
and
$$\left(\bigcup_{i=1}^{n} V_{x_i}\right) \cap \left(\bigcap_{i=1}^{n} U_{x_i}\right) = \phi. \blacksquare$$

1.9 Theorem *Let (X, T) be a Hausdorff space.*
Let A be T-compact.
Then,
A is T-closed.

Proof Let $y \in X - A$. By Lemma 1, there is a T-nghd. U of y such that $U \cap A = \phi$. Then, $y \in \text{Int}_T (X - A)$, implying that $X - A \in T$. ∎

Lemma 1 is actually a separation theorem—compact sets can be separated from points in Hausdorff spaces. This suggests that compact Hausdorff spaces might have very strong separation properties. In fact, any compact Hausdorff space is normal. Compactness is essential here, as there are Hausdorff spaces which are neither regular nor normal.

1.10 Theorem *Let (X, T) be a compact, Hausdorff space.*
Then,
T is a normal topology.

Proof Let A and B be disjoint, T-closed subsets of X. By Corollary 2, A and B are T-compact. If $y \in A$, then by Lemma 1 there are disjoint, T-open sets U_y and V_y with $y \in U_y$ and $B \subset V_y$. Now, $A \subset \bigcup_{y \in A} U_y$. Since A is T-compact, there are elements $y_1, ..., y_n$ of A such that $A \subset \bigcup_{i=1}^{n} U_{y_i} \in T$. Further, $B \subset V_{y_i}$ for $i = 1, ..., n$, so $B \subset \bigcap_{i=1}^{n} V_{y_i} \in T$. All that remains is to observe that

$$\left(\bigcup_{i=1}^{n} U_{y_i}\right) \cap \left(\bigcap_{i=1}^{n} V_{y_i}\right) = \phi. \quad \blacksquare$$

Just for the record, we state the following.

3 Corollary *Let (X, T) be a compact, regular space.*
Then, T is a normal topology.

Proof Immediate by Theorem 1.10, as a regular space is Hausdorff. ∎

There is a clever proof of Theorem 1.10 in Bourbaki using filterbases. The reader should attempt a proof using this approach, or consult Bourbaki.

We shall apply Theorem 1.10 to prove another separation theorem which will be needed later (in the proof of Theorem 2.8).

1.11 Theorem *Let X be a T-compact Hausdorff space.*
Let F be T-closed.
Let U be T-open and $F \subset U$.
Then,
for some V, V is T-open and $F \subset V \subset \bar{V} \subset U$ and $\bar{V}$ is T-compact.

Proof By Theorem 1.10, (X, T) is normal.

By Theorem 3.2.V, there is a T-open set V with $F \subset V \subset \bar{V} \subset U$.

Since $\bar{V}$ is a T-closed subset of a compact space, then $\bar{V}$ is T-compact by Corollary 2. ∎

We shall now prove the Tychonov Theorem, which may be the most important theorem in set topology. It says that a product space is compact exactly when each coordinate space is compact. The proof is quite short, as all the machinery has already been developed.

1.12 THEOREM (Tychonov) *Let $A \neq \phi$.*
Let T_α be a topology on X_α for each $\alpha \in A$.
Let $\mathbb{P}$ be the product topology on $\prod_{\alpha \in A} X_\alpha$.
Then,
$\prod_{\alpha \in A} X_\alpha$ *is $\mathbb{P}$-compact*
if and only if
X_α is T_α-compact for each $\alpha \in A$.

Proof Suppose first that $\prod_{\alpha \in A} X_\alpha$ is $\mathbb{P}$-compact. If $\beta \in A$, then p_β:
$\prod_{\alpha \in A} X_\alpha \to X_\beta$ is a $(\mathbb{P}, T_\beta)$ continuous surjection, hence X is T_β-compact by Theorem 1.7.

Conversely, suppose that X_α is T_α-compact for each $\alpha \in A$. Let $\mathcal{M}$ be an ultrafilterbase on $\prod_{\alpha \in A} X_\alpha$. By Theorem 5.5.IV, $p_\beta(\mathcal{M})$ is an ultrafilterbase on X_β for each $\beta \in A$.

Since X_β is T-compact if $\beta \in A$, then $p_\beta(\mathcal{M})$ is T_β-convergent, say to f_β, by Theorem 1.6.

By Theorem 3.6.IV, $\mathcal{M}$ is $\mathbb{P}$-convergent to f.

By Theorem 1.6, $\prod_{\alpha \in A} X_\alpha$ is $\mathbb{P}$-compact. ∎

Our proof of the Tychonov Theorem is quite like Bourbaki's with filterbases replacing filters. Kelley (p. 143) has an interesting proof using Alexander's Theorem, which says that one can replace open covers with subbasic open covers in the definition of compactness. Otherwise, a direct appeal to the definition of compactness in proving Tychonov's Theorem is too messy, as the open sets in the product topology are very cumbersome.

There is also a proof in Hocking and Young and in Singer and Thorpe using the finite intersection property.

The proof we have given of Tychonov's Theorem is quite deep, depending as it does upon properties of ultrafilterbases, which in turn depend for their existence upon Zorn's Lemma. John Kelley* proved in 1950 that in fact Tychonov's Theorem is equivalent to Zorn's Lemma in the form of the Axiom of Choice, thus verifying a conjecture made some years before by Kakutani. Kelley's argument goes as follows. Suppose that A is a nonempty set, and $X_a \neq \phi$ for each $a \in A$. Assuming the Tychonov Theorem, we wish to show that $\prod_{\alpha \in A} X_\alpha \neq \phi$.

Choose some object λ such that $\lambda \notin X_a$ for each $a \in A$ $\left(\text{e.g. } \lambda = \bigcup_{\alpha \in A} X_\alpha\right)$. Let T_a consist of ϕ, $\{\lambda\}$ and all complements of finite subsets of $Y_a = X_a \cup \{\lambda\}$ for each $a \in A$. Then, T_a is a topology on Y_a, and (Y_a, T_a) is compact. Hence $\left(\prod_{\alpha \in A} Y_\alpha, \mathbb{P}\right)$ is compact.

Now, if $b \in A$, let $z_b = \left(\prod_{\alpha \in A} Y_\alpha\right) \cap \{f | f(b) \in X_b\}$. Alternatively, $z_b = p_b^{-1}(X_b)$. Since $\{\lambda\} \in T_b$, then X_b is T_b-closed, and continuity of p_b implies that z_b is closed in $\prod_{\alpha \in A} Y_\alpha$. Thus $\{z_b | b \in A\}$ is a set of closed subsets of $\prod_{\alpha \in A} Y_\alpha$. Further, if $b_1, \ldots, b_r \in A$, then $\bigcap_{j=1}^{r} z_{b_j} \neq \phi$ by a simple mathematical induction argument. Then $\{z_b | b \in A\}$ has the finite intersection property, hence $\bigcap_{b \in A} z_b \neq \phi$.

To complete the proof, check that $\bigcap_{b \in A} z_b = \prod_{b \in A} X_b$.

We shall pause here to consider one of the important examples of set topology, the Tychonov plank. We shall use it to produce a non-normal subspace of a normal space, as promised in Chapter V.

1.12 EXAMPLE Let $\Omega^* = \Omega \cup \{\Omega\}$ and $\omega^* = \omega \cup \{\omega\}$. Then, Ω^* and ω^* are non-limit ordinals. Give each the order topology. The product space $\Omega^* \times \omega^*$ is called the Tychonov plank.

We shall prove first that the Tychonov plank is normal. Since Ω^* and ω^* are Hausdorff, then so is $\Omega^* \times \omega^*$. Further, by Example 1.8, Ω^* and ω^* are both compact, hence $\Omega^* \times \omega^*$ is compact. Finally, any compact Hausdorff space is normal.

One interesting feature of the Tychonov plank is that it has a non-normal subspace. Let $S = (\Omega^* \times \omega^*) - \{(\Omega, \omega)\}$. Then, S is an open subset of $\Omega^* \times \omega^*$ obtained by removing a corner from the plank (Figure 1.1). We claim that S (as a subspace of $\Omega^* \times \omega^*$) is not normal.

---

* John L. Kelley, "The Tychonov Theorem Implies the Axiom of Choice", *Fund. Math.*, 37, 1950, pp. 75–76.

150   FUNDAMENTAL CONCEPTS OF TOPOLOGY

The proof is by exhibiting disjoint, closed subsets of S having no disjoint neighborhoods.

Let $A = \{(\Omega, y) | y < \omega\}$ and $B = \{(x, \omega) | x < \Omega\}$. Then, $A \subset S$ and $B \subset S$. Immediately $A \cap B = \phi$, and the reader can easily check that A and B are closed in S.

Suppose that A and B have disjoint neighborhoods in S; say G_A and G_B are disjoint sets, open in S, and $A \subset G_A$ and $B \subset G_B$.

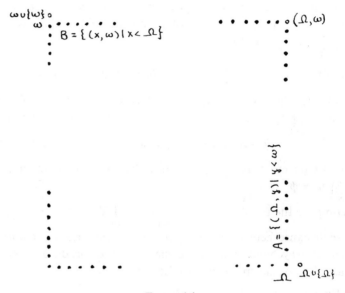

FIGURE 1.1

Now, for some U and V open in $\Omega^* \times \omega^*$, $G_A = S \cap U$ and $G_B = S \cap V$. As S is open in $\Omega^* \times \omega^*$, so are G_A and G_B.

Let $(\Omega, x) \in A$. As G_A is open in $\Omega^* \times \omega^*$, there are open sets U_x in Ω^* and V_x in ω^* with $(\Omega, x) \in U_x \times V_x \subset G_A$. Now, since $\Omega \in U_x$ open in Ω^*, then for some $\xi_x \in \Omega$, $]\xi_x, \Omega] \subset U_x$. Choose some $\zeta_x \in]\xi_x, \Omega]$. Then, since $x \in V_x$ and $U_x \times V_x \subset G_A$, we have $(\zeta_x, x) \in G_A$.

Note that $\{\zeta_x | x < \omega\}$ is countable. Hence for some $\zeta \in \Omega$, $\zeta_x < \zeta$ for each $x < \omega$ (e.g. take $\zeta = \bigcup_{x \in \omega} \{\zeta_x\}$). Then, $\zeta \in]\xi_x, \Omega] \subset U_x$ for each $x < \omega$. Then also $(\zeta, x) \in G_A$ for each $x < \omega$.

Now, $(\zeta, \omega) \in B \subset G_B$. Reasoning as before, produce W open in Ω^*, M open in ω^*, with $(\zeta, \omega) \in W \times M \subset G_B$.

Since ω is a limit ordinal, and $\omega \in M$ and M is open in ω^*, then there is a basic neighborhood $]\alpha, \omega] \subset M$ for some $\alpha \in \omega$. Then, $(\zeta, \alpha \cup \{\alpha\}) \in W \times M \subset G_B$. But also $(\zeta, \alpha \cup \{\alpha\}) \in G_A$, as $\alpha \cup \{\alpha\} \in \omega$. Thus $G_A \cap G_B \neq \phi$, a contradiction, implying that S is not normal. ∎

As a first application of the Tychonov Theorem, we shall extend Theorem 1.2 to Euclidean n-space. A second application will be the proof of Theorem 2.5, next section.

1.13 THEOREM *Let $A \subset E^n$.*
Then,
A is compact
if and only if
A is closed and bounded.

Proof If A is compact, then A is closed and bounded by an argument similar to that used in the proof of Theorem 1.2.

Conversely, suppose that A is closed and bounded. Let $\mathcal{O}$ be any open cover of A. Since A is bounded, there are real numbers a_{11}, a_{12}, a_{21}, a_{22}, ..., a_{n1}, a_{n2} such that

$$A \subset [a_{11}, a_{12}] \times [a_{21}, a_{22}] \times [a_{31}, a_{32}] \times \cdots \times [a_{n1}, a_{n2}].$$

Now, $\mathcal{O} \cup \{E^n - A\}$ is an open cover of $[a_{11}, a_{12}] \times \cdots \times [a_{n1}, a_{n2}]$, which is compact by Theorem 1.2 and Tychonov's Theorem. Hence there is a finite collection $\{V_1, ..., V_k\}$ of sets in $\mathcal{O}$ such that $\{V_1, ..., V_k\} \cup \{E^n - A\}$ covers $[a_{11}, a_{12}] \times \cdots \times [a_{n1}, a_{n2}]$. Then, $\{V_1, ..., V_k\}$ is a reduction of $\mathcal{O}$ to a finite cover of A, hence A is compact. ∎

In view of the fact that compact spaces enjoy a number of properties not generally shared by non-compact spaces, it is natural to ask the following question: can a given space be embedded homeomorphically in a compact space?

The answer to this is always yes. In fact, there are many ways of doing it, among them the Wallman, Stone-Čech and Alexandroff compactifications. We shall consider only the Alexandroff method, as this will suffice for our purposes (see Dugundji for a complete discussion of the other two).

Alexandroff's method is to adjoin a new point, often denoted ∞, to X to form a new set $X \cup \{\infty\}$. A topology $\mathcal{O}(T)$ is defined on $X \cup \{\infty\}$ by specifying the $\mathcal{O}(T)$ open sets to be all the T-open sets (which of course do not contain ∞), together with each subset A of $X \cup \{\infty\}$ which contains

∞ and has the additional two properties that $A \cap X \,(= A - \{\infty\})$ is T-open and $X - A \,(= (X \cup \{\infty\}) - A)$ is T-compact.

The resulting space $(X \cup \{\infty\}, \mathcal{O}(T))$ is compact, and the map $\varphi\colon X \to X \cup \{\infty\}$ defined by $\varphi(x) = x$ is an homeomorphism of X into $X \cup \{\infty\}$. Further, if the original space is not compact, then in addition $\overline{\varphi(X)} = X \cup \{\infty\}$, so that ∞ can be approximated as closely as we like (in the sense of the topology) by elements of X. Finally, the relative topology $\mathcal{O}(T)_X$ coincides with T, so that in a sense (X, T) has been left undisturbed. The space $(X \cup \{\infty\}, \mathcal{O}(T))$ is called an Alexandroff one-point compactification of (X, T).

1.14 Theorem (Alexandroff) *Suppose that (X, T) is not compact.*
Let $z \notin X$.
Let $Y = X \cup \{z\}$.
Let $\mathcal{O}(T) = T \cup \{A | z \in A \subset Y$ and $A \cap X \in T$ and $X - A$ is T-compact$\}$.
Then,

(1) $\mathcal{O}(T)$ is a topology on Y.
(2) $\mathcal{O}(T)_X = T$.
(3) Y is $\mathcal{O}(T)$ compact.
(4) Let $\varphi\colon X \to Y$ be defined by $\varphi(x) = x$ for each $x \in X$.

   Then, φ is a $(T, \mathcal{O}(T)_X)$ homeomorphism, and $\overline{\varphi(X)} = Y$.

Proof of (1) This is a routine (though lengthy) calculation, and is left to the reader.

Proof of (2) Left as an exercise.

Proof of (3) Let $\mathscr{C}$ be an $\mathcal{O}(T)$-cover of Y. We must produce a finite subset B of $\mathscr{C}$ such that $\cup B = Y$.

Since $\mathscr{C}$ covers Y, then $z \in W$ for some $W \in \mathscr{C}$. Now, $\{c \cap X | c \in \mathscr{C}\}$ is a T-cover of X, hence also of $X - W$. But, $X - W$ is T-compact. Then, there are finitely many elements $c_1, \ldots, c_n$ of $\mathscr{C}$ such that $c_1 \cap X, \ldots, c_n \cap X$ covers $X - W$. Then, $\{c_1, \ldots, c_n, W\}$ covers Y, and we may choose $B = \{c_1, \ldots, c_n, W\}$.

Proof of (4) It is easy to check that φ is a $(T, \mathcal{O}(T)_X)$ homeomorphism. To show that $\overline{\varphi(X)} = Y$, where $\overline{\varphi(X)}$ denotes $\mathrm{Cl}_{\mathcal{O}(T)}(\varphi(X))$, it suffices to show that $z \in \overline{\varphi(X)}$. Let W be an $\mathcal{O}(T)$-nghd. of z. If $W \cap X = \phi$, then $W = \{z\}$. Then, $X - W = X$ is T-compact, a contradiction. Thus, $W \cap X \neq \phi$, and $z \in \overline{\varphi(X)}$. This completes the proof. ∎

In practice, the space $(Y, \mathcal{O}(T))$ of Alexandroff's Theorem can often be realized more concretely through an additional homeomorphism, say $(Y, \mathcal{O}(T)) \cong (P, M)$. In this event, the space (P, M) is also called a one-point compactification of (X, T).

1.6 DEFINITION *Let (X, T) be a non-compact space. Let z, Y and $\mathcal{O}(T)$ be as in Theorem 1.14. Then, any space homeomorphic to $(Y, \mathcal{O}(T))$ is an Alexandroff one-point compactification of (X, T).*

Note that the assumption that (X, T) is not itself compact is used only in showing that $\overline{\varphi(X)} = Y$. In practice, of course, one does not consider compactifications of compact spaces anyway, so that this assumption is not very restrictive.

1.13 EXAMPLE We know that E^1 is not compact. To produce a compactification of E^1, let $Y = X \cup \{\infty\}$. The open sets in Y are:

(1) open sets in E^1 (that is, unions of open intervals),

(2) each subset A of Y with $\infty \in A$, and $A \cap E^1$ open in E^1, and $E^1 - A$ closed and bounded.

Thus, for example, $\{x|x < 1\} \cup \{x|x > 2\} \cup \{\infty\}$ is open in Y, since $\{x|x < 1\} \cup \{x|x > 2\}$ is open in E^1 and $X - (\{x|x < 1\} \cup \{x|x > 2\} \cup \{\infty\}) = [1, 2]$ is compact in E^1.

A more concrete representation of this compactification can be obtained as follows. We know that $E^1 \cong S^1 - \{(0, 1)\}$ (see Example 2.1.II). Compactify $S^1 - \{(0, 1)\}$ by putting back the point $(0, 1)$. Then, $S^1 \cong (Y, \mathcal{O}(T))$ as defined above, and has the advantage of being a more familiar space to visualize and work with.

In one higher dimension, $E^2 \cong S^2 - \{(0, 0, 1)\}$, the stereographic projection being a homeomorphism. A one point compactification of E^2 may then be thought of as S^2, the unit sphere in E^3. This is standard procedure in complex analysis, where one often works on the unit sphere (which is compact) instead of in the complex plane. See, for example, Einar Hille's Introduction to Complex Analysis, Vol. I. ∎

1.14 EXAMPLE Let $X = Z^+$, with T the topology of Example 1.4. This space is not compact. A one point compactification is obtained by adjoining ∞ to Z^+ and defining the open sets as: ϕ, Z^+, all sets $Z^+ \cap [1, n]$ for $n \in Z^+$, and $Z^+ \cup \{\infty\}$.

To see this, note that, if $\infty \in A \subset Z^+ \cup \{\infty\}$, and $A \cap Z^+$ is open, then

$$A \cap Z^+ = \begin{cases} Z^+ \cap [1, n] \text{ for some } n \in Z^+, \\ \text{or } \phi, \\ \text{or } Z^+. \end{cases}$$

If $A \cap Z^+ = \phi$ or $Z^+ \cap [1, n]$, then $Z^+ - A$ is not finite, hence not compact (see Example 1.6). Then $A \cap Z^+ = Z^+$, so that $A = Z^+ \cup \{\infty\}$ is necessarily the case if A is to be open in $Z^+ \cup \{\infty\}$. ∎

The use of one-point compactifications will be illustrated in Theorems 2.6 through 2.9 of the next section.

7.2 Local compactness

Euclidean n-space is not compact, but each point of E^n has a neighborhood whose closure is compact. Such a space is called locally compact.

2.1 Definition X is T-locally compact
if and only if
if $x \in X$, then there is a T-nghd. V of x such that $\bar{V}$ is T-compact.

2.1 Example E^n is locally compact; $\overline{B_{\rho_n}(x, 1)}$ is compact for each $x \in E^n$. ∎

2.2 Example Any compact space is locally compact—take $V = X$ in the definition. This is stated as Theorem 2.1 for reference. ∎

2.1 Theorem *If X is T-compact, then X is T-locally compact.*

Proof Immediate. ∎

2.3 Example Z^+, with the topology of Example 1.4, is not locally compact and not compact. If $n \in Z^+$, then $Z^+ \cap [1, n]$ is a nghd. of n which is finite, hence compact, but $\overline{Z^+ \cap [1, n]} = Z^+$. ∎

2.4 Example Let $X_\alpha = E^1$ for each $\alpha \in E^1$. Then the product space $\prod_{\alpha \in E^1} X_\alpha$ is not locally compact. The reader might try proving this directly. It will follow quite easily later from Theorem 2.5. ∎

Local compactness is not preserved by continuous surjections. For example, if (X, T) is any non-locally compact space, map $f: X \to X$, the identity map. Then f is a (D, T) continuous surjection, where D is the discrete topology on X. But, while (X, D) is locally compact, (X, T) is not by choice.

However, a continuous, open map onto a Hausdorff space does preserve local compactness.

2.2 THEOREM *Let (Y, M) be a Hausdorff space.*
Let $f: X \to Y$ be a (T, M) continuous, open surjection.
Let X be T-locally compact.
Then,
Y is M-locally compact.

Proof Let $y \in Y$. For some $x \in X$, $y = f(x)$. There is a T-nghd. V of x such that $\bar{V}$ is T-compact. Since f is an open map, $f(V)$ is an M-nghd. of y. Then, by Theorem 1.8.III, and Theorem 1.10.III, $f|\bar{V}: \bar{V} \to f(\bar{V})$ is a $(T_{\bar{V}}, M_{f(\bar{V})})$ continuous surjection.

By Theorem 1.7, $f(\bar{V})$ is M-compact.

Now, by Theorem 1.1 (4).II, $f(\bar{V}) \subset \overline{f(V)}$, as f is (T, M) continuous. Since $V \subset \bar{V}$, then $\overline{f(V)} \subset \overline{f(\bar{V})} \subset \overline{\overline{f(V)}} = \overline{f(V)}$ by Theorem 2.3 (6).I.
Then, $\overline{f(V)} = \overline{f(\bar{V})}$.

But, M is a Hausdorff topology, and $f(\bar{V})$ is M-compact, so $f(\bar{V})$ is M-closed by Theorem 1.9.

Then, $f(\bar{V}) = \overline{f(\bar{V})} = \overline{f(V)}$ by Theorem 2.3 (5).I. Since $f(\bar{V})$ is M-compact, then so is $\overline{f(V)}$. Then $f(V)$ is a nghd. of $f(x)$ whose closure is compact, and Y is M-locally compact. ∎

2.3 THEOREM *Let $(X, T) \cong (Y, M)$.*
Then,
X is T-locally compact $\Leftrightarrow$ Y is M-locally compact.

Proof Left as an exercise (caution: don't use Theorem 2.2). ∎

Local compactness of a subset of X is defined in the obvious way.

2.2 DEFINITION *Let $A \subset X$.*
Then,
A is T-locally compact
if and only if
A is T_A-locally compact.

Thus, if $A \subset X$, we say that A is T-locally compact if, given $x \in A$, there is a T-nghd. V of x with $\mathrm{Cl}_{T_A}(A \cap V)$ T_A-compact. In view of Theorem 1.5.III, this is equivalent to saying that $A \cap \text{Cl}_T (A \cap V)$ is T-compact. This sketches the proof of the next theorem, which casts Definition 2.2 in more convenient terms.

2.4 Theorem *Let $A \subset X$.*
Then,
A is T-locally compact
if and only if
if $x \in A$, then there is a T-nghd. V of x such that $A \cap \overline{(A \cap V)}$ is T-compact.

Proof Left as an exercise. ∎

2.5 Example $[0, 1[$ is a locally compact subset of E^1. To see this, choose $x \in [0, 1[$ and consider two cases:

1) $x \neq 0$. Then choose $\delta > 0$ such that $]x - \delta, x + \delta[\subset]0, 1[$. Then, $]x - \delta, x + \delta[$ is a T-nghd. of x, and

$$[0, 1[\cap \overline{([0, 1[\cap]x - \delta, x + \delta[)} = [0, 1[\cap \overline{([x - \delta, x + \delta])}$$
$$= [x - \delta, x + \delta] \text{ is compact in } E^1.$$

2) $x = 0$. Use $]-\tfrac{1}{2}, \tfrac{1}{2}[$ for V in Theorem 2.3. We have

$$[0, 1[\cap \overline{([0, 1[\cap]-\tfrac{1}{2}, \tfrac{1}{2}[)} = [0, 1[\cap \overline{[0, \tfrac{1}{2}[} = [0, 1[\cap [0, \tfrac{1}{2}] = [0, \tfrac{1}{2}],$$

and this is compact in E^1. ∎

Note that this example also shows that a locally compact subset of a Hausdorff space need not be closed.

2.6 Example A subspace of a locally compact space is not necessarily locally compact. For example, the set $\mathscr{I}$ of irrationals is not locally compact in E^1, although E^1 is locally compact.

To see this, let V be any nghd. of π. If $\mathscr{I} \cap \overline{(\mathscr{I} \cap V)}$ is compact, then $\mathscr{I} \cap \overline{(\mathscr{I} \cap V)}$ is bounded and closed. For some $\varepsilon > 0$, $]\pi - \varepsilon, \pi + \varepsilon[\subset V$. Choose any rational y with $\pi - \varepsilon < y < \pi + \varepsilon$. Then, y is a cluster point of $\mathscr{I} \cap \overline{(\mathscr{I} \cap V)}$, but $y \notin \mathscr{I} \cap \overline{(\mathscr{I} \cap V)}$, as $y \notin \mathscr{I}$. This means that $\mathscr{I} \cap \overline{(\mathscr{I} \cap V)}$ is not closed after all, a contradiction. Then, $\mathscr{I} \cap \overline{(\mathscr{I} \cap V)}$ cannot be compact, so $\mathscr{I}$ is not locally compact.

Of course, there is nothing special about π in this argument—any irrational will do. ∎

As with subspaces, a product of locally compact spaces need not be locally compact. If, however, the coordinate spaces are Hausdorff, and if enough of them are compact, then the product will be locally compact.

COMPACTNESS AND COVERING THEOREMS   157

2.5 Theorem *Let $A \neq \phi$.*
Let T_α be a Hausdorff topology on X_α for each $\alpha \in A$.
Let $\mathbb{P}$ be the product topology on $\prod_{\alpha \in A} X_\alpha$.
Then,
$\prod_{\alpha \in A} X_\alpha$ *is $\mathbb{P}$-locally compact*
if and only if
X_β is T_β-locally compact for each $\beta \in A$ and $A \cap \{\alpha | X_\alpha$ is not T_α-compact$\}$ is finite.

Proof Suppose first that $\prod_{\alpha \in A} X_\alpha$ is $\mathbb{P}$-locally compact. Let $\beta \in A$. Then, p_β is a $(\mathbb{P}, T_\beta)$ continuous, open surjection $\prod_{\alpha \in A} X_\alpha \to X_\beta$ by Theorem 2.5.III. By Theorem 1.7.V and Theorem 2.2, X_β is T_β-locally compact.

We must now show that $A \cap \{\alpha | X_\alpha$ is not T_α-compact$\}$ is finite.

Let $f \in \prod_{\alpha \in A} X_\alpha$. Since $\prod_{\alpha \in A} X_\alpha$ is $\mathbb{P}$-locally compact, there is a $\mathbb{P}$-nghd. V of f such that $\bar{V}$ is $\mathbb{P}$-compact. Choose a basic $\mathbb{P}$-open nghd. $\bigcap_{i=1}^{n} p_{\alpha_i}^{-1}(U_{\alpha_i})$ of f such that $\bigcap_{i=1}^{n} p_{\alpha_i}^{-1}(U_{\alpha_i}) \subset V$.

Now note that $p_\alpha(\bar{V})$ is T_α-compact for each $\alpha \in A$, by Theorem 1.7, Theorem 2.5 (1).III, and Theorem 1.8.III.

Further, if $\beta \in A$, then

$$p_\beta\left(\bigcap_{i=1}^{n} p_{\alpha_i}^{-1}(U_{\alpha_i})\right) \subset p_\beta(V) \subset p_\beta(\bar{V}).$$

But,

$$p_\beta\left(\bigcap_{i=1}^{n} p_{\alpha_i}^{-1}(U_{\alpha_i})\right) = X_\beta \quad \text{if} \quad \beta \neq \alpha_1, ..., \alpha_n.$$

Then, $p_\beta(\bar{V}) = X_\beta$ for $\beta \neq \alpha_1, ..., \alpha_n$, so X_β is T_β-compact for at least each $\beta \in A$ with $\beta \neq \alpha_1, ..., \alpha_n$.

Conversely, suppose that X_β is T_β-locally compact for each $\beta \in A$ and that $A \cap \{\alpha | X_\alpha$ is not T_α-compact$\}$ is finite.

If $\{\alpha | X_\alpha$ is not T_α-compact$\} = \phi$, then $\prod_{\alpha \in A} X_\alpha$ is $\mathbb{P}$-compact by Tychonov's Theorem, hence $\mathbb{P}$-locally compact.

Suppose then that $A \cap \{\alpha | X_\alpha$ is not T_α-compact$\} = \{\alpha_1, ..., \alpha_n\}$.

Let $g \in \prod_{\alpha \in A} X_\alpha$. For $i = 1, ..., n$, $g_{\alpha_i} \in X_{\alpha_i}$, and X_{α_i} is T_{α_i}-locally compact, so there is some T_{α_i}-nghd. V_{α_i} of g_{α_i} such that $\bar{V}_{\alpha_i}$ is T_{α_i}-compact.

Now, $g \in \bigcap_{i=1}^{n} p_{\alpha_i}^{-1}(V_{\alpha_i}) \in \mathbb{P}$. Further, by Theorem 2.2.III, $\text{Cl}_{\mathbb{P}}\left(\bigcap_{i=1}^{n} p_{\alpha_i}^{-1}(V_{x_i})\right)$
$= \bigcap_{i=1}^{n} p_{\alpha_i}^{-1}(\bar{V}_{\alpha_i}) = \prod_{\alpha \in A} Y_\alpha$, where

$$Y_\alpha = \begin{cases} \bar{V}_{\alpha_i} & \text{if } \alpha = \alpha_i, i = 1, \ldots, n, \\ X_\alpha & \text{if } \alpha \in A \text{ and } \alpha \neq \alpha_i, i = 1, \ldots, n. \end{cases}$$

Now, Y_α is T_α-compact for each $\alpha \in A$. By Tychonov's Theorem, $\prod_{\alpha \in A} Y_\alpha$ is $\mathbb{P}$-compact, and the theorem is proved. ∎

As an immediate application of Theorem 2.5, the space of Example 2.4 is now easily seen to be not locally compact.

There is an important connection between local compactness and Alexandroff's Theorem which we shall have occasion to exploit a number of times. A one point compactification of a non-compact space is Hausdorff exactly when the space is both Hausdorff and locally compact.

2.6 THEOREM *Let (X, T) be a non-compact space.*
Let (Y, M) be an Alexandroff one-pont compactification of (X, T).
Then,
(Y, M) is a Hausdorff space
if and only if
(X, T) is Hausdorff and locally compact.

Proof For simplicity, let $z \notin X$, let $Y = X \cup \{z\}$, and let $M = \mathcal{O}(T)$, as in Theorem 1.14.

Suppose first that (Y, M) is a Hausdorff space.

Since $T = M_X$, it is immediate by Theorem 1.6.V, that (X, T) is Hausdorff.

Now let $x \in X$. Since M is Hausdorff and $z \neq x$, there are disjoint M-nghds. V and W of x and z respectively. Now, (Y, M) is a compact space, and $\text{Cl}_M(V)$ is M-closed, so $\text{Cl}_M(V)$ is M-compact by Corollary 2, sec. 1.

Since $z \notin V$, then $V = V \cap X \in T$. Further, $\text{Cl}_T(V) = \text{Cl}_{M_X}(V) = X \cap \text{Cl}_M(V)$ by Theorem 1.5 (2).III.

Now, $\text{Cl}_M(V) \subset X$, since $z \in W \in M$ and $W \cap V = \phi$. Then, $\text{Cl}_T(V) = \text{Cl}_M(V)$.

Finally, if $\mathcal{O}$ is a T-cover of $\text{Cl}_T(V)$, then $\mathcal{O}$ is an M-cover of $\text{Cl}_M(V)$, which is M-compact, hence has a finite reduction.

Then, V is a T-nghd. of x and $\text{Cl}_T(V)$ is T-compact, implying that X is T-locally compact.

Conversely, suppose that (X, T) is Hausdorff and locally compact.

To show that (Y, M) is Hausdorff, let x and y be distinct points of Y.

If $x \in X$ and $y \in X$, then there are disjoint T (hence M) neighborhoods of x and y, as T is Hausdorff.

Thus, we may suppose that, say, $y = z$. Now, X is T-locally compact, so there is a T-nghd. V of x such that $\text{Cl}_T(V)$ is T-compact.

Then, $z \in Y - \text{Cl}_T(V)$. But, $X \cap (Y - \text{Cl}_T(V)) = X - \text{Cl}_T(V) \in T$, and $X - (Y - \text{Cl}_T(V)) = \text{Cl}_T(V)$ is T-compact, so $Y - \text{Cl}_T(V) \in M$.

Finally, $\text{Cl}_T(V) \cap (Y - \text{Cl}_T(V)) = \phi$ so that M is Hausdorff. ∎

We shall apply the last theorem immediately to the proof of three important theorems, both as an illustration of the use of one point compactifications and for later reference.

The first theorem concerns the problem of existence of continuous functions, considered in Chapter V in connection with Urysohn's Lemma.

2.7 THEOREM *Let (X, T) be a locally compact, Hausdorff space.*
Let F be a T-compact subset of X.
Let $V \in T$ such that $V \neq X$ and $F \subset V$.
Then,
there is a continuous $f: X \to [0, 1]$ such that $f(X - V) = \{0\}$ and $f(F) = \{1\}$.

Proof If (X, T) is compact, then (X, T) is normal by Theorem 1.10, and the conclusion follows from Urysohn's Lemma and the fact that F and $X - V$ are disjoint, T-closed sets.

Thus suppose that X is not T-compact. Let $z \notin X$, and let $Y = X \cup \{z\}$. Let M denote the Alexandroff topology on Y of Theorem 1.14.

Since $V \in T \subset M$, then $V \in M$, so $Y - V$ is M-closed. Also, F is M-closed, since $z \in Y - F$ and $X \cap (Y - F) = X - F \in T$ and $X - (Y - F)$ is T-compact, implying that $Y - F \in M$.

Now, (Y, M) is a compact Hausdorff space, hence is normal. By Urysohn's Lemma, there is a continuous $g: Y \to [0, 1]$ with $g(Y - V) = \{0\}$ and $g(F) = \{1\}$.

Let $f = g|X$. Since $T = M_X$, then $f: X \to [0, 1]$ is continuous by Theorem 1.8.III. Finally, $f(X - V) = \{0\}$ and $f(F) = \{1\}$. ∎

As a second application of Theorem 2.6, we shall **extend** Theorem 1.11 to locally compact spaces. The details of the proof are similar to those of the last theorem, so some will be omitted.

2.8 THEOREM *Let (X, T) be a locally compact Hausdorff space.
Let F be T-closed.
Let $U \in T$ and $F \subset U$.
Then,
for some V, $V \in T$ and $F \subset V \subset \bar{V} \subset U$ and $\bar{V}$ is T-compact.*

Proof If (X, T) is compact, then the result is immediate by Theorem 1.11.

Suppose then that (X, T) is not T-compact. Let (Y, M) be an Alexandroff one point compactification of (X, T). As usual, we may for convenience suppose that $X \subset Y$ and $T = M_X$.

Now, F is M-closed. Further, (Y, M) is a compact Hausdorff space by Theorem 2.6, and $U \in M$ such that $F \subset U$. By Theorem 1.11, there is some M-open V such that $F \subset V \subset \text{Cl}_M(V) \subset U$, and $\text{Cl}_M(V)$ is M-compact.

Finally, $V \subset U \subset X$ and $V \in T$, and $\text{Cl}_T(V) = \text{Cl}_M(V)$ is M-compact, hence T-compact. ∎

The final application of Theorem 2.6 for now is to the proof of a separation theorem: every locally compact Hausdorff space is completely regular.

2.9 THEOREM *Let (X, T) be a locally compact Hausdorff space.
Then,
T is completely regular.*

Proof If X is T-compact, then T is normal by Theorem 1.10, hence also completely regular.

Suppose then that X is not T-compact. Let (Y, M) be a one point compactification of (X, T), with $X \subset Y$ and $T = M_X$.

Since (Y, M) is a compact Hausdorff space, then M is a normal topology. Then (X, T) is a subspace of a normal space, hence is completely regular by Corollary 1, Ch. V, sec. 4. ∎

We have already noted, in Example 2.6, that a subspace of a locally compact space need not be locally compact. We shall conclude this section with an application of Theorem 2.9 to a derivation of sufficient conditions for a subspace of a locally compact space to be locally compact.

2.10 THEOREM *Let (X, T) be a locally compact Hausdorff space.
Let A be T-open or T-closed.
Then,
A is T-locally compact.*

Proof Suppose first that A is T-open. Let $x \in A$. By Theorem 2.9, T is completely regular, hence regular. For some V, V is a T-nghd. of x and $\bar{V} \subset A$ and $\bar{V}$ is T-compact. Now, $V \subset \bar{V} \subset A$, so $V \subset A$. Then, $A \cap \overline{(A \cap V)} = \bar{V}$ is T-compact, and A is T-locally compact by Theorem 2.4.

If A is T-closed, let $x \in A$ and note that, for some V, V is a T-nghd. of x and $\bar{V}$ is T-compact. Now, $A \cap \overline{(A \cap V)}$ is a T-closed subset of a T-compact set $\bar{V}$, hence is T-compact by Theorem 1.8. By Theorem 2.4 again, A is T-locally compact. ∎

Actually, the proof in the case that A is closed does not make use of the assumption that (X, T) is Hausdorff, as does the case that A is open. In fact, any closed subset of any locally compact space is locally compact.

7.3 Countable compactness

The definition of compactness requires that each open cover be reducible to a finite cover. If this is relaxed to the reduction of just countable open covers, then we have a weaker condition called countable compactness.

3.1 DEFINITION *X is T-countably compact*
if and only if
each countable T-cover of X can be reduced to a finite T-cover of X.

For convenience, we shall abbreviate "X is T-countably compact" to "X is T-c.c.".

It is obvious that compactness implies countable compactness.

3.1 THEOREM *If X is T-compact, then X is T-c.c.*

Proof Immediate. ∎

There are spaces which are countably compact, but not compact. We shall produce an example after we have developed some more theory.

The next theorem gives a condition equivalent to countable compactness which is very like the finite intersection form of compactness.

3.2 THEOREM *X is T-c.c.*
if and only if
if F is a countable set of T-closed subsets of X, and if $\cap F = \phi$, then, for some finite, non-empty subset B of F, $\cap B = \phi$.

Proof The proof is like that of Theorem 1.4, and is left to the reader. ∎

A second reformulation of the notion of countable compactness is analogous to Theorem 1.6. However, where ultrafilterbases are used to characterize compactness, sequences suffice for countable compactness, and not by convergence, but by a weaker condition. A sequence S to X has x as a limit point if every nghd. of x contains elements S_n for arbitrarily large n. Often this is paraphrased: S is frequently in every nghd. of x. This does not mean that S is eventually in every (or any) nghd. of x. For example, let $S_n = (-1)^n$ for $n = 1, 2, \ldots$ Then, -1 and 1 are limit points of S, as S is frequently in each nghd. of 1 or of -1, but S has no limit.

3.2 Definition *Let S be a sequence to X.*
Let $x \in X$.
Then,
x is a T-limit point of S
if and only if
if V is a T-nghd. of x, and $K \in Z^+$, then there is some $n \in Z^+$ such that $n \geq K$ and $S_n \in V$.

3.3 Theorem *X is T-c.c.*
if and only if
each sequence to X has a T-limit point in X.

Proof Suppose first that X is T-c.c. Let S be a sequence to X. Let $F_n = \text{Cl}_T \{S_j | j \in Z^+ \text{ and } j \geq n\}$ for each $n \in Z^+$. Then, F_n is T-closed, and $F_{n+1} \subset F_n$ for each $n \in Z^+$. If $i_1, \ldots, i_n$ are positive integers, then $\bigcap_{j=1}^{n} F_{i_j} = F_\alpha$, where $\alpha = \max \{i_1, \ldots, i_n\}$, hence $\bigcap_{j=1}^{n} F_{i_j} \neq \phi$. By Theorem 3.2, $\bigcap_{n=1}^{\infty} F_n \neq \phi$. Let $x \in \bigcap_{n=1}^{\infty} F_n$. We shall show that x is a T-limit point of S.

For, let $x \in V \in T$, and let K be a positive integer. Then, $x \in F_K$, so $V \cap \{S_j | j \in Z^+ \text{ and } j \geq K\} \neq \phi$. Then, for some n, $n \in Z^+$ and $n \geq K$ and $S_n \in V$.

Conversely, suppose that each sequence to X has a T-limit point in X. If X is not T-c.c., then there is a countable T-cover $\{G_n | n \in Z^+\}$ of X which has no finite reduction. Then, for each $n \in Z^+$, $\bigcup_{i=1}^{n} G_i \neq X$. Choose $S_n \in X - \bigcup_{i=1}^{n} G_i$ for each $n \in Z^+$. This defines a sequence S to X.

Now, S has no T-limit point in X. For, suppose that $z \in X$. Then, $z \in G_K$ for some $K \in Z^+$. But, $S_n \notin G_K$ for $n \geq K$, so that z is not a T-limit point of S.

This contradiction implies that $X - \bigcup_{i=1}^{k} G_i = \phi$ for some $K \in Z^+$. Hence $\{G_1, ..., G_K\}$ is a finite reduction of the cover $\{G_i | i \in Z^+\}$. ∎

The last theorem can be phrased in terms of nets and filterbases, but this would not be of any value to us.

Our final equivalent formulation of countable compactness is similar in form to the Bolzano-Weierstrass Theorem of classical analysis.

3.4 Theorem *X is T-c.c.*
if and only if
if A is an infinite subset of X, then A has a T-cluster point in X.

Proof Suppose first that X is T-c.c. Let A be an infinite subset of X. Then, A has a countable subset, say $\{S_i | i \in Z^+\}$, where $S_i \neq S_j$ for $i \neq j$. By Theorem 3.3, S has a T-limit point λ in X. It is easy to check that λ is a T-cluster point of A.

Conversely, suppose that each infinite subset of X has a T-cluster point in X. Let S be a sequence to X.

If $\{S_i | i \in Z^+\}$ is finite, then for some $K \in Z^+$, $S_i = S_K$ for infinitely many values of i. Then, S_K is a T-limit point of S.

If $\{S_i | i \in Z^+\}$ is infinite, then, for some $\lambda \in X$, λ is a T-cluster point of $\{S_i | i \in Z^+\}$. Then, λ is a T-limit point of S. ∎

We now give an example of a countably compact, non-compact space.

3.1 Example Ω is a countably compact, non-compact space.

That Ω is not compact is easy to show. One way is to observe that $\{[0, \alpha[| \alpha \in \Omega\}$ is an open cover which cannot be reduced to a finite cover. A second method is to note that $[\alpha, \Omega[$ is closed for each $\alpha \in \Omega$, and $\bigcap_{\alpha \in \Omega} [\alpha, \Omega[= \phi$, but $\bigcap_{i=1}^{n} [\alpha_i, \Omega[\neq \phi$ for any $\alpha_1, ..., \alpha_n \in \Omega$.

To show that Ω is countably compact, let $A \subset \Omega$ and suppose that A is infinite. It suffices to show that A has a cluster point in Ω.

Now, A has a countable subset, say $\{a_i | i \in \omega\}$. Since Ω is uncountable, then $\phi \neq \Omega - \{a_i | i \in \omega\} \subset \Omega$. Then $\Omega - \{a_i | i \in \omega\}$ has a least element, say L. We claim that L must be a cluster point of A. Consider two cases.

i) L is a limit ordinal.

Choose any basic neighborhood of L, say $]\alpha, L]$, where $\alpha \in \Omega$ and $\alpha < L$. If $]\alpha, L] \cap \{a_i | i \in \omega\} = \phi$, then $\alpha \cup \{\alpha\} \in \Omega - \{a_i | i \in \omega\}$. But, $\alpha \cup \{\alpha\} < L$, contradicting the choice of L as the least element of $\Omega - \{a_i | i \in \omega\}$.

ii) L is a non-limit ordinal.

Then, $[0, L[$ is compact (Example 1.8). Since $\{a_i | i \in \omega\}$ is an infinite subset of the compact, hence countably compact, $[0, L[$, then $\{a_i | i \in \omega\}$ has a cluster point c in $[0, L[$. It is now easy to check that c is also a cluster point of A in Ω. ∎

We shall return to the notion of countable compactness in the next section and again in Chapter VIII, where we prove that compactness and countable compactness are equivalent in metric spaces. We conclude this section with the fact that countable compactness is a topological invariant.

3.5 THEOREM *Let* $(X, T) \cong (Y, M)$.
Then,
X is T-c.c. if and only if Y is M-c.c.

Proof Left as an exercise. ∎

7.4 Other kinds of compactness and 1°-countable spaces

In this section we briefly sketch two more kinds of compactness and examine how they stand in relation to those already developed.

4.1 DEFINITION *X is T-Bolzano-Weierstrass compact
if and only if
each sequence to X has a limit point in X.*

Bolzano-Weierstrass compactness is often abbreviated to B-W compactness. Actually, this concept was treated last section, although at that time we did not assign a name to it. We do so now simply because the term is used in the literature.

4.1 THEOREM *X is T-B-W compact $\Leftrightarrow$ X is T-c.c.*

Proof Immediate by Theorem 3.3. ∎

Finally, there is sequential compactness. A space is sequentially compact if each sequence has a convergent subsequence. The reader is probably familiar with subsequences from calculus, but we can quickly sketch the idea with an example. Let $S_n = 1/n$ for $n = 1, 2, \ldots$ Then, S is a sequence to E^1. Let $V_n = 1/2n$ for $n = 1, 2, \ldots$ Then, V is also a sequence to E^1, but may be considered as a subsequence of S in the sense that $V_n = S_{2n}$. That is, $V = S \circ \varphi$, where φ is an increasing function on Z^+ to Z^+. This motivates the following definition.

4.2 DEFINITION *Let S be a sequence to X.*
Then,
V is a subsequence of S
if and only if
for some $\varphi: Z^+ \to Z^+$, we have:
$\varphi(n) < \varphi(m)$ if $n < m$,
and,
$V = S \circ \varphi$.

4.3 DEFINITION *X is T-sequentially compact*
if and only if
each sequence to X has a T-convergent subsequence.

4.2 THEOREM *If X is T-sequentially compact, then X is T-B-W compact.*

Proof Suppose that X is T-sequentially compact. Let S be a sequence to X. Let V be a T-convergent subsequence of S, say $V = S \circ \varphi$ T-converges to $y \in X$. We leave it for the reader to show that y is a T-limit point of S. ∎

Thus far we have the following table of implications:

$$\text{compactness} \Rightarrow \text{countable compactness}$$
$$\Updownarrow$$
$$\text{sequential compactness} \Rightarrow \text{B-W compactness}$$

It is not true that compactness or B-W (or countable) compactness implies sequential compactness. For example, let $X_\alpha = [0, 1]$, considered as a subspace of E^1, for each $\alpha \in [0, 1]$. Then, $\prod_{\alpha \in [0,1]} X_\alpha$ is compact in the product topology, hence also B-W and countably compact. The reader can show as a substantial exercise, however, that $\prod_{\alpha \in [0,1]} X_\alpha$ is not sequentially compact. An easier exercise is to check that Ω is sequentially compact, but not compact.

In some spaces, countable compactness does imply sequential compactness. A space is said to be 1°-countable (or satisfy the first axiom of countability) if, given any point x, there is a countable collection B_x of neighborhoods of x such that each nghd. of x contains at least one set from B_x. Such a collection is often called a base for the neighborhood system at x (by analogy with Theorem 3.1.I). It is obvious that any metric space is 1°-countable (see Theorem 6.1.VIII). However, the reader can show that the space of Example 1.7.I is not 1°-countable.

4.4 DEFINITION X is $T - 1°$-countable
if and only if
if $x \in X$, then there is a countable set B_x of T-nghds. of x such that, if $x \in V \in T$, then $b \subset V$ for some $b \in B_x$.

With the help of a preliminary Lemma, we shall show that sequential and countable compactness are equivalent in $1°$-countable spaces.

1 LEMMA *Let X be $T - 1°$-countable.*
Let S be a sequence to X.
Let x be a T-limit point of S.
Then,
there is a subsequence of S which T-converges to x.

Proof Let $B_x = \{V_i | i \in Z^+\}$ be a countable set of T-nghds. of x such that, if $x \in V \in T$, then $x \in V_j \subset V$ for some positive integer j.

Let $U_i = \bigcap_{j=1}^{i} V_j$ for $i = 1, 2, \ldots$

Then, $\{U_i | i \in Z^+\}$ is a set of T-nghds. of x, and $U_{i+1} \subset U_i$ for each $i \in Z^+$. Further, if $x \in V \in T$, then $U_n \subset V$ for some $n \in Z^+$.

Define a subsequence s of S inductively as follows.

Let $\varphi_1 = \min \{j | j \in Z^+ \text{ and } S_j \in U_1\}$.

Let $\varphi_2 = \min \{j | j \in Z^+ \text{ and } j > \varphi_1 \text{ and } S_j \in U_2\}$.

In general, if $\varphi_1, \ldots, \varphi_n$ have been chosen, let $\varphi_{n+1} = \min \{j | j \in Z^+ \text{ and } j > \varphi_n \text{ and } S_j \in U_{n+1}\}$. Then, $s = S \circ \varphi$ is a subsequence of S, and it is easy to check that s is T-convergent to x. ∎

4.3 THEOREM *Let X be $T - 1°$-countable.*
Then,
X is T-sequentially compact
if and only if
X is T-c.c.

Proof By Theorem 4.1 and Theorem 4.2, if X is T-sequentially compact, then X is T-c.c.

Conversely, suppose that X is T-c.c. Let S be a sequence to X. By Theorem 4.1, X is T-B-W- compact. Hence S has a T-limit point x in X. By Lemma 1, some subsequence of S is T-convergent to x. Then, X is T-sequentially compact. ∎

Our chart now reads:

compactness $\Rightarrow$ countable compactness

$\Updownarrow$

sequential compactness $\underset{1°\text{-spaces}}{\rightleftarrows}$ B-W compactness

The next theorem is by way of summarizing previous results.

4.4 Theorem *Let X be $T - 1°$-countable.*
Then,
X is T-c.c.
if and only if
X is T-B-W compact
if and only if
X is T-sequentially compact.

Proof Immediate by Theorems 4.1 and 4.3. ∎

Finally, all the topological concepts defined in this section are invariants.

4.5 Theorem *Let $(X, T) \cong (Y, M)$.*
Then,
(1) X is T-B-W compact $\Leftrightarrow$ Y is M-B-W compact.
(2) X is T-sequentially compact $\Leftrightarrow$ Y is M-sequentially compact.
(3) X is $T - 1°$-countable $\Leftrightarrow$ Y is $M - 1°$-countable.

Proof Left as an exercise. ∎

7.5   2°-countable, separable and Lindelöf spaces

In this section we shall briefly sketch three additional concepts which play an important role in set topology.

5.1 Definition *X is $T - 2°$-countable*
if and only if
T has a countable base.

5.1 Theorem *If X is $T - 2°$-countable, then X is $T - 1°$-countable.*

Proof Immediate by Theorem 3.1.I. ∎

We shall see in Chapter VIII that every metric space is $1°$-countable, but not necessarily $2°$-countable. In fact, a metric space is $2°$-countable exactly when it has a countable subset whose closure is the whole space. The terminology used in this connection is quite standard, and is given in the following two definitions.

5.2 DEFINITION *A is T-dense in X
if and only if
$\bar{A} = X$.*

5.3 DEFINITION *X is T-separable
if and only if
there is a countable, T-dense subset of X.*

If A is dense in X, then each element of X can be approximated arbitrarily closely (in the sense of the topology) by elements of A. For example, the rationals are dense in E^1, and the polynomials are dense in the sup-norm space $C([a, b])$. This is the Weierstrass Approximation Theorem. A noncompact space is also dense in its one point compactification.

We shall now show that $2°$-countability implies separability.

5.2 THEOREM *Let X be $T - 2°$-countable.
Then,
X is T-separable.*

Proof Let $\mathscr{B}$ be a countable base for T. We may assume that $\phi \notin \mathscr{B}$.
If $b \in \mathscr{B}$, choose $x_b \in b$. Let $A = \{x_b | b \in \mathscr{B}\}$. Then, $A \subset X$ and A is countable. Further, $\bar{A} = X$. For suppose that $y \in X$ and $y \in V \in T$. For some b, $b \in \mathscr{B}$ and $y \in b \subset V$ by Theorem 3.1.I. Then, $x_b \in A \cap V$, so $y \in \bar{A}$ by Theorem 2.3 (2).I. ∎

It is not true in general that separable implies $2°$-countable.

5.1 EXAMPLE Let R be the set of real numbers and T the topology consisting of ϕ and all sets $R - A$, A finite. Then, (R, T) is separable (for example, Z^+ is a countable dense subset).

Now suppose that $\{U_i | i \in Z^+\}$ is a countable set of T-open sets. For convenience, we may suppose that each U_i is not empty. Then, there is a finite subset S_i of R with $U_i = R - S_i$ for each positive integer i. Now, $\bigcap_{i=1}^{\infty} U_i = \bigcap_{i=1}^{\infty} (R - S_i) = R - \bigcup_{i=1}^{\infty} S_i \neq \phi$. Let $x \in \bigcap_{i=1}^{\infty} U_i$. Then, $R - \{x\}$ is T-open, but does not have any U_i as a subset, so that $\{U_i | i \in Z^+\}$ cannot be a base for T. ∎

The notion of a Lindelöf space is in a sense complementary to that of countable compactness. A space is Lindelöf if each open cover has a countable reduction.

5.4 DEFINITION *X is T-Lindelöf if and only if each T-cover of X can be reduced to a countable T-cover.*

It is obvious that each compact space is Lindelöf and that each countably compact Lindelöf space is compact.

5.3 THEOREM
(1) If X is T-compact, then X is T-Lindelöf.
(2) If X is T-Lindelöf and T-c.c., then X is T-compact.

Proof Immediate from the definitions. ∎

There are separable spaces which are not Lindelöf, and Lindelöf spaces which are not separable, so that these two notions are independent.

5.2 EXAMPLE Give the set R of real numbers the topology T generated by the half-open intervals $]a, b]$. Then, $\{(x, y) | x$ and y are rational$\}$ is a countable, dense subset, so that $R \times R$ is separable in the product topology. However, $R \times R$ is not Lindelöf.

To see this, suppose $R \times R$ is Lindelöf. It is easy to check that each closed subspace would then also be Lindelöf (the proof is like that of Theorem 1.8). Now, $\{(r, -r) | r$ is irrational$\}$ is a closed subspace of $R \times R$, so would be Lindelöf. But, for any irrational number r, $]r - 1, r] \times]-r - 1, -r] \cap \{(y, -y) | y$ is irrational$\} = \{(r, -r)\}$ is a nghd. of $(r, -r)$, so that the subspace $\{(r, -r) | r$ is irrational$\}$ is discrete. And obviously an uncountable, discrete space cannot be Lindelöf (take an open cover of singletons). ∎

5.3 EXAMPLE The ordinal space $\Omega \cup \{\Omega\}$ is compact (as $\Omega \cup \{\Omega\}$ is a non-limit ordinal), hence Lindelöf. But $\Omega \cup \{\Omega\}$ is not separable. For, let C be a countable subset of $\Omega \cup \{\Omega\}$. Then, $C - \{\Omega\}$ is countable, hence has an upper bound γ in Ω. Now, $\{\gamma \cup \{\gamma\}\}$ is a neighborhood of $\gamma \cup \{\gamma\}$ which does not intersect C, so that C cannot be dense in $\Omega \cup \{\Omega\}$. ∎

It is also possible for a space to be Lindelöf and neither compact, countably compact, nor $2°$-countable.

5.4 EXAMPLE Let R be the set of real numbers, and T the topology generated by the sets $]-\infty, x]$. It is very easy to check that (R, T) is Lindelöf, but neither compact, countably compact, nor $2°$-countable. ∎

However, every $2°$-countable space is Lindelöf.

5.4 THEOREM *Let X be $T - 2°$-countable. Then, X is T-Lindelöf.*

Proof Let $\mathcal{O}$ be a T-cover of X. We must product a countable subset of $\mathcal{O}$ which also covers X.

Let $\mathcal{B}$ be a countable base for T. If $x \in \alpha \in \mathcal{O}$, then there is some $V(\alpha, x) \in \mathcal{B}$ with $x \in V(\alpha, x) \subset \alpha$ by Theorem 3.1.I. Now, $\{V(\alpha, x) | x \in \alpha \in \mathcal{O}\} \subset \mathcal{B}$, hence $\{V(\alpha,x) | x \in \alpha \in \mathcal{O}\}$ is countable. For convenience, write $\{V(\alpha,x) | x \in \alpha \in \mathcal{O}\}$ as $\{V_i | i \in Z^+\}$. For each $i \in Z^+$, choose some $\beta_i \in \mathcal{O}$ with $V_i \subset \beta_i$. Now, if $\alpha \in \mathcal{O}$, then $\alpha = \bigcup_{x \in \alpha} V(\alpha, x)$, so $\bigcup_{\alpha \in \mathcal{O}} \alpha = X = \bigcup_{x \in \alpha \in \mathcal{O}} V(\alpha, x) = \bigcup_{i=1}^{\infty} V_i \subset \bigcup_{i=1}^{\infty} \beta_i$ implies that $\bigcup_{i=1}^{\infty} \beta_i = X$. Then, $\{\beta_i | i \in Z^+\}$ is a countable subset of $\mathcal{O}$ which covers X, and the proof is complete. ∎

Summary chart

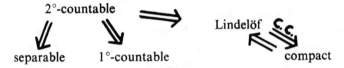

We conclude this section with a separation theorem which typifies how the concepts of this section can strengthen separation properties, and an invariance theorem.

5.5 THEOREM *Let (X, T) be a regular, Lindelöf space. Then, (X, T) is normal.*

Proof Let A and B be disjoint, T-closed sets. If $a \in A$, then by Theorem 2.2 (3), Ch. V, there is a T-nghd. U_a of a such that $\bar{U}_a \cap B = \phi$. Similarly, if $b \in B$, there is a T-nghd. V_b of b such that $\bar{V}_b \cap A = \phi$.

Then, $\{X - (A \cup B)\} \cup \{U_a | a \in A\} \cup \{V_b | b \in B\}$ is a T-cover of X.

Now, X is T-Lindelöf, hence there are countably many elements a_i of A, b_i of B such that $\{X - (A \cup B)\} \cup \{U_{a_i} | i \in Z^+\} \cup \{V_{b_i} | i \in Z^+\}$ covers X.

For each positive integer n, let

$$W_n = U_{a_n} - \bigcup_{j=1}^{n} \bar{V}_{b_j},$$

$$S_n = V_{b_n} - \bigcup_{j=1}^{n} \bar{U}_{a_j}.$$

Then, by Theorem 2.5 (2) and Theorem 1.2 (2), Ch. I, W_n and S_n are T-open for each positive integer n. Then, $\bigcup_{n=1}^{\infty} W_n$ and $\bigcup_{n=1}^{\infty} S_n$ are T-open. Further, $A \subset \bigcup_{n=1}^{\infty} W_n$ and $B \subset \bigcup_{n=1}^{\infty} S_n$.

There remains to show that $\left(\bigcup_{n=1}^{\infty} W_n\right) \cap \left(\bigcup_{n=1}^{\infty} S_n\right) = \phi$.

Suppose that $x \in \left(\bigcup_{n=1}^{\infty} W_n\right) \cap \left(\bigcup_{n=1}^{\infty} S_n\right)$.

For some integers m and k, $x \in W_m \cap S_k$.

If $m \leq k$, then $x \in V_{b_k} - \bigcup_{j=1}^{k} \bar{U}_{a_j}$ implies that $x \notin U_{a_m}$ a contradiction.

If $k \leq m$, then $x \in U_{a_m} - \bigcup_{j=1}^{m} \bar{V}_{b_j}$ implies $x \notin V_{b_k}$, a contradiction.

Then, $\left(\bigcup_{n=1}^{\infty} W_n\right) \cap \left(\bigcup_{n=1}^{\infty} S_n\right) = \phi$, and the proof is complete. ∎

1 COROLLARY *Let (X, T) be a regular, $2°$-countable space. Then, (X, T) is normal.*

Proof Immediate by Theorems 5.4 and 5.5. ∎

5.6 THEOREM *Let $(X, T) \cong (Y, M)$. Then,*
(1) X is $T - 2°$-countable $\Leftrightarrow$ Y is $M - 2°$-countable.
(2) X is T-separable $\Leftrightarrow$ Y is M-separable.
(3) X is T-Lindelöf $\Leftrightarrow$ Y is M-Lindelöf.

Proof Left as an exercise. ∎

Problems

1) Prove that the Cantor set in E^1 is compact.
2) Show that, in a Hausdorff space, disjoint compact sets have disjoint open neighborhoods.

3) Show that an intersection of compact subsets of a space need not be compact. However, an intersection of compact subsets of a Hausdorff space is compact.

4) The closure of a compact subset of a space may not be compact.

5) Let R be the set of real numbers. Examine the space (R, T) in each of the following cases with regard to compactness, local, countable and sequential compactness, separation properties, separability, $1°$- and $2°$-countability, and the Lindelöf property.
   (a) T consists of ϕ and all sets $R - S$, with S finite.
   (b) T is the discrete topology on R.
   (c) T consists of ϕ and all sets $R - S$, with S countable.

*6) Let $X = [0, 1] \times [0, 1]$, with the order topology $T_<$ of Example 3.7.I. Examine the space $(X, T_<)$ for compactness, local, countable, and sequential compactness, separation properties, separability, $1°$- and $2°$-countability, and the Lindelöf property.

7) Does $1°$-countable imply separable?

8) A continuous bijection of a compact space to a Hausdorff space is an homeomorphism.

9) If X is a space in which each compact subspace is closed, then sequential limits in X are unique.
   Thus show:
   a $1°$-countable space is Hausdorff if and only if its compact subsets are closed.

10) Let $f: X \to Y$. Let X be Hausdorff and Y compact. Then,
    f is continuous
    if and only if
    $\{(x, f(x)) | x \in X\}$ is closed in $X \times Y$.

11) Let $f: X \to Y$ be a continuous, closed surjection. Let X be a compact, Hausdorff space. Then, Y is Hausdorff. If also X is $2°$-countable, then Y is $2°$-countable.

*12) Let X consist of all real-valued sequences x such that $\sum_{i=1}^{\infty} |x_i|^2$ converges. Let $\varrho(x, y) = \left[\sum_{i=1}^{\infty} (x_i - y_i)^2 \right]^{1/2}$ for each $x, y \in X$.
    Examine the space (X, T_0) for compactness, local compactness, and countable compactness. Is (X, T_0) sequentially compact?

13) Let X be $1°$-countable. Let $x \in X$ and $A \subset X$. Then, $x \in \bar{A}$ if and only if there is a sequence to A which converges to x.

14) Let X be separable. Then, any set of pairwise-disjoint, open subsets of X is countable.

*15) A dense, locally compact subspace of a Hausdorff space is an open set.

*16) Prove Alexander's Theorem:
Let S be a subbase for T. Then,
(X, T) is compact if and only if each T-cover of X by sets in S has a finite subcover.

*17) Use Alexander's Theorem to prove the Tychonov Theorem.

18) X is T-compact if and only if each universal net to X is T-convergent.

19) X is T-compact if and only if each net to X has a T-convergent subnet.

20) X is T-compact if and only if each ultrafilter on X is T-convergent.

21) Give a topological proof of the following theorem from calculus:
Let $f: A \to E^1$ be continuous, where A is a closed, bounded subset of E^1. Then, there are points a and b in A such that $f(a) \leq f(t) \leq f(b)$ for each $t \in A$.

22) Give a topological proof of the Maximum Modulus Theorem from complex analysis:
If f is analytic on a closed, bounded subset A of E^2, then f has its maximum modulus on the boundary of A. (Hint: show that $|f|$ is an homeomorphism).

23) $\Omega \cup \{\Omega\}$ is a one point compactification of Ω.

24) $\prod_{i=1}^{\infty} X_i$ is $1°$-($2°$-)countable if and only if each X_i is $1°$-($2°$-)countable.

25) $\prod_{i=1}^{\infty} X_i$ is separable if and only if each X_i is separable.

26) A subspace of a separable space need not be separable. But an open subspace of a separable space is separable.

27) Every subspace of a $2°$-countable space is separable.

28) Let X be compact and $\sim$ an equivalence relation on X. Then, $X/\sim$ is compact in the quotient topology.

29) Let X be countably compact and $\sim$ an equivalence relation on X. Then, $X/\sim$ is countably compact.

30) Let X be T-locally compact. Let $(\hat{X}, \hat{T})$ and $(\hat{Y}, \hat{M})$ be compact spaces. Suppose that (X, T) is homeomorphic to proper dense subspaces $(\hat{X} - \{x_0\}, \hat{T}_{\hat{X}-\{x_0\}})$ and $(\hat{Y} - \{y_0\}, \hat{M}_{\hat{Y}-\{y_0\}})$. Then, $(\hat{X}, \hat{T}) \cong (\hat{Y}, \hat{M})$.

Thus, the Alexandrov one-point compactification of (X, T) is unique up to homeomorphism.

31) Let X be T-locally compact and suppose that T is Hausdorff. Then, T is the weak topology determined by the T-compact subsets of X (i.e., $A \in T \Leftrightarrow A \cap C \in T_C$ for each compact subset C of X).

*32) This exercise deals with the notion of paracompactness, which is especially useful in differential geometry.

If $\mathscr{T}$ and $\mathscr{T}^*$ are covers of X by subsets of X, then $\mathscr{T}^*$ is a refinement of $\mathscr{T}$ if, given $g^* \in \mathscr{T}^*$, there is some $g \in \mathscr{T}$ with $g^* \subset g$.

(a) X is T-compact if and only if each T-cover of X has a finite refinement of T-open sets which covers X.

(b) X is T-c.c. if and only if each countable T-cover of X has a finite refinement of T-open sets which covers X.

If $\mathscr{T} \subset \mathscr{P}(X)$, then $\mathscr{T}$ is T-locally finite if each point in X has a T-nghd. having a non-empty intersection with at most finitely many sets in $\mathscr{T}$.

(c) If $\mathscr{T}$ is a locally finite set of subsets of X, then $\overline{\cup \mathscr{T}} = \cup\{\bar{g}/g \in \mathscr{T}\}$. X is T-paracompact if and only if T is Hausdorff and each T-cover of X has a locally finite refinement of open sets.

(d) Compact $\Rightarrow$ paracompact.

(e) Paracompact does not imply compact.

(f) Ω is not paracompact.

(g) Each closed subset of a paracompact space is paracompact.

(h) Paracompact implies normal (first show that paracompact implies regular).

(i) Paracompact and separable imply Lindelöf.

(j) Is paracompactness preserved by continuous surjections?—by homeomorphism?

(k) Research project: prove Stone's Theorem: Every metric space is paracompact.

*33) This exercise is a brief introduction to function spaces. Let (X, T) and (Y, M) be spaces, and let Y^X denote the set of all (T, M) continuous functions on X to Y.

If A is T-compact, and $V \in M$, let $[A, V] = Y^X \cap \{f | f(A) \subset V\}$. The compact-open topology $c(X, Y)$ is that generated by the sets $[A, V]$, with A T-compact and $V \in M$.

(a) Let (X, T) be compact. Define a sup-norm metric ϱ on $(E^1)^X$ by $\varrho(f, g) = \sup \{|f(x) - g(x)| \mid x \in X\}$ for f, g continuous on X to E^1. Compare T_ϱ and $c(X, E^1)$.

(b) Let $\mathbb{P}$ be the product topology on $(E^1)^{E^1}$, considered as a subspace of $\prod_{\alpha \in E^1} X_\alpha$ (each $X_\alpha = E^1$). Define a metric ϱ on $(E^1)^{E^1}$ by:
$\varrho(f, g) = 1$ if $|f(x) - g(x)| \geq 1$ for some $x \in E^1$; $\varrho(f, g) = \sup \{|f(x) - g(x)| \mid x \in E^1\}$ if $|f(x) - g(x)| \leq 1$ for each $x \in E^1$. Compare $\mathbb{P}$, T_ϱ and $c(E^1, E^1)$.

(c) If $(X, T) \cong (\hat{X}, \hat{T})$, and $(Y, M) \cong (\hat{Y}, \hat{M})$, then $(Y^X, c(X, Y)) \cong (\hat{Y}^{\hat{X}}, c(\hat{X}, \hat{Y}))$.

(d) (Y, M) is homeomorphic to a subspace of $(Y^X, c(X, Y))$.

(e) $c(X, Y)$ is Hausdorff (regular) if and only if M is Hausdorff (regular).

(f) Let (S, K) be any space. Map $\varphi: Y^X \times S^Y \to S^X$ by $\varphi(f, g) = g \circ f$ for $f \in Y^X$ and $g \in S^Y$. Then,

i) The map $\varphi_f: S^Y \to S^X$ defined by $\varphi_f(g) = \varphi(f, g)$ for a given $f \in Y^X$, is $(c(Y, S), c(X, S))$ continuous.

ii) If T and K are Hausdorff and Y is M-locally compact, then φ is continuous (in the product topology on $Y^X \times S^Y$).

(g) If Y is M-locally compact, then the evaluation map $e: S^Y \times Y \to S$ given by $e(f, y) = f(y)$ for $f \in S^Y$ and $y \in Y$, is continuous.

(h) Let (S, ϱ) be a metric space. Let $\{f_n\}_{n=1}^\infty$ be a sequence to S^Y. Let $g \in S^Y$. Then,
$f_n \to g$ in the $c(Y, S)$ topology on S^Y
if and only if
$f_n \to g$ uniformly on each compact subset of Y (i.e. if A is an M-compact subset of Y, and $\varepsilon > 0$, then there is an integer K such that $\varrho(f_n(a), g(a)) < \varepsilon$ for each $n \geq K$ and $a \in A$.

CHAPTER VIII

A closer look at metric spaces

8.1 Invariance and hereditary properties

HISTORICALLY, METRIC spaces were the first topological spaces to be studied, being the most natural generalization of Euclidean n-space. Today they are still of paramount importance, not only in set topology but also in analysis. In this chapter, we shall tie together many of the concepts of the first seven chapters into a coherent package as they relate to metric spaces. We shall also develop a number of important concepts which are peculiar to the metric space setting.

For the remainder of this chapter, ϱ is a metric on X. We have already seen (Chapter V) that metric spaces are Hausdorff, regular, completely regular and normal. It has not been formally stated yet, but the property of being a metric space is a topological invariant. That is, if $(X, T_\varrho) \cong (Y, M)$, then there is a metric σ on Y which induces the given topology M in the sense that $M = T_\sigma$. The metric σ is concocted in the obvious way in the proof of the first theorem.

1.1 THEOREM *Let $(X, T_\varrho) \cong (Y, M)$.*
Then,
there is a metric σ on Y such that $M = T_\sigma$.

Proof Let $\varphi \colon X \to Y$ be a (T_ϱ, M) homeomorphism. If $x, y \in Y$, define $\sigma(x, y) = \varrho((\operatorname{inv} \varphi)(x), (\operatorname{inv} \varphi)(y))$. We leave it for the reader to verify that σ is a metric on Y. To show that $M = T_\sigma$, use Theorem 3.4.I, and Definition 4.2.I. ∎

It is also the case that each subspace A of a metric space is a metric space, the relative topology $(T_\varrho)_A$ being induced by the restriction of ϱ to $A \times A$.

1.2 THEOREM *Let $A \subset X$.*
Then,
$(T_\varrho)_A = T_{\varrho|(A \times A)}.$

Proof Left to the reader. ∎

8.2 Equivalent metrics

It is not true that each metric space has a unique metric which induces the metric topology of the space. Many metrics on X may give rise to exactly the same topology. Such metrics are called equivalent.

2.1 DEFINITION *Let σ and τ be metrics on X. Then,*
$$\sigma \sim \tau$$
if and only if
$$T_\sigma = T_\tau.$$

It is obvious that $\sim$ is an equivalence relation on the class of metrics on X.

2.1 THEOREM
(1) $\varrho \sim \varrho$
(2) $\varrho \sim \sigma \Leftrightarrow \sigma \sim \varrho$
(3) If $\varrho \sim \sigma$ and $\sigma \sim \tau$, then $\varrho \sim \tau$.

Proof Left to the reader. ∎

2.1 EXAMPLE Let $\tau(x, y) = |x - y|$ and $\sigma(x, y) = K|x - y|$ for any $K > 0$ and real numbers x, y. For each K, τ and σ are equivalent metrics on the reals, inducing the Euclidean topology. ∎

2.2 EXAMPLE Let $\tau(x, y) = \dfrac{\varrho(x, y)}{1 + \varrho(x, y)}$. Then, $\tau \sim \varrho$. An interesting thing about this example is that $0 \leq \tau(x, y) \leq 1$ for each $x, y \in X$. Thus, boundedness is not a topological invariant. ∎

2.3 EXAMPLE Let $\sigma((x, y), (a, b)) = |x - a| + |y - b|$ for $(x, y), (a, b)$ in the plane. Then, $\sigma \sim \varrho_2$, the Euclidean metric on E^2. ∎

2.4 EXAMPLE Let $\varrho(x, y) = \int_0^1 |x - y|$ and $\sigma(x, y) = \sup_{0 \leq t \leq 1} |x(t) - y(t)|$ for $x, y \in C([0, 1])$. The reader can show that the metrics ϱ and σ are not equivalent. ∎

One condition for the equivalence of two metrics ϱ and σ is a simple consequence of Theorem 3.4, Chapter I, and the fact that the ϱ-spheres form a base for T_ϱ, and the σ-spheres for T_σ. If can we fit a ϱ-sphere about x inside each σ-sphere about x, then $T_\sigma \subset T_\varrho$. If we can do it the other way around as well, then $T_\varrho \subset T_\sigma$, and then $\sigma \sim \varrho$.

2.2 Theorem *Let σ be a metric on X.*
Then,
$$\sigma \sim \varrho$$
if and only if
the following two conditions are satisfied:

(1) If $x \in X$ and $\varepsilon > 0$, then $B_\varrho(x, \delta) \subset B_\sigma(x, \varepsilon)$ for some $\delta > 0$.
(2) If $x \in X$ and $\varepsilon > 0$, then $B_\sigma(x, \delta) \subset B_\varrho(x, \varepsilon)$ for some $\delta > 0$.

Proof Use Theorem 3.4.I. ■

For example, an application of Theorem 2.2 to the metrics σ and ϱ_2 of Example 2.3 leads to the conclusion that $\sigma \sim \varrho_2$ because inside each circle about (x, y) can be fit a square about (x, y), and conversely.

8.3 Continuity and uniform continuity

We noted in the introductory remarks of section 1, Ch. II, that the usual ε, δ-definition of continuity in calculus generalizes immediately to metric spaces. With the general concept of continuity in arbitrary spaces now behind us, we shall for completeness record the exact sense in which this is true.

3.1 Theorem *Let σ be a metric on Y.*
Let $f: X \to Y$ and let $x \in X$.
Then,
f is (T_ϱ, T_σ) continuous at x
if and only if
if $\varepsilon > 0$, there is some $\delta > 0$ such that $\sigma(f(x), f(y)) < \varepsilon$ whenever $y \in X$ and $\varrho(x, y) < \delta$.

Proof Immediate by Theorem 4.1.I, and Theorem 1.2.II. ■

1 Corollary *Let σ be a metric on Y.*
Let $f: X \to Y$.
Then,
f is (T_ϱ, T_σ) continuous
if and only if
if $x \in X$ and $\varepsilon > 0$, there is some $\delta > 0$ such that $\sigma(f(x), f(y)) < \varepsilon$ whenever $y \in X$ and $\varrho(x, y) < \delta$.

Proof Immediate by Theorem 3.1. ■

The reader is probably familiar with the notion of uniform continuity (at least of real valued functions from sets in E^1) from calculus. The key to the concept is implicit in Corollary 1. If $f: X \to Y$ is (T_ϱ, T_σ) continuous, Corollary 1 insures the existence of an appropriate δ, *given x and ε*. This means that δ probably depends both upon x and ε, and the same δ will usually not work for x and, say, a different point z. If the same δ can be used for each point of X (though it may still depend upon ε), then f is said to be uniformly continuous on X.

3.1 DEFINITION *Let σ be a metric on Y.*
Let $f: X \to Y$.
Then,
f is (T_ϱ, T_σ) uniformly continuous
if and only if
given $\varepsilon > 0$, there is some $\delta > 0$ such that $\sigma(f(x), f(y)) < \varepsilon$ whenever $x, y \in X$ and $\varrho(x, y) < \delta$.

Put another way, given $\varepsilon > 0$, there must be some $\delta > 0$ such that $f(B_\varrho(x, \delta)) \subset B_\sigma(f(x), \varepsilon)$ for each $x \in X$.

Clearly uniform continuity implies continuity.

3.2 THEOREM *Let σ be a metric on Y.*
Let $f: X \to Y$ be (T_ϱ, T_σ) uniformly continuous.
Then,
f is (T_ϱ, T_σ) continuous.

Proof Immediate by Corollary 1. ∎

However, continuity does not in general imply uniform continuity. For example, let $f(x) = 1/x$ for $0 < x \leq 1$.

A classical theorem of calculus is that continuity on a closed, bounded subset A of E^n implies uniform continuity of $f: A \to E^1$. Theorem 1.13 of Chapter VII suggests a generalization of this to arbitrary metric spaces, with "compactness" replacing "closed and bounded" (actually, closed and bounded does make sense in any metric space, but it is not always synonymous with compactness).

The theorem is preceded by the Lebesgue Covering Lemma, which has many applications in both topology and analysis (in particular, real analysis). The Lemma says that, if X is a compact metric space, and $\mathcal{O}$ an open cover of X, then there is a positive number $\lambda(\mathcal{O})$, called a Lebesgue number of the cover, such that each open sphere of radius $\lambda(\mathcal{O})$ is contained in some set of the cover.

3.3 THEOREM (Lebesgue Covering Lemma) *Let X be T_ϱ-compact.*
Let $\mathcal{O}$ be a T_ϱ-cover of X.
Then,
there is a positive number $\lambda(\mathcal{O})$ such that, if $x \in X$, then there is some $V_x \in \mathcal{O}$ such that $B_\varrho(x, \lambda(\mathcal{O})) \subset V_x$.

Proof Note that $\cup \mathcal{O} = X$. If $x \in X$, then, there is some $U_x \in \mathcal{O}$ such that $x \in U_x$.

Since $\{B_\varrho(y, \varepsilon) | y \in X \text{ and } \varepsilon > 0\}$ is a base for T_ϱ, then $B_\varrho(x, \varepsilon_x) \subset U_x$ for some $\varepsilon_x > 0$, for each $x \in X$.

Now, $\{B_\varrho(x, \tfrac{1}{2}\varepsilon_x) | x \in X\}$ is a T_ϱ-cover of X. Since X is T_ϱ-compact, there are points $x_1, \ldots, x_n$ of X such that $X = \bigcup_{i=1}^{n} B_\varrho(x_i, \tfrac{1}{2}\varepsilon_{x_i})$. Let $\lambda(\mathcal{O}) = \min_{1 \leq i \leq n} \tfrac{1}{2}\varepsilon_{x_i}$.

Then, $\lambda(\mathcal{O}) > 0$. To show that $\lambda(\mathcal{O})$ has the desired property, let $z \in X$. We seek $V_z \in \mathcal{O}$ such that $B_\varrho(z, \lambda(\mathcal{O})) \subset V_z$.

First, for some j, $1 \leq j \leq n$ and $z \in B_\varrho(x_j, \tfrac{1}{2}\varepsilon_{x_j})$. Further, $B_\varrho(x_j, \varepsilon_{x_j}) \subset U_{x_j}$. We shall show that $B_\varrho(z, \lambda(\mathcal{O})) \subset B_\varrho(x_j, \varepsilon_{x_j})$.

Let $t \in B_\varrho(z, \lambda(\mathcal{O}))$. Then, $\varrho(t, x_j) \leq \varrho(t, z) + \varrho(z, x_j)$. Now, $\varrho(t, z) < \lambda(\mathcal{O}) \leq \tfrac{1}{2}\varepsilon_{x_j}$, and $\varrho(z, x_j) < \tfrac{1}{2}\varepsilon_{x_j}$, so $\varrho(t, x_j) < \varepsilon_{x_j}$. Then, $B_\varrho(z, \lambda(\mathcal{O})) \subset B_\varrho(x_j, \varepsilon_{x_j}) \subset U_{x_j}$, and we may choose $V_z = U_{x_j}$. ∎

3.4 THEOREM *Let σ be a metric on Y.*
Let $f: X \to Y$ be (T_ϱ, T_σ) continuous.
Let X be T_ϱ-compact.
Then,
f is (T_ϱ, T_σ) uniformly continuous.

Proof Let $\varepsilon > 0$. Then, $\{B_\sigma(y, \varepsilon/2) | y \in Y\}$ is a T_σ-cover of Y. Since f is (T_ϱ, T_σ) continuous, then $\{f^{-1}(B_\sigma(y, \varepsilon/2)) | y \in Y\}$ is a T_ϱ-cover of X.

Let λ be a Lebesgue number for this cover. We claim:

$$\sigma(f(x), f(y)) < \varepsilon \quad \text{if} \quad x, y \in X \quad \text{and} \quad \varrho(x, y) < \lambda.$$

To show this, let $x \in X$ and note that, by the Lebesgue Covering Lemma, there is some $y \in Y$ such that $B_\varrho(x, \lambda) \subset f^{-1}(B_\sigma(y, \varepsilon/2))$. Then, if $\varrho(x, z) < \lambda$, we have $f(z) \in B_\sigma(y, \varepsilon/2)$. Then, $\sigma(f(x), f(z)) \leq \sigma(f(z), y) + \sigma(y, f(x)) < \varepsilon/2 + \varepsilon/2 = \varepsilon$ so that f is (T_ϱ, T_σ) uniformly continuous. ∎

Note that the condition of Theorem 3.4 is sufficient, but not necessary. For example, let $f(x) = 1/(1 + x^2)$ for $x \in E^1$. Then, $f: E^1 \to E^1$ is uniformly continuous (given $\varepsilon > 0$, choose $\delta = \varepsilon/2$), but E^1 is not compact.

8.4 Isometries

Given metric spaces (X, ϱ) and (Y, σ), a map $f: X \to Y$ is said to be distance-preserving if $\sigma(f(x), f(y)) = \varrho(x, y)$ whenever $x, y \in X$. A distance-preserving bijection is called an isometry.

4.1 DEFINITION f is a (ϱ, σ) isometry
if and only if
$f: X \to Y$ is a bijection and $\varrho(x, y) = \sigma(f(x), f(y))$ whenever $x, y \in X$.

Isometries play an important role in the theory of normed linear spaces (see, for example, Taylor or Yosida).

It is easy to see that an isometry is an homeomorphism. Of course, not every homeomorphism between metric spaces is an isometry. For example, the map $f: E^1 \to E^1$ given by $f(x) = \frac{1}{2}x$ shrinks distances, but is a homeomorphism.

4.1 THEOREM *Let σ be a metric on Y.*
Let f be a (ϱ, σ) isometry.
Then,
f is a (T_ϱ, T_σ) homeomorphism.

Proof Left as an exercise. ∎

It is also a simple consequence of the definition that any isometry is uniformly continuous (choose $\delta = \varepsilon$), regardless of whether or not X is compact.

4.2 THEOREM *Let σ be a metric on Y.*
Let $f: X \to Y$ be a (ϱ, σ) isometry.
Then,
f is (T_ϱ, T_σ) uniformly continuous.

Proof Immediate. ∎

We shall need the following fact about isometries in proving the completion theorem of section 7.

4.3 THEOREM *Let σ be a metric on Y.*
Let $f: X \to Y$ be a (ϱ, σ) isometry.
Then,
inv $f: Y \to X$ is a (σ, ϱ) isometry.

Proof Left as an exercise. ∎

8.5 Adequacy of sequences

When metric spaces were generalized to arbitrary topological spaces, mathematicians were surprised by the failure of the sequential theory of convergence in the new setting. The reason for this is that in metric spaces sequences were always sufficient to describe the objects of interest—closure points, cluster points, continuity, and so on. In this section we shall describe exactly the sense in which sequences provide an adequate theory of convergence in metric spaces.

First of all, sequential convergence characterizes the closed (hence open) sets relative to the topology T_ϱ. This has already been proved (Theorem 1.2.IV). For completeness, we shall also state the sequential characterization of cluster points which forms the metric space analogue of Theorems 2.2 and 3.3.IV.

5.1 THEOREM *Let $A \subset X$ and $x \in X$.*
Then,
x is a T_ϱ-cluster point of A
if and only if
there is a sequence to $A - \{x\}$ which T_ϱ-converges to x.

Proof Suppose first that x is a T-cluster point of A. If $n \in Z^+$, then $B_\varrho(x, 1/n) \cap (A - \{x\}) \neq \phi$. Define a sequence S to $A - \{x\}$ by choosing $S_n \in B_\varrho(x, 1/n) \cap (A - \{x\})$ for each positive integer n. It is easy to check that S is T_ϱ-convergent to x.

Conversely, suppose that there is a sequence to $A - \{x\}$ which T_ϱ-converges to x. Then, x is a T_ϱ-cluster point of A by Theorem 2.2.IV, since a sequence is a net. ∎

Continuity can also be tested sequentially in metric spaces.

5.2 THEOREM *Let σ be a metric on Y.*
Let $f: X \to Y$.
Let $x \in X$.
Then,
f is (T_ϱ, T_σ) continuous at x
if and only if
if S is a sequence to X and S T_ϱ-converges to x, then $f \circ S$ T_σ-converges to $f(x)$.

Proof Suppose first that f is (T_ϱ, T_σ) continuous at x. Let S be a sequence to X which T_ϱ-converges to x. Then, $f \circ S$ is T_σ-convergent to $f(x)$ by Theorem 2.4.IV.

Conversely, suppose that, if S is a sequence to X which T_ϱ-converges to x, then $f \circ S$ T_σ-converges to $f(x)$.

Let $f(x) \in V \in T_\sigma$. We must produce a T_ϱ-nghd. U of x such that $f(U) \subset V$.

Suppose there is no such T_ϱ-nghd. of x. Then, for each positive integer n, there is some $S_n \in B_\varrho(x, 1/n)$ such that $f(S_n) \notin V$. This defines a sequence S to X such that S T_ϱ-converges to x.

But $f(S_n) \notin V$ for each $n \in Z^+$, so that $f \circ S$ does not T_σ-converge to $f(x)$, a contradiction. This completes the proof. ∎

1 COROLLARY *Let σ be a metric on Y.*
Let $f: X \to Y$.
Then,
f is (T_ϱ, T_σ) continuous
if and only if
if $x \in X$ and S is a sequence to X which T_ϱ-converges to x, then $f \circ S$ T_σ-converges to $f(x)$.

Proof Immediate by Theorem 5.2. ∎

Actually, we used the metric ϱ on X in proving Theorem 5.2, but not the metric σ on Y. The theorem would go through for any topology on Y. Note also that ϱ is needed only in proving sufficiency—necessity follows from the corresponding theorem for nets, which does not depend upon properties of metrics.

8.6 Compactness and covering theorems in metric spaces

In this section we shall investigate the various kinds of compactness, and the notions of 1°- and 2°-countable, separable, and Lindelöf spaces as they apply to metric spaces.

We begin by showing that every metric space is 1°-countable.

6.1 THEOREM *X is T_ϱ-1°-countable.*

Proof Let $x \in X$. It is a simple matter to show that $\{B_\varrho(x, 1/n) | n \in Z^+\}$ satisfies the requirements of the set B_x of Definition 4.4.VII. ∎

By virtue of the implication chart on p. 167, the notions of countable compactness, B-W compactness and sequential compactness are all equivalent in metric spaces. We would like to improve upon this result and conclude that these are all in fact equivalent to compactness in metric spaces.

The proof of this is not so easy. As a means of attack, we shall first show that each countably compact metric space is 2°-countable. This in itself requires an involved argument, se we precede it with a simplifying Lemma of some interest in its own right. The Lemma states that each countably compact space can be covered by finitely many spheres of arbitrarily small radius. Such a space is often called totally bounded in the literature. As this term will also be used later (section 7), we shall define it now.

6.1 DEFINITION *X is ϱ-totally bounded
if and only if
if $\varepsilon > 0$, there is some $n \in Z^+$ and there are points $x_1, \ldots, x_n \in X$ such that*
$$X = \bigcup_{i=1}^{n} B_{\varrho}(x_i, \varepsilon).$$

1 LEMMA *Let X be T_{ϱ}-c.c.
Then,
X is ϱ-totally bounded.*

Proof The proof is by contradiction. Suppose that X is not ϱ-totally bounded.

Then, there is some $\varepsilon > 0$ such that, given any finite set of points $x_1, \ldots, x_n$ of X, then $\bigcup_{i=1}^{n} B_{\varrho}(x_i, \varepsilon) \neq X$. Construct a sequence S to X inductively as follows.

Choose $S_1 \in X$,
$$S_2 \in X - B_{\varrho}(S_1, \varepsilon),$$
and, in general, for any positive integer $n \geq 1$,
$$S_{n+1} \in X - \bigcup_{i=1}^{n} B_{\varrho}(S_i, \varepsilon).$$

Now consider two cases.

i) $\bigcup_{i=1}^{\infty} B_{\varrho}(S_i, \varepsilon) = X$.

Then, $\{B_{\varrho}(S_i, \varepsilon) | i \in Z^+\}$ is a countable T_{ϱ}-cover of X.
But this cover cannot be reduced, since, for any $j \in Z^+$, $S_j \in X - \bigcup_{\substack{i=1 \\ i \neq j}}^{\infty} B_{\varrho}(S_i, \varepsilon)$.
This contradicts the countable compactness of the space (X, T_{ϱ}).

ii) $X \neq \bigcup_{i=1}^{\infty} B_{\varrho}(S_i, \varepsilon)$.

Let $V = \cup B_{\varrho}(x, \varepsilon)$, the union being over all $x \in X - \bigcup_{i=1}^{\infty} B_{\varrho}(S_i, \varepsilon)$.

Then, $\{B_{\varrho}(S_i, \varepsilon) | i \in Z^+\} \cup \{V\}$ is a T_{ϱ}-cover of X. This cover cannot be reduced either. For suppose that $X = V \cup \left(\bigcup_{j=1}^{k} B_{\varrho}(S_{i_j}, \varepsilon) \right)$. Choose a positive integer r with $r > \max_{i \leq j \leq k} i_j$. We shall derive a contradiction by showing that $S_r \notin V \cup \left(\bigcup_{j=1}^{k} B_\varrho(S_{i_j}, \varepsilon) \right)$.

Suppose $S_r \in V$. Then, $S_r \in B_\varrho(x, \varepsilon)$ for some $x \in X - \bigcup_{i=1}^{\infty} B_\varrho(S_i, \varepsilon)$. Since $\varrho(x, S_r) < \varepsilon$, then $x \in B_\varrho(S_r, \varepsilon)$, a contradiction.

Finally, $S_r \notin \bigcup_{j=1}^{k} (B_\varrho(S_{i_j}, \varepsilon))$ by definition of r, and the proof is complete. ∎

6.2 THEOREM *Let X be T_ϱ-c.c. Then, X is T_ϱ-$2°$-countable.*

Proof Given any positive integer n, there exist, by virtue of Lemma 1, elements $x_{n1}, \ldots, x_{nr_n}$ of X such that $X = \bigcup_{i=1}^{r_n} B_\varrho(x_{ni}, 1/n)$.

Let $\mathscr{B} = \{B_\varrho(x_{ni}, 1/n) | n \in Z^+ \text{ and } i \in Z^+ \text{ and } 1 \leq i \leq r_n\}$.
Then, $\mathscr{B}$ is a countable subset of T_ϱ.
We shall prove that $\mathscr{B}$ is a base for T_ϱ, using Theorem 3.1.I.
Let $y \in V \in T_\varrho$. For some $\varepsilon > 0$, $y \in B_\varrho(y, \varepsilon) \subset V$. Choose a positive integer K such that $1/K < \varepsilon/2$. For some elements $x_{K1}, \ldots, x_{Kr_K}$ of X, $X = \bigcup_{i=1}^{r_K} B_\varrho(x_{Ki}, 1/K)$. Then, $y \in B_\varrho(x_{Kj}, 1/K)$ for some j, $1 \leq j \leq r_K$. We claim that $B_\varrho(x_{Kj}, 1/K) \subset B_\varrho(y, \varepsilon)$.
For, let $t \in B_\varrho(x_{Kj}, 1/K)$. Then,

$$\varrho(y, t) \leq \varrho(y, x_{Kj}) + \varrho(x_{Kj}, t) < 1/K + 1/K < \varepsilon,$$

by choice of K.
This implies that $\mathscr{B}$ is a base for T_ϱ. ∎

With this accumulated machinery, it is now easy to show that compactness and countable compactness are equivalent in metric spaces.

6.3 THEOREM *X is T_ϱ-compact $\Leftrightarrow$ X is T_ϱ-c.c.*

Proof If X is T_ϱ-compact, then X is T_ϱ-c.c. by Theorem 3.1.VII.
Conversely, suppose that X is T_ϱ-c.c. By Theorem 6.2, X is T_ϱ-$2°$-countable. By Theorem 5.4.VII, X is T_ϱ-Lindelöf. Finally, by Theorem 5.3.VII, X is T_ϱ-compact. ∎

Summary chart

$$\begin{matrix} & \text{compact} & \Leftrightarrow & \text{countably compact} \\ \text{for} & \Updownarrow & & \Updownarrow \\ \text{metric} & \text{sequentially} & \Leftrightarrow & \text{B-W compact} \\ \text{spaces} & \text{compact} & & \end{matrix}$$

With the hindsight afforded by historical perspective, it is now not very surprising that many different definitions of compactness were tried when topology was in its formative stages. The main attempts are all equivalent in the spaces which were best understood at the time.

In proving Theorem 6.3, we used the fact that any $2°$-countable space is Lindelöf. If the space is metric, the reverse implication can be proved: a Lindelöf metric space is $2°$-countable. We shall not prove this directly, but by means of another metrically strengthened theorem. Specifically, we shall show that Theorem 5.2 of Chapter VII is reversible in metric spaces. If we can then show that Lindelöf implies separable in metric spaces, then we will have a nicely completed chain of implications.

6.4 THEOREM X is T_ϱ-$2°$-countable $\Leftrightarrow X$ is T_ϱ-separable.

Proof In view of Theorem 5.2.VII, if X is T_ϱ-$2°$-countable, then X is T_ϱ-separable.

Thus, suppose that X is T_ϱ-separable. Let A be a countable dense subset of X.

Let $\mathscr{B} = \{B_\varrho(x, \varepsilon) | x \in A$ and $\varepsilon > 0$ and ε is rational$\}$. Then, $\mathscr{B}$ is countable. We shall use Theorem 3.1.1 to show that $\mathscr{B}$ is a base for T.

Let $y \in V \in T$. Choose $\varepsilon > 0$ such that $y \in B_\varrho(y, \varepsilon) \subset V$. Choose a rational number δ such that $0 < \delta \leq \varepsilon$. Then, $B_\varrho(y, \delta) \subset V$. Since A is dense in X, then for some $a \in A$, $a \in B_\varrho(y, \delta/3)$. We claim that $B_\varrho(a, \delta/3) \subset V$.

For, let $t \in B_\varrho(a, \delta/3)$. Then,

$$\varrho(t, y) \leq \varrho(t, a) + \varrho(a, y) < \delta/3 + \delta/3 = \tfrac{2}{3}\delta < \delta.$$

Then, $y \in B_\varrho(a, \delta/3) \subset V$ and $B_\varrho(a, \delta/3) \in \mathscr{B}$, so that $\mathscr{B}$ is a base for T_ϱ. ∎

6.5 THEOREM X is T_ϱ-Lindelöf if and only if X is T_ϱ-separable.

Proof If X is T_ϱ-separable, then X is T_ϱ-$2°$-countable, hence T_ϱ-Lindelöf by Theorem 6.4 and by Theorem 5.4.VII.

Conversely, suppose that X is T_ϱ-Lindelöf. We must produce countable $A \subset X$ such that $\bar{A} = X$.

If n is a positive integer, let $C_n = \{B_\varrho(x, 1/n) | x \in X\}$. Then, C_n is a T_ϱ-cover of X, hence has a countable reduction, say $B_n = \{B_\varrho(x_{ni}, 1/n) | i \in Z^+\}$.
Let $F_n = \{x_{ni} | B_\varrho(x_{ni}, 1/n) \in B_n\}$ for each $n \in Z^+$. Then, F_n is countable.
Let $A = \bigcup_{n=1}^{\infty} F_n$.
Then, A is countable.
To show that $\bar{A} = X$, let $y \in V \in T$.
For some $\varepsilon > 0$, $y \in B_\varrho(y, \varepsilon) \subset V$. Choose some positive integer K such that $1/K < \varepsilon$. Since B_K covers X, then $y \in B_\varrho(x_{Kj}, 1/K)$ for some $x_{Kj} \in F_K$. Then, $\varrho(y, x_{Kj}) < 1/K < \varepsilon$, so $x_{Kj} \in A \cap V$, implying that $y \in \bar{A}$. ∎

It is now a simple matter of bookkeeping to show the equivalence of the Lindelöf condition with 2°-countability in metric spaces.

6.6 Theorem *X is T_ϱ-Lindelöf $\Leftrightarrow$ X is T_ϱ-2°-countable.*

Proof If X is T_ϱ-2°-countable, then X is T_ϱ-Lindelöf by Theorem 5.4.VII. Conversely, suppose that X is T_ϱ-Lindelöf. By Theorem 6.5, X is T_ϱ-separable, hence T_ϱ-2°-countable by Theorem 6.4. ∎

We can now fill in some missing links in the implication chart of p. 170.

For metric spaces

2°-countable $\Rightarrow$ 1°-countable
$\Updownarrow$
separable $\Leftrightarrow$ Lindelöf

We conclude this section by showing that any compact metric space is 2°-countable (hence separable and Lindelöf).

6.7 Theorem *Let (X, T_ϱ) be compact.*
Then,
X is T_ϱ-2°-countable.

Proof Since X is T_ϱ-compact, then X is T_ϱ-c.c. by Theorem 6.3, hence T_ϱ-2°-countable by Theorem 6.2. ∎

Summary chart for metric spaces

compact $\Leftrightarrow$ countably compact   2°-countable $\Rightarrow$ 1°-countable
$\Updownarrow$ $\Updownarrow$ $\Rightarrow$ $\Updownarrow$
sequentially $\Leftrightarrow$ B-W compact   separable $\Leftrightarrow$ Lindelöf
compact

8.7 Complete metric spaces

The Cauchy criterion for convergence of real sequences says that $S: Z^+ \to E^1$ converges if and only if, given $\varepsilon > 0$, then $|S_n - S_m| < \varepsilon$ for m and n sufficiently large. Any sequence satisfying this condition is called a Cauchy sequence, and the theorem may be rephrased: a real sequence converges if and only if it is a Cauchy sequence.

The notion of a Cauchy sequence is easily generalized to arbitrary metric spaces—simply replace E^1 by X and $|S_n - S_m|$ by $\varrho(S_n, S_m)$.

7.1 DEFINITION *S is a ϱ-Cauchy sequence to X
if and only if
$S: Z^+ \to X$ and, given $\varepsilon > 0$, there exists some positive integer K such that $\varrho(S_n, S_m) < \varepsilon$ whenever $n, m \in Z^+$ and $n \geq K$ and $m \geq K$.*

It is obvious that every convergent sequence is Cauchy.

7.1 THEOREM *Let S be a T_ϱ-convergent sequence to X.
Then,
S is a ϱ-Cauchy sequence.*

Proof Left as an exercise. ∎

However, the Cauchy theorem from E^1 does not carry over into metric spaces in general. For example, let X be the set of rationals in E^1. Any sequence of rationals converging to an irrational limit (e.g. $S_n = (1 + 1/n)^n$) is a Cauchy sequence to X having no limit in X.

A space in which each Cauchy sequence converges is called complete. This concept forms the basis of the theory of Hilbert and Banach spaces, which play a fundamental role in modern analysis.

7.2 DEFINITION *X is ϱ-complete
if and only if
each ϱ-Cauchy sequence to X is T_ϱ-convergent.*

7.1 EXAMPLE E^n is complete. ∎

7.2 EXAMPLE Let $X = \{p \mid p \text{ is a polynomial on } [a, b]\}$, and consider X as a subspace of the sup norm space $C([a, b])$. Then, X is not complete. For, define $S_n(x) = \sum_{j=0}^{n} x^j/j!$ for $a \leq x \leq b$, $n = 1, 2, \ldots$ Then S is a sequence to X which converges to the exponential function in $C([a, b])$, hence is Cauchy, but does not converge in X.

Note that $C([a, b])$ is complete. As we have seen (Example 1.4, Ch. IV), convergence in the sup norm topology is equivalent to ordinary uniform convergence on $[a, b]$. Completeness of $C([a, b])$ now follows easily from the fact that a uniform limit of continuous functions is again continuous. ∎

7.3 EXAMPLE The rationals form an incomplete subspace of the reals. ∎

Observe that the only incomplete spaces we have seen so far have been subspaces of complete spaces. This is always the case. Any incomplete metric space can be embedded isometrically as a dense subset of a complete space, which is called a completion of the original space.

The method of doing this is entirely analogous to a construction of the reals from the rationals, the reals forming a completion of the rationals. We briefly recall the steps in outline:

Given the set Q of rationals,

(1) Let K denote the set of all Cauchy sequences (convergent or not in Q) to Q,

(2) If $x, y \in K$, define $x \sim y$ to mean that $\lim_{n \to \infty} |x_n - y_n| = 0$. This is an equivalence relation on K.

(3) Let $K/\sim$ denote the set of equivalence classes of elements of K. Given two such classes A and B, define $\sigma(A, B) = \lim_{n \to \infty} |x_n - y_n|$ for any $x \in A$ and $y \in B$.

(4) $K/\sim$ is complete in the metric σ.

(5) Q is isometric to a dense subset of $K/\sim$.

The proof of the general theorem on existence of completions is an almost exact rewording of the above construction, with Q replaced by X and $|x_n - y_n|$ by $\varrho(x_n, y_n)$. The argument is quite lengthy because we shall prove all the assertions made so easily above. However, an occasional referral back to the above construction should keep the reader from foundering in the details.

To help organize the argument, we precede the theorem with a sequence of five preparatory lemmas. Notation will be allowed to accumulate from lemmas the to the completion theorem in order to avoid pointless repetition.

In the following, K denotes the set of Cauchy sequences to X.

1 LEMMA *If $f, g \in K$, define $f \sim g$ to mean $\lim_{n \to \infty} \varrho(f_n, g_n) = 0$. Then,*

$\sim$ is an equivalence relation on K.

Proof Left as an exercise. ∎

2 LEMMA *If $x, x', y, y' \in X$, then*
$$|\varrho(x, x') - \varrho(y, y')| \leq \varrho(x, y) + \varrho(x', y').$$

Proof We consider two cases.

1) If $\varrho(x, x') - \varrho(y, y') \geq 0$, then
$$\varrho(x, x') - \varrho(y, y') \leq \varrho(x, y) + \varrho(y, x') - \varrho(y, y')$$
$$\leq \varrho(x, y) + \varrho(y, y') + \varrho(y', x') - \varrho(y, y')$$
$$= \varrho(x, y) + \varrho(x', y').$$

2) If $\varrho(x, x') - \varrho(y, y') \leq 0$, then
$$|\varrho(x, x') - \varrho(y, y')| = \varrho(y, y') - \varrho(x, x') \leq \varrho(x, y) + \varrho(x', y')$$
by the argument of 1). ∎

3 LEMMA *Let $A, B \in K/\sim$.*
Let $x, x' \in A$ and $y, y' \in B$.
Then,

(1) $\{\varrho(x_n, y_n)\}_{n=1}^{\infty}$ converges in E^1 to a non-negative limit.
(2) $\lim_{n \to \infty} \varrho(x_n, y_n) = \lim_{n \to \infty} \varrho(x'_n, y'_n)$.

Proof of (1) For any positive integers m and n,
$$|\varrho(x_n, y_n) - \varrho(x_m, y_m)| \leq \varrho(x_n, x_m) + \varrho(y_n, y_m)$$
by Lemma 2.

The conclusion now follows easily from the fact that
$$\varrho(x_n, x_m) \to 0 \text{ and } \varrho(y_n, y_m) \to 0 \text{ as } n, m \to \infty.$$

Since $\{\varrho(x_n, y_n)\}_{n=1}^{\infty}$ is a Cauchy sequence of non-negative real numbers, then $\{\varrho(x_n, y_n)\}_{n=1}^{\infty}$ converges in E^1 to a non-negative limit.

Proof of (2) By Lemma 2,
$$|\varrho(x_n, y_n) - \varrho(x'_n, y'_n)| \leq \varrho(x_n, x'_n) + \varrho(y_n, y'_n).$$

Now, $x \sim x'$, so $\lim_{n \to \infty} \varrho(x_n, x'_n) = 0$. Similarly, $y \sim y'$, so $\lim_{n \to \infty} \varrho(y_n, y'_n) = 0$.
Then, $|\varrho(x_n, y_n) - \varrho(x'_n, y'_n)| \to 0$ as $n \to \infty$. ∎

4 LEMMA *If $A \in K/\sim$ and $B \in K/\sim$, define $\sigma(A, B) = \lim_{n \to \infty} \varrho(x_n, y_n)$ for any $x \in A$ and $y \in B$.*
Then,
σ is a metric on $K/\sim$.

Proof By Lemma 3, σ is a well-defined function on $(K/\sim) \times (K/\sim)$ to $\{x | x \text{ is real and } x \geq 0\}$.

There remains to verify that σ is a metric. The details are left to the reader. ∎

5 LEMMA *Let $D = (K/\sim) \cap \{A | \text{for some } x \in A, x_i = x_j \text{ for all } i, j \in Z^+\}$. Then,*

(1) $\text{Cl}_{T_\sigma}(D) = K/\sim$.

(2) (X, ϱ) is isometric to $(D, \sigma|(D \times D))$.

Proof of (1) Note first that $D \subset K/\sim$. Let $A \in K/\sim$ and let $\varepsilon > 0$. It suffices to show that $B_\sigma(A, \varepsilon) \cap D \neq \phi$. Let $x \in A$. Then, $x \in K$, so for some positive integer M, $\varrho(x_n, x_m) < \varepsilon/2$ if $n, m \geq M$. In particular, $\varrho(x_n, x_M) < \varepsilon/2$ if $n \geq M$. Let $s: Z^+ \to X$ be defined by $s_n = x_M$ for each $n \in Z^+$. Then, $s \in K$. Let B be the equivalence class of s in $K/\sim$. Then, $B \in D$. Further, $\sigma(A, B) = \lim_{n \to \infty} \varrho(x_n, s_n) = \lim_{n \to \infty} \varrho(x_n, x_M) \leq \varepsilon/2 < \varepsilon$. Then, $B \in B_\sigma(A, \varepsilon) \cap D$.

Proof of (2) Define $\varphi: X \to D$ by letting, for $x \in X$, $\varphi(x)$ be the equivalence class of $K/\sim$ determined by the sequence $\{\varphi(x)_n\}_{n=1}^\infty$ where $(\varphi(x))_n = x$ for each $n \in Z^+$.

We leave it for the reader to show that $\varphi: X \to D$ is an isometry. ∎

We can now state and prove the completion theorem.

7.2 THEOREM *There exists a complete metric space (Y, δ) such that (X, ϱ) is isometric to a dense subspace of (Y, δ).*

Proof If X is ϱ-complete, choose $Y = X$ and $\delta = \varrho$.

Thus, suppose that X is not ϱ-complete. By the previous Lemma, it suffices to show that $K/\sim$ is σ-complete.

To do this, let S be a σ-Cauchy sequence to $K/\sim$. We must produce $L \in K/\sim$ such that S is σ-convergent to L.

Now, if $n \in Z^+$, then $B_\sigma(S_n, 1/n) \cap D \neq \phi$, as $K/\sim = \bar{D}$. Choose $B_n \in B_\sigma(S_n, 1/n) \cap D$ for each positive integer n. This defines a sequence B to D.

We claim that B is a σ-Cauchy sequence. For, let $\varepsilon > 0$ and note that

$$\sigma(B_n, B_m) \leq \sigma(B_n, S_n) + \sigma(S_n, B_m)$$
$$\leq \sigma(B_n, S_n) + \sigma(S_n, S_m) + \sigma(S_m, B_m)$$
$$< 1/n + \sigma(S_n, S_m) + 1/m \to 0 \text{ as } n, m \to \infty,$$

since S is a σ-Cauchy sequence.

Now, for each $n \in Z^+$, let $x_n = (\text{inv } \varphi)(B_n)$. This defines a sequence x to X. Since φ is an isometry: $X \to D$, then $\text{inv } \varphi$ is an isometry: $D \to X$ by Theorem 4.3. Since B is a σ-Cauchy sequence to D, it is immediate that x is a ϱ-Cauchy sequence to X.

Let L be the equivalence class of x. Then, $L \in K/\sim$. We shall prove that S is T_σ-convergent to L.

First, if n is a positive integer, then $\sigma(S_n, L) \leq \sigma(S_n, B_n) + \sigma(B_n, L)$. Now, $\sigma(S_n, B_n) \leq 1/n$ by the choice of B_n. Further, $\sigma(B_n, L) = \lim_{k \to \infty} \varrho(t_k, x_k)$, by definition of σ, where t is any element of B_n.

But, $\varphi(x_n) = B_n$, so the sequence α given by $\alpha_k = x_n$ for each $k \in Z^+$ is an element of B_n, by definition of φ.

Then, x is a ϱ-Cauchy sequence to X. We may therefore choose a positive integer r such that $\varrho(x_i, x_j) < \varepsilon/2$ for $i, j \geq r$.

Choose $n \geq r$. Then, $\lim_{k \to \infty} \varrho(x_n, x_k) \leq \varepsilon/2$.

Then, for $n \in Z^+$ and $n > \max(r, 2/\varepsilon)$, we have

$$\sigma(S_n, L) < 1/n + \lim_{k \to \infty} (x_n, x_k) \leq \varepsilon/2 + \varepsilon/2 = \varepsilon.$$

Then, S is T_σ-convergent to L, and the theorem is proved. ∎

While the completion theorem is the most important theorem of this section, there remain other questions of some interest. One concerns the hereditary properties of completeness. Of course, subspace of complete spaces need not be complete (this would make the completion theorem quite pointless). However, a closed subspace of a complete space is complete. This is important because it sheds some light upon how a completion of any subspace of a complete space is formed—simply attach all the cluster points of the subspace.

7.3 THEOREM *Let X be ϱ-complete.*
Let A be T_ϱ-closed.
Then,
A is ϱ-complete.

Proof Let S be a ϱ-Cauchy sequence to A. Since X is ϱ-complete, then S is T_ϱ-convergent to some $x \in X$. By Theorem 1.2.IV $x \in \bar{A}$, hence $x \in A$. ∎

Very strictly speaking, we should say in Theorem 7.3 that A is $\varrho|(A \times A)$ complete. But this is an unnecessarily fine point at this stage.

The property of being Cauchy is not in general preserved by continuous maps. It is, however, by uniformly continuous maps.

7.4 THEOREM *Let σ be a metric on Y.*
Let $f: X \to Y$ be a (T_ϱ, T_σ) uniformly continuous.
Let S be a ϱ-Cauchy sequence to X.
Then,
$f \circ S$ is a σ-Cauchy sequence to Y.

Proof Let $\varepsilon > 0$. For some $\delta > 0$, $\sigma(f(x), f(y)) < \varepsilon$ if $x, y \in X$ and $\varrho(x, y) < \delta$.

Choose $M \in Z^+$ such that $\varrho(S_n, S_m) < \delta$ if $n, m \geq M$. Then, $\sigma(f(S_n), f(S_m)) < \varepsilon$ if $n, m \geq M$. ∎

In view of the last theorem and Theorem 4.2, isometries also preserve Cauchy sequences. We regarded this fact as sufficiently obvious by itself, to use it in the proof of Theorem 7.2. The reader who is upset by this can switch the order of the theorems to put 7.4 before 7.2 without incurring any logical difficulties, as the two theorems are in no way dependent upon one another.

There is an alternative formulation of completeness which is often useful (see, for example, the proof of Baire's Category Theorem in the next section). If $A \subset X$, the diameter of A, as measured by ϱ, is the supremum of distances $\varrho(x, y)$ between points x and y in A. Generally, one considers only diameters of bounded sets. For completeness, we include a definition of boundedness of subsets of X, which is a direct generalization from the usual definition in E^1. Theorem 7.5 then says that a space is complete exactly when each descending sequence of closed, bounded sets whose diameters tend to zero has an intersection consisting of a single point.

7.3 DEFINITION *Let $A \subset X$.*
Then,
A is ϱ-bounded
if and only if
$A \subset B_\varrho(x, \varepsilon)$ for some $\varepsilon > 0$ and $x \in X$.

7.4 DEFINITION *Let $\phi \neq A \subset X$ and suppose that A is ϱ-bounded.*
Then,
diam $(A) = \sup \{\varrho(x, y) | x \in A$ and $y \in A\}$.

7.5 THEOREM *The following two conditions are equivalent:*
(1) X is ϱ-complete.
(2) If F_i is a non-empty, T_ϱ-closed, ϱ-bounded subset of X for each $i \in Z^+$, and if $F_{i+1} \subset F_i$, and if $\lim_{n \to \infty}$ diam $(F_n) = 0$, then, for some $x \in X$, $\bigcap_{n=1}^{\infty} F_n = \{x\}$.

Proof (1) ⇒ (2).

Assume (1). Let F be a sequence of non-empty, T_ϱ-closed, ϱ-bounded sets with $F_{i+1} \subset F_i$ and $\lim\limits_{n \to \infty} \text{diam}(F_n) = 0$.

For each $n \in Z^+$, choose some $S_n \in F_n$. This defines a sequence S to X. Since diam $(F_n) \to 0$ as $n \to \infty$, it is easy to check that S is a ϱ-Cauchy sequence. Since X is ϱ-complete, S converges, say to y, in X. By Theorem 1.2.IV, it is easy to check that $y \in \bigcap\limits_{n=1}^{\infty} F_n$. Finally, use the assumption that diam $(F_n) \to 0$ as $n \to \infty$ to show that $\bigcap\limits_{n=1}^{\infty} F_n = \{y\}$.

(2) ⇒ (1):

Assume (2). Let S be a ϱ-Cauchy sequence to X. Let $F_n = \text{Cl}_{T_\varrho}(\{S_j | j \geq n\})$ for each positive integer n.

We leave it for the reader to show that each F_n is T_ϱ-closed, non-empty, ϱ-bounded, that $F_{n+1} \subset F_n$, and that $\lim\limits_{n \to \infty} \text{diam}(F_n) = 0$.

For some $x \in X$, $\bigcap\limits_{n=1}^{\infty} F_n = \{x\}$ by (2). It is easy to check that S is T_ϱ-convergent to x. ∎

Theorem 7.5 is often referred to as Cantor's Theorem, after Georg Cantor, who used a statement like (2) in laying the foundations for the theory of transfinite numbers.

We conclude this section with a theorem displaying a relationship between compactness, completeness, and total boundedness. A preliminary Lemma will be used in the proof.

6 LEMMA

(1) If $A \subset B \subset X$, and B is ϱ-bounded, then
diam $(A) \leq$ diam $(B) =$ diam $(\bar{B})$.

(2) If $x \in X$ and $\varepsilon > 0$, then
diam $(B_\varrho(x, \varepsilon)) \leq 2\varepsilon$.

Proof Left to the reader. ∎

7.6 THEOREM *X is T_ϱ-compact if and only if X is ϱ-complete and ϱ-totally bounded.*

Proof Suppose first that X is T_ϱ-compact. Let $\varepsilon > 0$. Then, $\{B_\varrho(x, \varepsilon) | x \in X\}$ is a T_ϱ-cover of X, and so has a finite reduction. Then, there are points $x_1, \ldots, x_n$ of X such that $X = \bigcup_{i=1}^{n} B_\varrho(x_i, \varepsilon)$, implying that X is ϱ-totally bounded.

To show that X is ϱ-complete, let S be a ϱ-Cauchy sequence to X. Since X is T_ϱ-compact, then X is T_ϱ-sequentially compact. Thus, there is a subsequence V of S which is T_ϱ-convergent, say, to $x \in X$.

We shall show that S is T_ϱ-convergent to x.

Let $\varepsilon > 0$.

Write $V = S \circ \varphi$ for some non-decreasing $\varphi \colon Z^+ \to Z^+$. For some $K > 0$, $V_j \in B_\varrho(x, \varepsilon/2)$ for $j \in Z^+$ and $j \geq K$. For some $M > 0$, $\varrho(S_i, S_j) < \varepsilon/2$ for $i, j \geq M$. Choose some integer $N \geq M + K$. Then, for $i \geq N$, $\varrho(S_i, x) \leq \varrho(S_i, V_i) + \varrho(V_i, x)$. But, $\varrho(S_i, V_i) = \varrho(S_i, S(\varphi(i))) < \varepsilon/2$, as $\varphi(i) \geq i \geq N \geq M$. And, $\varrho(V_i, x) < \varepsilon/2$, as $i \geq N \geq K$. Then, $\varrho(S_i, x) < \varepsilon$ for $i \geq N$, hence S is T_ϱ-convergent to x.

Conversely, suppose that X is ϱ-complete and ϱ-totally bounded. Let $\mathcal{M}$ be an ultrafilterbase on X. By Theorem 1.6.VII, it suffices to show that $\mathcal{M}$ is T_ϱ-convergent.

We do this in two steps.

i) If $\varepsilon > 0$, then there is some $m \in \mathcal{M}$ such that diam $(m) < \varepsilon$.

To prove this, let $\varepsilon > 0$. Since X is ϱ-totally bounded, there are points $x_1, \ldots, x_n \in X$ such that $X = \bigcup_{i=1}^{n} B_\varrho(x_i, \varepsilon/4)$. If $1 \leq i \leq n$, then, for some $m_i \in \mathcal{M}$, either $m_i \subset B_\varrho(x_i, \varepsilon/4)$ or $m_i \subset X - B_\varrho(x_i, \varepsilon/4)$ by Theorem 5.3.IV. Suppose that $m_i \subset X - B_\varrho(x_i, \varepsilon/4)$ for $i = 1, \ldots, n$. By Theorem 3.1.IV, there is some $m \in \mathcal{M}$ such that $m \subset \bigcap_{i=1}^{n} m_i$. Then, $m \subset \bigcap_{i=1}^{n} (X - B_\varrho(x_i, \varepsilon/4))$ $= X - \bigcup_{i=1}^{n} B_\varrho(x_i, \varepsilon/4) = \phi$, which is impossible.

Hence for some i, $m_i \subset B_\varrho(x_i, \varepsilon/4)$. By Lemma 6,

$$\text{diam}(m_i) \leq \text{diam}(B_\varrho(x_i, \varepsilon/4)) \leq \varepsilon/2 < \varepsilon.$$

ii) $\mathcal{M}$ is T_ϱ-convergent.

If $n \in Z^+$, choose some $f_n \in \mathcal{M}$ such that diam $(f_n) < 1/n$.

Let $F_n = \bigcap_{i=1}^{n} f_i$ for $n = 1, 2, \ldots$

Then, F_n is non-empty, and F_n is T_ϱ-closed and ϱ-bounded. Further, $F_{n+1} \subset F_n$ for each positive integer n.

Finally, diam $(F_n) \leq 1/n \to 0$ as $n \to \infty$, by Lemma 6 (1).

Since X is ϱ-complete, then by Theorem 7.5, there is some $x \in X$ such that $\bigcap_{i=1}^{\infty} F_i = \{x\}$. We claim that $\mathscr{M}$ is T_ϱ-convergent to x.

To show this, let $\varepsilon > 0$. Choose $k \in Z^+$ such that $1/k < \varepsilon$. Then, $F_k \subset B_\varrho(x, \varepsilon)$, since, if $y \in F_k$, then $\varrho(x, y) \leq \text{diam}(F_k) \leq 1/k < \varepsilon$.

Then, $\bigcap_{i=1}^{k} f_i \subset B_\varrho(x, \varepsilon)$.

For some g, $g \in \mathscr{M}$ and $g \subset \bigcap_{i=1}^{k} f_i$, by Theorem 3.1.IV.

Then, $g \in \mathscr{M}$ and $g \subset B_\varrho(x, \varepsilon)$, hence $\mathscr{M}$ is T_ϱ-convergent to x, and the proof is complete. ∎

8.8 The Baire category theorem

In this section we shall prove a theorem due to Baire which has wide application in topology and such areas of analysis as real function theory, measure and integration, Fourier analysis and functional analysis.

The traditional statement of the theorem requires the introduction of some terminology, which is actually valid in non-metric as well as metric spaces.

8.1 DEFINITION *Let T be a topology on X.*
Let $A \subset X$.
Then,
A is T-nowhere dense
if and only if
$\text{Int}_T(\bar{A}) = \phi$.

For example, the Cantor set (sec. 2, Ch. III) is nowhere dense in $[0, 1]$.

8.2 DEFINITION *Let T be a topology on X and let $A \subset X$.*
Then,
(1) A is of T-first category
if and only if
A is a countable union of T-nowhere dense sets.
(2) A is of T-second category
if and only if
A is not of T-first category.

Baire's Category Theorem says that a complete metric space must be of second category. To clarify the proof, we state two rather obvious lemmas.

1 LEMMA *If $x \in X$ and $\varepsilon > 0$, then $\overline{B_\varrho(x, \varepsilon)} \subset B_\varrho(x, 2\varepsilon)$.*

Proof Let $z \in \overline{B_\varrho(x, \varepsilon)}$. Then, $B_\varrho(x, \varepsilon) \cap B_\varrho(z, \varepsilon) \neq \phi$. Let $t \in B_\varrho(x, \varepsilon) \cap B_\varrho(z, \varepsilon)$. Then, $\varrho(z, x) \leq \varrho(z, t) + \varrho(t, x) < \varepsilon + \varepsilon = 2\varepsilon$. ∎

2 LEMMA *If $x \in B_\varrho(z, \varepsilon)$, then for some $\delta > 0$, $\overline{B_\varrho(x, \delta)} \subset B_\varrho(z, \varepsilon)$.*

Proof Since $x \in B_\varrho(z, \varepsilon) \in T_\varrho$, then $B_\varrho(x, \xi) \subset B_\varrho(z, \varepsilon)$ for some $\xi > 0$. Choose $\delta = \xi/2$. By Lemma 1, $\overline{B_\varrho(x, \delta)} \subset B_\varrho(x, \xi) \subset B(z, \varepsilon)$. ∎

8.1 THEOREM (Baire Category Theorem) *Let X be ϱ-complete. Then, X is of T_ϱ-second category.*

Proof Suppose instead that X is of T_ϱ-first category. Then, $X = \bigcup_{i=1}^{\infty} D_i$, where each D_i is T_ϱ-nowhere dense in X.

Now, $\text{Int}_{T_\varrho}(\bar{D}_1) = \phi$, so $\bar{D}_1 \neq X$. Choose $S_1 \in X - \bar{D}_1$.

Since $S_1 \notin \bar{D}_1$, there is some $\varepsilon_1 > 0$ such that $B_\varrho(S_1, 2\varepsilon_1) \cap D_1 = \phi$. Now, $\overline{B_\varrho(S_1, \varepsilon_1)} \subset B_\varrho(S_1, 2\varepsilon_1)$ by Lemma 1. Let $F_1 = \overline{B_\varrho(S_1, \varepsilon_1)}$. Then, $F_1 \cap \bar{D}_1 = \phi$. Next, D_2 is T_ϱ-nowhere dense, so $\text{Int}_{T_\varrho}(\bar{D}_2) = \phi$. Then, $B_\varrho(S_1, \varepsilon_1) \cap (X - \bar{D}_2) = B_\varrho(S_1, \varepsilon_1) - \bar{D}_2 \neq \phi$. Choose $S_2 \in B_\varrho(S_1, \varepsilon_1) - \bar{D}_2$, and, using Lemma 2, choose ε_2 such that $0 < \varepsilon_2 < \frac{1}{2}\varepsilon_1$ and $\overline{B_\varrho(S_2, \varepsilon_2)} \subset B_\varrho(S_1, \varepsilon_1) - \bar{D}_2$.

Let $F_2 = \overline{B_\varrho(S_2, \varepsilon_2)}$. Then, $F_2 \cap \bar{D}_2 = \phi$, and $F_2 \subset F_1$.

Now proceed by induction. For each positive integer n, choose $S_{n+1} \in B_\varrho(S_n, \varepsilon_n) - \bar{D}_{n+1}$ and choose ε_{n+1} such that $0 < \varepsilon_{n+1} < \varepsilon_n/2$ and $\overline{B_\varrho(S_{n+1}, \varepsilon_{n+1})} \subset B_\varrho(S_n, \varepsilon_n) - \bar{D}_{n+1}$. Let $F_{n+1} = \overline{B_\varrho(S_{n+1}, \varepsilon_{n+1})}$.

In this way, generate a sequence F of T_ϱ-closed, ϱ-bounded sets such that $F_{n+1} \subset F_n$. Further, by Lemma 6, section 7, $\text{diam}(F_{n+1}) \leq 2\varepsilon_{n+1} \leq \varepsilon_1/2^{n-1} \to 0$ as $n \to \infty$.

By Cantor's Theorem (Th. 7.5), $\bigcap_{n=1}^{\infty} F_n = \{t\}$ for some $t \in X$.

But, for each positive integer n, $F_n \subset X - \bar{D}_n$. Then, $t \in \bigcap_{n=1}^{\infty} (X - \bar{D}_n) = X - \bigcup_{n=1}^{\infty} \bar{D}_n \subset X - \bigcup_{n=1}^{\infty} D_n = \phi$, a contradiction, and the proof is complete. ∎

Proofs using Baire's Category Theorem are often called category arguments. We shall give three examples from analysis.

8.1 EXAMPLE As a first example of a category argument, we shall prove the existence of a real-valued, nowhere differentiable, continuous function on [0, 1].

For each positive integer n, let

$$D_n = C([0, 1]) \cap \left\{ f \,\middle|\, \begin{array}{l} \text{for some } x \in [0, 1], \left|\dfrac{f(x + h) - f(x)}{h}\right| \leq n \\ \text{whenever } h \neq 0 \text{ and } 0 \leq x + h \leq 1 \end{array} \right\}.$$

The idea is to show that each D_n is nowhere dense in $C([0, 1])$. Since $C([0, 1])$ is complete in the sup-norm metric, we can then conclude that $\bigcup_{n=1}^{\infty} D_n \neq C([0, 1])$ by Baire's Theorem. Then we can choose some $F \in C([0, 1])$ such that $F \notin D_n$ for $n = 1, 2, \ldots$ Such an F can have a derivative at no point of [0, 1].

To see this, suppose $0 \leq z \leq 1$ and $F'(z) = L$. Since $F \notin D_n$, there is some $h_n \neq 0$ such that $0 \leq z + h_n \leq 1$ and $\left|\dfrac{F(z + h_n) - F(z)}{h_n}\right| > n$.

Then, $|F(z + h_n) - F(z)| \geq n|h_n|$. Now, F is continuous on [0, 1], hence bounded. Let $M < 0$ such that $|F(z + h_n) - F(z)| < M$ for each positive integer n. This can happen only if $h_n \to 0$ as $n \to \infty$. Now, since $F'(z)$ exists by assumption, then $|F'(z)| = \lim_{n \to \infty} \left|\dfrac{F(z + h_n) - F(z)}{h_n}\right| \geq \lim_{n \to \infty} n = \infty$, a contradiction, implying that F has no derivative on [0, 1].

This is the outline of the argument. To make it work, we must show that each D_n is nowhere dense in $C([0, 1])$. That is, Int $(\bar{D}_n) = \phi$ for $n = 1, 2, \ldots$ We shall outline the main steps of the argument, leaving some of the details to the reader.

1) $\bar{D}_n = D_n$ for each positive integer n.

Let $g \in \bar{D}_n$. Produce a sequence S to D_n which converges to g. For convenience, continuously extend g and each S_j to E^1 (recall E^1 is normal). Since $S_j \in D_n$, there is some $z_j \in [0, 1]$ such that $\left|\dfrac{S_j(z_j + h) - S_j(z_j)}{h}\right| \leq n$ if $h \neq 0$ and $0 \leq z_j + h \leq 1$. This defines a sequence z to [0, 1]. Since [0, 1] is compact, z has a subsequence $\zeta = z \circ \varphi$ converging to some L

in [0, 1]. Let $P = S \circ \varphi$. Then, for $h \neq 0$, we have:

$$\left| \frac{g(L+h) - g(L)}{h} \right| \leq \left| \frac{g(L+h) - g(\zeta_j + h)}{h} \right| + \left| \frac{g(\zeta_j + h) - P_j(\zeta_j + h)}{h} \right|$$

$$+ \left| \frac{P_j(\zeta_j + h) - P_j(\zeta_j)}{h} \right| + \left| \frac{P_j(\zeta_j) - g(\zeta_j)}{h} \right|$$

$$+ \left| \frac{g(\zeta_j) - g(L)}{h} \right|.$$

Thus show that, given $\varepsilon > 0$, $\left| \frac{g(L+h) - g(L)}{h} \right| \leq n + \varepsilon$ by taking j sufficiently large.

Hence conclude that $g \in D_n$.

2) D_n has no interior points.

Let $g \in D_n$ and let $\varepsilon > 0$. We shall produce $p \in C([0, 1])$ such that $\sup |p(x) - g(x)| < \varepsilon$ and $p \notin D_n$.

To do this, note that g is uniformly continuous on $[0, 1]$. Produce $\delta_1 > 0$ such that $|g(x) - g(y)| < \varepsilon/2$ if $|x - y| < \delta_1$. Let $\delta = \min(\delta_1, \varepsilon/2n)$.
Partition $[0, 1]$ by points $0 = t_0 < t_1 < t_2 < \cdots < t_{k-1} < t_k = 1$, where $|t_{j+1} - t_j| < \delta$ for $j = 1, \ldots, k - 1$.

In each $[t_j, t_{j+1}]$, define p as the function whose graph is the straight line from $(t_j, g(t_j) + \varepsilon/8)$ to $(t_{j+1}, g(t_{j+1}) - \varepsilon/8)$ if j is odd, and from $(t_j, g(t_j) - \varepsilon/8)$ to $(t_{j+1}, g(t_{j+1}) + \varepsilon/8)$ if j is even. Thus, for $t_j \leq x \leq t_{j+1}$,

$$p(x) = \begin{cases} g(t_j) + \varepsilon/8 - \left[\dfrac{g(t_j) - g(t_{j+1}) + \varepsilon/4}{t_{j+1} - t_j} \right] (x - t_j), & j \text{ odd} \\[2ex] g(t_{j+1}) + \varepsilon/8 + \left[\dfrac{g(t_{j+1}) - g(t_j) + \varepsilon/4}{t_{j+1} - t_j} \right] (x - t_{j+1}), & j \text{ even} \end{cases}$$

(see Figure 8.1).

Finally, show that $p \notin D_n$ and that $\sup_{0 \leq x \leq 1} |p(x) - g(x)| < \varepsilon$. This completes the proof. ∎

8.2 EXAMPLE A subset A of E^1 is an F_σ set if A is a countable union of closed sets; A is G_δ if A is a countable intersection of open sets. In measure theory, F_σ and G_δ sets play an important role. See, for example, Royden, p. 52, for an F_σ, G_δ characterization of Lebesgue measurable sets, or Munroe, p. 65, for the Borel classification scheme, of which F_σ and G_δ are a part.

In this example, we show by a category argument that the set Q of rationals is not a G_δ set.

If Q is a G_δ set, then $Q = \bigcap_{n=1}^{\infty} V_n$, where each V_n is open in E^1.

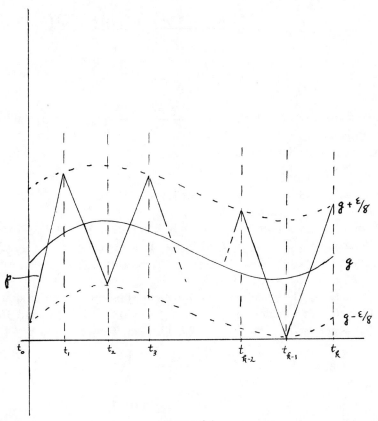

FIGURE 8.1

Since the rationals are countable, there is a bijection $f: Z^+ \to Q$. Since $Q = \bigcup_{n=1}^{\infty} \{f(n)\}$, then $E^1 = \bigcup_{n=1}^{\infty} ((E^1 - V_n) \cup \{f(n)\})$.

Now, $(E^1 - V_n) \cup \{f(n)\}$ is nowhere dense in E^1. To see this, note that $E^1 - Q = E^1 - \bigcap_{i=1}^{\infty} V_i = \bigcup_{i=1}^{\infty} (E^1 - V_i)$, so that $(E^1 - Q) \cup \{f(n)\}$ $= \{f(n)\} \cup \bigcup_{i=1}^{\infty} (E^1 - V_i) = \bigcup_{i=1}^{\infty} ((E^1 - V_i) \cup \{f(n)\})$. Any interior point of $(E^1 - V_n) \cup \{f(n)\}$ is also an interior point of $(E^1 - Q) \cup \{f(n)\}$, an impossibility, as any interval on the real line about an irrational will contain rationals as well.

Then, E^1 is of the first category, contradicting Baire's Theorem, as E^1 is complete. ∎

8.3 EXAMPLE As a final example, we shall prove the Principle of Uniform Boundedness for the class $C(X)$ of continuous, real-valued functions on a complete metric space (X, ϱ). This theorem is similar in form to the important Banach-Steinhaus Theorem of functional analysis (see, for example, Horvath, Vol. I).

THEOREM (*Principle of Uniform Boundedness*) *Let X be ϱ-complete. Let $B \subset C(X)$.*
Suppose, given $x \in X$, there is some $M_x > 0$ such that $|f(x)| \leq M_x$ for each $f \in B$.
Then,
for some non-empty, T_ϱ-open set U, and some $K > 0$,
$|f(y)| \leq K$ for each $f \in B$ and $y \in U$.

Proof If n is a positive integer and $f \in B$, let

$$D(n, f) = X \cap \{y \mid |f(y)| \leq n\}.$$

We claim that $D(n, f)$ is T_ϱ-closed. For suppose that S is a sequence to $D(n, f)$ which T_ϱ-converges to $z \in X$. Then, $f \circ S$ converges to $f(z)$. Since $|f(S_j)| \leq n$ for each $j \in Z^+$, then $|f(z)| \leq n$ by continuity of f. Then, $z \in D(n, f)$, so $D(n, f)$ is closed.

Let $D_n = \bigcap_{f \in B} D(n, f)$ for each positive integer n.

Then, D_n is closed also.

Now, if $x \in X$, then $|f(x)| \leq n$ for all $f \in B$, and $n \geq M_x$. If $n \geq M_x$, then, $x \in D(n, f)$ for all $f \in B$, so $x \in D_n$.

Thus, $X = \bigcup_{n=1}^{\infty} D_n$.

Now, X is complete, so for some positive integer K, D_K is not nowhere dense.

Then, $\text{Int}(\bar{D}_K) = \text{Int}(D_K) \neq \phi$. Let $x \in \text{Int}(D_K)$.

For some T_ϱ-open U, $x \in U \subset D_K$.

Finally, note that, if $y \in U$, and $f \in B$, then $y \in D(K, f)$, so $|f(y)| \leq K$, and the proof is complete. ∎

If the reader is interested in Fourier analysis, Gaal has a proof (p. 293) using the Uniform Boundedness Principle, of the following fact: given any countable set D of real numbers, there is a coutinuous function of period 2π whose Fourier series diverges at each point in D.

8.9 Ascoli's theorem

Ascoli's Theorem (sometimes called Arzela's Theorem) is an excellent example of a characterization theorem, distinguishing the compact subsets in the sup norm space $C([a, b])$. In so doing, it achieves a purely topological goal, but the theorem also has many applications in analysis.

The usual statement requires the notions of equicontinuous and uniformly bounded families of functions.

9.1 DEFINITION *Let $F \subset C([a, b])$.*
Then,
F is equicontinuous
if and only if
given $\varepsilon > 0$, there is some $\delta > 0$ such that $|f(x) - f(y)| < \varepsilon$ whenever $f \in F$ and $x, y \in [a, b]$ and $|x - y| < \delta$.

9.2 DEFINITION *Let $F \subset C([a, b])$.*
Then,
F is uniformly bounded
if and only if
for some $M > 0$, $|f(x)| \leq M$ for each $f \in F$ and $x \in [a, b]$.

Thus, the Principle of Uniform Boundedness of Example 8.3 says that the subset B of $C(X)$ is uniformly bounded on some open set, under suitable conditions on X and B.

9.1 THEOREM (*Ascoli's Theorem*) *Let $F \subset C([a, b])$.*
Then,
F is compact
if and only if
F is closed, equicontinuous and uniformly bounded.

Proof Suppose first that F is compact.
i) F is closed.
Since $C([a, b])$ is a metric space, hence Hausdorff, then F is closed by Theorem 1.9.VII.

ii) F is equicontinuous.

Suppose that F is not equicontinuous. Then, there is some $\varepsilon > 0$ such that: given $n \in Z^+$, there is some $f_n \in F$, and there are points x_n and y_n of $[a, b]$ such that $|x_n - y_n| < 1/n$ and $|f_n(x_n) - f_n(y_n)| \geq \varepsilon$.

This defines a sequence f to F. Since F is compact, then f has a convergent subsequence, say $g = f \circ \varphi$, with g convergent to $h \in F$. Now, h is continuous on $[a, b]$, hence uniformly continuous on $[a, b]$ by Theorem 3.4. Then, for some $\delta > 0$, $|h(x) - h(y)| < \varepsilon/3$ whenever $x, y \in [a, b]$ and $|x - y| < \delta$. Choose a positive integer K such that $1/K < \delta$ and such that $\sup_{a \leq t \leq b}|g_i(t) - h(t)| < \varepsilon/3$ for $i \in Z^+$ with $i \geq K$. Now, we have $|f_{\varphi_K}(x_{\varphi_K}) - f_{\varphi_K}(y_{\varphi_K})|$
$= |g_K(x_{\varphi_K}) - g_K(y_{\varphi_K})| \geq \varepsilon$ by choice of the sequences x and y.

Since $|x_{\varphi_K} - y_{\varphi_K}| \leq 1/\varphi_K < 1/K < \delta$, a repeated application of the triangle inequality yields:

$$|g_K(x_{\varphi_K}) - g_K(y_{\varphi_K})| \leq |g_K(x_{\varphi_K}) - h(x_{\varphi_K})| + |g_K(y_{\varphi_K}) - h(y_{\varphi_K})|$$
$$+ |h(x_{\varphi_K}) - h(y_{\varphi_K})|$$
$$< \varepsilon/3 + \varepsilon/3 + \varepsilon/3 = \varepsilon,$$

a contradiction.

Thus, F is equicontinuous.

iii) F is uniformly bounded.

If F is not uniformly bounded, then, given any positive integer n, there is an $f_n \in F$ and some $x_n \in [a, b]$ such that $|f_n(x_n)| \geq n$. This defines a sequence f to F. Since F is compact, f has a convergent subsequence. The reader can now derive a contradiction quite like that used in ii).

Conversely, suppose that F is closed, equicontinuous and uniformly bounded.

For a straightforward, but somewhat involved, argument that F is compact, see Goffman, p. 151–152, MAA Studies in Analysis. However, having filled many lines with down to earth analysis, we shall give instead a more elegant topological argument.

Let $\mathcal{M}$ be an ultrafilterbase on F. By Theorem 1.6.VII, it suffices to show that $\mathcal{M}$ converges in F.

If $x \in [a, b]$, let $\varphi_x: C([a, b]) \to E^1$ be defined by $\varphi_x(f) = f(x)$ for each $f \in C([a, b])$. It is easy to check that φ_x is continuous for each $x \in [a, b]$. Then, $\varphi_x(\mathcal{M})$ is an ultrafilterbase on $\varphi_x(F)$ by Theorem 5.5.IV and, by the uniform boundedness of F, $\varphi_x(F)$ is a bounded subset of E^1, hence $\overline{\varphi_x(F)}$ is compact in E^1. Then, $\varphi_x(\mathcal{M})$ converges, say to $g(x)$, in $\overline{\varphi_x(F)}$, by Theorem 1.6.VII. To complete the proof, the reader can check that $g \in F$ and that $\mathcal{M}$ converges to g. ∎

For alternate and, in some cases, more general statements of the Ascoli (or Ascoli-Arzela) Theorem, see Royden, Dugundji, Kelley, Gaal, Simmons, or Kolmogorov and Fomin, Vol. I, to name just a few references. Goffman has some examples of its applications, among them a proof of the existence of solutions of the differential equation $y'(x) = f(x, y)$, $y(x_0) = y_0$ (under suitable conditions on f, of course) and a discussion of normal families in complex analysis.

8.10 Metrization

Since metric spaces enjoy a number of properties not shared by topological spaces in general, it is natural to ask: when is a given space a metric space? That is, given (Y, M), is there a metric σ on Y such that $M = T_\sigma$? This is among the most difficult problems in set topology. If there is such a σ, then (Y, M) is said to be metrizable; if not, then (Y, M) is non-metrizable.

10.1 DEFINITION (Y, M) *is metrizable*
if and only if
there is a metric σ on Y such that $M = T_\sigma$.

It is easy to produce examples of nonmetrizable spaces. For example, if a space is not Hausdorff (or regular, or completely regular, or normal), then it certainly cannot be a metric space. Note that any set has at least one metric topology on it—but this does not mean that any metric on the set induces the given topology of the space.

We shall now consider conditions on a space sufficient to insure metrizability.

As a beginning, it is easy to show that a finite product is a metric space exactly when each coordinate space is metric.

10.1 THEOREM *Let T_i be a topology on X_i, $i = 1, ..., n$.*
Let $\mathbb{P}$ denote the product topology on $\prod_{i=1}^{n} X_i$.
Then,
$\left(\prod_{i=1}^{n} X_i, \mathbb{P} \right)$ *is metrizable*
if and only if
(X_i, T_i) is metrizable for $i = 1, ..., n$.

Proof In view of Theorem 2.1.III, and Theorem 1.1, the theorem is trivial if $n = 1$, so we may assume that $n \geq 2$.

Suppose first that $\left(\prod_{i=1}^{n} X_i, \mathbb{P}\right)$ is metrizable. Then, there is a metric σ on $\prod_{i=1}^{n} X_i$ such that $\mathbb{P} = T_\sigma$.

Let $f \in \prod_{i=1}^{n} X_i$ and let $j \in Z^+$ with $1 \leq j \leq n$. Let

$$Y = \left(\prod_{i=1}^{n} X_i\right) \cap \{g | g_i = f_i \text{ for } 1 \leq i \leq n \text{ and } i \neq j\}.$$

Then, $(Y, \mathbb{P}_Y) \cong (X_j, T_j)$, by Theorem 2.6.III.
By Theorem 1.2, $(Y, \mathbb{P}_Y)$ is metrizable.
By Theorem 1.1, (X_j, T_j) is metrizable.
Conversely:
Suppose that (X_i, T_i) is metrizable for $i = 1, \ldots, n$. Let ϱ_i be a metric on X_i such that $T_i = T_{\varrho_i}$ for $i = 1, \ldots, n$.

If $f, g \in \prod_{i=1}^{n} X_i$, define $\sigma(f, g) = \max_{1 \leq i \leq n} \varrho_i(f_i, g_i)$.

We leave it for the reader to check that $\mathbb{P} = T_\sigma$. Hint: use Theorem 3.4.I. ∎

The problem is not so easily solved when infinite Cartesian products are involved. An uncountable Cartesian product space, in which uncountably many factors have at least two points, is never metrizable (see Dugundji). However, a countable product of metrizable spaces is always metrizable.

10.2 THEOREM *Let ϱ_i be a metric on X_i for each $i \in Z^+$.*
Let $\mathbb{P}$ be the product topology on $\prod_{i=1}^{\infty} X_i$ determined by the topologies T_{ϱ_i} on X_i, $i = 1, \ldots$
Let $\sigma(f, g) = \sum_{i=1}^{\infty} \dfrac{\varrho_i(f_i, g_i)}{2^i(1 + \varrho_i(f_i, g_i))}$
for $f, g \in \prod_{i=1}^{\infty} X_i$.
Then,
(1) σ is a metric on $\prod_{i=1}^{\infty} X_i$.
(2) $\mathbb{P} = T_\sigma$.

Proof (1) is a routine calculation, and is omitted. To prove (2), we use Theorem 3.4.I.

First, let $\varepsilon > 0$ and $f \in \prod_{i=1}^{\infty} X_i$. We seek a $\mathbb{P}$-basic nghd. V of f such that $V \subset B_\sigma(f, \varepsilon)$.

Since $\sum_{i=1}^{\infty} 1/2^i$ converges, we can choose $N \in Z^+$ such that $\sum_{i=N+1}^{\infty} 1/2^i < \varepsilon/2$.

Let $V = \bigcap_{i=1}^{N} p_i^{-1}(B_{\varrho_i}(f_i, \varepsilon/2N))$. Then, $f \in V \in \mathbb{P}$.

To show that $V \subset B_\sigma(f, \varepsilon)$, let $g \in V$. Then, $p_i(g) = g_i \in B_{\varrho_i}(f_i, \varepsilon/2N)$ for $i = 1, \ldots, N$, so $\varrho_i(g_i, f_i) < \varepsilon/2N$. Then,

$$\sigma(f, g) = \sum_{i=1}^{\infty} \frac{1}{2^i} \frac{\varrho_i(f_i, g_i)}{1 + \varrho_i(f_i, g_i)} = \sum_{i=1}^{N} \frac{1}{2^i} \frac{\varrho_i(f_i, g_i)}{1 + \varrho_i(f_i, g_i)}$$

$$+ \sum_{i=N+1}^{\infty} \frac{1}{2^i} \frac{\varrho_i(f_i, g_i)}{1 + \varrho_i(f_i, g_i)} < N\left(\frac{\varepsilon}{2N}\right) + \frac{\varepsilon}{2} = \varepsilon.$$

This shows that $T_\sigma \subset \mathbb{P}$.

Now let $\bigcap_{i=1}^{n} p_{\alpha_i}^{-1}(U_{\alpha_i})$ be a $\mathbb{P}$-basic nghd. of f. We seek $W \in T_\sigma$ such that $f \in W \subset \bigcap_{i=1}^{n} p_{\alpha_i}^{-1}(U_{\alpha_i})$.

We shall produce $\varepsilon > 0$ such that $B_\sigma(f, \varepsilon) \subset \bigcap_{i=1}^{n} p_{\alpha_i}^{-1}(U_{\alpha_i})$.

Since $f_{\alpha_i} \in U_{\alpha_i} \in T_{\varrho_i}$ for $i = 1, \ldots, n$, then, for some $\delta_{\alpha_i} > 0$, $f_{\alpha_i} \in B_{\varrho_i}(f_{\alpha_i}, \delta_{\alpha_i}) \subset U_{\alpha_i}$.

Choose $\varepsilon > 0$ such that $\varepsilon < \max_{1 \leq i \leq n} \left(\dfrac{\delta_{\alpha_i}}{2^{\alpha_i}(1 + \delta_{\alpha_i})}\right)$.

We claim that $B_\sigma(f, \varepsilon) \subset \bigcap_{i=1}^{n} p_{\alpha_i}^{-1}(U_{\alpha_i})$.

For, let $g \in B_\sigma(f, \varepsilon)$. For $i = 1, \ldots, n$, we have

$$\sigma(f, g) < \varepsilon \leq \frac{\delta_{\alpha_i}}{2^{\alpha_i}(1 + \delta_{\alpha_i})},$$

so

$$\frac{\varrho_{\alpha_i}(f_{\alpha_i}, g_{\alpha_i})}{2^{\alpha_i}(1 + \varrho_{\alpha_i}(f_{\alpha_i}, g_{\alpha_i}))} \leq \sum_{j=1}^{\infty} \frac{\varrho_j(f_j, g_j)}{2^j(1 + \varrho_j(f_j, g_j))} < \varepsilon \leq \frac{\delta_{\alpha_i}}{2^{\alpha_i}(1 + \delta_{\alpha_i})}.$$

Then, $\varrho_{\alpha_i}(f_{\alpha_i}, g_{\alpha_i}) < \delta_{\alpha_i}$, so $g \in p_{\alpha_i}^{-1}(B_{\varrho_{\alpha_i}}(f_{\alpha_i}, \delta_{\alpha_i}))$.

Then, $B_\sigma(f, \varepsilon) \subset \bigcap_{i=1}^{n} p_{\alpha_i}^{-1}(U_{\alpha_i}))$, so that $\mathbb{P} \subset T_\sigma$, and the proof is complete. ∎

Of course, the "converse" of Theorem 10.2 is also true. Using the slice method of Theorem 2.6.III, it is easy to show that each (X_i, T_i) is metrizable if $\left(\prod_{i=1}^{\infty} X_i, \mathbb{P}\right)$ is metrizable.

We shall conclude this section with the classical Urysohn Metrization Theorem, which was historically the first significant result on the general metrization problem. The proof is very instructive, the idea being to embed the given space homeomorphically in a space which is known to be metrizable.

10.3 THEOREM (Urysohn Metrization Theorem) *Let (X, T) be a regular, $2°$-countable space.*
Then,
(X, T) is metrizable.

Proof We begin with a preliminary observation. Suppose that $\phi \neq V \in T$. By the regularity of T (see Theorem 2.2.V), there is some $U \in T$ with $\phi \neq U \subset \overline{U} \subset V$. Since a $2°$-countable, regular space is normal (Corollary 1, section 5, Ch. VII), there is by Urysohn's Lemma a continuous $f: X \to [0, 1]$ such that $f(\overline{U}) = \{0\}$ and $f(X - V) = \{1\}$.

Now let $\mathscr{B}$ be a countable base for T.
Define
$$M = \{(U, V) | U \in \mathscr{B} \text{ and } V \in \mathscr{B} \text{ and } \phi \neq U \subset \overline{U} \subset V\}.$$

Then, M is also countable.

Let $Y_{(U,V)} = [0, 1]$, considered as a subspace of E^1, for each $(U, V) \in M$, and let $\mathbb{P}$ be the product topology on $\prod_{(U,V) \in M} Y_{(U,V)}$.

By Theorems 1.2 and 10.2, $\left(\prod_{(U,V) \in M} Y_{(U,V)}, \mathbb{P} \right)$ is metrizable.

To prove our theorem, it now suffices, by Theorems 1.2 and 1.1, to show that (X, T) is homeomorphic to a subspace of $\left(\prod_{(U,V) \in M} Y_{(U,V)}, \mathbb{P} \right)$.

If $(U, V) \in M$, then by our initial observation there is some $f_{(U,V)}: X \to Y_{(U,V)}$ such that $f_{(U,V)}(\overline{U}) = \{0\}$ and $f_{(U,V)}(X - V) = \{1\}$.
Define
$$\varphi: X \to \prod_{(U,V) \in M} Y_{(U,V)}$$
by specifying, if $x \in X$, then
$$(\varphi(x))(U, V) = f_{(U,V)}(x) \text{ for each } (U, V) \in M.$$

We now proceed in steps.

i) φ is an injection.

Suppose that x and y are in X and $\varphi(x) = \varphi(y)$. If $x \neq y$, then, since T is a Hausdorff topology, there are disjoint, basic open sets b_1 and b_2 with $x \in b_1$ and $y \in b_2$.

For some U, $U \in \mathscr{B}$ and $x \in U \subset \bar{U} \subset b_1$.
Then, $(U, b_1) \in M$.
Now, $f_{(U,b_1)}(\bar{U}) = \{0\}$ and $f_{(U,b_1)}(X - b_1) = \{1\}$.
Since $x \in \bar{U}$, then $f_{(U,b_1)}(x) = 0$.
Since $y \in X - b_1$, then $f_{(U,b_1)}(y) = 1$.
But, then

$$(\varphi(x))(U, b_1) = f_{(U,b_1)}(x) = 0 = (\varphi(y))(U, b_1) = f_{(U,b_1)}(y) = 1,$$

a contradiction.
Hence, $x = y$.

ii) φ is $(T, \mathbb{P})$ continuous.

If $(U, V) \in M$, and $x \in X$, then

$$(p_{(U,V)} \circ \varphi)(x) = p_{(U,V)}(\varphi(x)) = \varphi(x)((U, V)) = f_{(U,V)}(x).$$

But then $p_{(U,V)} \circ \varphi = f_{(U,V)} : X \to [0, 1]$ is continuous.
Then, φ is continuous by Theorem 2.7, Ch. III.

iii) If $W \in T$, then $\varphi(W) \in \mathbb{P}_{\varphi(X)}$.

To show this, let $W \in T$, and $x \in W$. Let $b_1 \in \mathscr{B}$ such that $x \in b_1 \subset W$. Since $\varphi(b_1) \subset \varphi(W)$, it suffices to produce $L \in \mathbb{P}$ such that $\varphi(x) \in L \cap \varphi(X) \subset \varphi(b_1)$.

Now, for some $b_2 \in \mathscr{B}$, $x \in b_2 \subset \bar{b}_2 \subset b_1$, so $(b_2, b_1) \in M$.
Then, $f_{(b_2,b_1)} : X \to [0, 1]$ is continuous, and $f_{(b_2,b_1)}(x) = 0$ as $x \in b_2$ and $f_{(b_2,b_1)}(X - b_1) = \{1\}$.
Note that $\varphi(x) \in p_{(b_2,b_1)}^{-1}([0, 1[)$, as $\varphi(x)((b_2, b_1)) = f_{(b_2,b_1)}(x) = 0 \in [0, 1[$.
Further, $p_{(b_2,b_1)}^{-1}([0, 1[) \in \mathbb{P}$, as $[0, 1[$ is open in $[0, 1]$.
Let $L = p_{(b_2,b_1)}^{-1}([0, 1[)$.
There remains to show that $L \cap \varphi(X) \subset \varphi(b_1)$.
Let $h \in L \cap \varphi(X)$.
Then, $h \in \varphi(X)$, so $h = \varphi(y)$ for some $y \in X$. Then,

$$p_{(b_2,b_1)}(h) = p_{(b_2,b_1)}(\varphi(y)) = (\varphi(y))((b_2, b_1)) = f_{(b_2,b_1)}(y) \in [0, 1[.$$

Now, if $y \notin b_1$, then $y \in X - b_1$, hence $f_{(b_2,b_1)}(y) = 1 \notin [0, 1[$. Then, $y \in b_1$, so $h = \varphi(y) \in \varphi(b_1)$.

By steps i), ii) and iii), and by Theorem 2.1.II, φ is a $(T, \mathbb{P}_{\varphi(X)})$ homeomorphism, and the proof is complete. ∎

The proof of Urysohn's Metrization Theorem is as interesting as the theorem itself, because it characterizes regular, 2°-countable spaces as subspaces of cubes having countably many sides, which is by itself a worthwhile topological result. This is very like Theorem 4.6.V, which characterized completely regular spaces as exactly the subspaces of cubes having an arbitrary (not necessarily countable) number of sides $[0, 1]$.

In view of Theorem 6.4, a separable metric space is $2°$-countable, so that we have also obtained a characterization of separable metric spaces as subspaces of countably-many-sided cubes.

It is tempting to try to use Theorem 4.6.V to shorten the proof of Urysohn's Theorem, since a $2°$-countable, regular space is normal, hence completely regular. But the proof of Urysohn's Theorem used Theorem 10.2, which requires that the space be embedded in a *countable* product, something which Theorem 4.6.V fails to provide.

For additional theorems on metrization, see, for example, Kelley or Dugundji.

Problems

1) Reconsider Problem 26, Ch. I.
2) Let R be the set of real numbers, with the topology T consisting of ϕ and all sets $R - S$, with S finite. Is (R, T) metrizable?
3) Give R the topology T consisting of ϕ and all sets $R - S$, with S countable. Is (R, T) metrizable?
4) Let T be the topology of Example 5.2.VII. Is (R, T) metrizable?
5) Let $X = [0, 1] \times [0, 1]$, with the order topology $T_<$ generated by the order: $(a, b) < (c, d)$ if $a < c$ of $(a = c$ and $b < d)$. Is (X, T) metrizable? (See also Problem 6, Ch. VII).
6) Let $f: X \to Y$ be a continuous surjection. Let X be a compact metric space, and Y Hausdorff. Then, Y is a compact metric space.
7) In a metric space, every closed set is a G_δ (countable intersection of open sets), and every open set an F_σ (countable union of closed sets).
8) Let ϱ be a metric on X. If A and B are non-empty subsets of X, let $d(A, B) = \inf \{\varrho(x, y) | x \in A \text{ and } y \in B\}$.
   (a) $d(\bar{A}, \bar{B}) = d(A, B)$.
   (b) Let $\mathscr{B}$ consist of all non-empty closed subsets A of X with diam $(A) = \sup \{\varrho(x, y) | x \in A \text{ and } y \in A\}$ finite. Let

   $\xi(A, B) = \max (\sup \{d(\{x\}, B) | x \in A\}, \sup \{d(\{x\}, A) | x \in B\})$,

   for $A, B \in \mathscr{B}$. Then, ξ is a metric on $\mathscr{B}$.

9) Let (X, ϱ) be a metric space. If A is a non-empty, closed subset of X, and $\sup \{\varrho(x, y) | x \in A \text{ and } y \in A\}$ is finite, prove that there are points $x_0, y_0 \in A$ such that
$$\sup \{\varrho(x, y) | x \in A \text{ and } y \in A\} = \varrho(x_0, y_0).$$

10) A metric space (X, ϱ) is complete if and only if $\overline{B_\varrho(x, \varepsilon)}$ is compact for each $x \in X$ and $\varepsilon > 0$.

11) Give an example of a complete metric space (X, ϱ), with an equivalent metric τ, such that (X, τ) is not complete.
Prove: If (X, ϱ) is a compact, complete metric space, and $\tau \sim \varrho$, then (X, τ) is also complete.

12) Every compact metric space is complete, but not conversely.

13) A compact subset of a metric space is closed and bounded. However, a closed, bounded subset of a metric space need not be compact.

14) Each compact metric space is separable.

15) A countable product of complete metric spaces is complete.

16) Each separable, metrizable space is homeomorphic to a subspace of a produce space $\prod_{\alpha \in \mathscr{A}} X_\alpha$, where each $X_\alpha = E^1$.

17) Let (X, ϱ) be a complete metric space. Let $\mathcal{O}$ be a countable subset of T_ϱ and $\bar{A} = X$ for each $A \in \mathcal{O}$. Then, $\cap \mathcal{O} \neq \phi$. Use this to give an alternate proof of Baire's Category Theorem.

18) Show that Baire's Theorem is equivalent to:
If (X, ϱ) is a complete metric space, and $\{G_n\}_{n=1}^\infty$ is a sequence of open sets with $\bar{G}_n = X$ for each n, then $\bigcap_{n=1}^\infty G_n$ is dense in X.

*19) Prove the following Baire-type theorem. Let X be a compact Hausdorff space. Then, X cannot be written as a union $\bigcup_{i=1}^\infty A_i$, where each A_i is closed and $\text{Int}(A_i) = \phi$.
Hence show: a compact Hausdorff space cannot be both countable and connected.

*20) Prove the following generalization of Baire's Theorem: a countably compact, regular space is of second category.

21) Let (X, ϱ) be a complete metric space and A a subset of first category as a subspace of (X, ϱ). Then, $X - A$ is dense in X.

*22) Prove the following Baire-type theorem. Let (X, T) be a regular, locally countably compact (each point has a nghd. whose closure is countably compact) space. Then, X cannot be written as a union $\bigcup_{n=1}^{\infty} A_n$, where each A_n is T-closed and $\text{Int}_T(A_n) = \phi$.

*23) Research problem: Show that every compact metric space is the continuous image of the Cantor set in E^1.
Can this result be extended to compact Hausdorff spaces?

24) A filterbase $\mathscr{F}$ on a metric space (X, ϱ) is called a ϱ-Cauchy filterbase if, given $\varepsilon > 0$, there is some $A \in \mathscr{F}$ with diam (A) = sup $\{\varrho(x, y) | x, y \in A\} < \varepsilon$. (Recall the proof of Theorem 7.6.) Each ϱ-convergent filterbase is ϱ-Cauchy, but not conversely. However, X is ϱ-complete if and only if each ϱ-Cauchy filterbase on X is T_ϱ-convergent.

25) Let $l_2 = \{x | x : Z^+ \to E^1 \text{ and } \sum_{i=1}^{\infty} x_i^2 \text{ converges}\}$, with metric $\varrho(x, y)$ $= \left[\sum_{i=1}^{\infty} (x_i - y_i)^2\right]^{1/2}$.
Show that (l_2, ϱ) is separable.
Let $C = l_2 \cap \{x | 0 \leq x \leq 1/j \text{ for } j = 1, 2, \ldots\}$.
Show that C is compact.

*26) If $\alpha < \beta$ and $0 < t < 1$, let
$F_{\alpha, \beta}(t) = \{f | \alpha < f(t) < \beta \text{ and } f \text{ is continuous: }]0, 1[\to E^1\}$.
The sets $F_{\alpha, \beta}(t)$ generate a topology T on $(E^1)^{]0,1[}$.
Show:
(a) T is Hausdorff, but not $1°$-countable, hence not metrizable.
(b) If $\{f_n\}_{n=1}^{\infty}$ is a sequence to $(E^1)^{]0,1[}$, and $g \in (E^1)^{]0,1[}$, then
$\{f_n\}$ is T-convergent to g
if and only if
$\{f_n\}$ is pointwise convergent to g in E^1.
Hence conclude: pointwise convergence cannot be described by a metric.

*27) In this exercise, we sketch the notion of a uniformity, which allows a generalization of the concept of uniform continuity to non-metric spaces.
Let $X \neq \phi$. Then, $\mathscr{U}$ is a uniformity (or uniform structure) on X if:
1) $\phi \neq \mathscr{U} \subset \mathscr{P}(X \times X)$.
2) $A, B \in \mathscr{U}$ implies $A \cap B \in \mathscr{U}$.

3) $A \in \mathcal{U}$ and $A \subset B \subset X \times X$ implies $B \in \mathcal{U}$.
(Thus far, $\mathcal{U}$ is almost a filter on $X \times X$).
4) $\Delta \subset \cap \mathcal{U}(\Delta = \{(x, x)|x \in X\} =$ diagonal of X).
5) $A \in \mathcal{U}$ implies $A^{-1} = \{(y, x)|(x, y) \in A\} \in \mathcal{U}$.
6) If $A \in \mathcal{U}$, then there is some $B \in \mathcal{U}$ such that
$B \circ B = \{(x, y)|$for some $z, (x, z) \in B$ and $(z, y) \in B\} \subset A$.

The pair $(X, \mathcal{U})$, with $\mathcal{U}$ a uniformity on X, is called a uniform space. Bourbaki calls the elements of a uniformity, entourages.

a) If $x \in X$, let $\mathcal{B}_x$ consist of all sets $X \cap \{y|(x, y) \in V\}$, with $V \in \mathcal{U}$. Then, there is a unique topology $T(\mathcal{U})$ on X such that, for each $x \in X$, $\mathcal{B}_x$ consists exactly of all subsets of X containing a $T(\mathcal{U})$-open nghd. of x (see problem 14, Ch. I). $T(\mathcal{U})$ is the topology induced by $\mathcal{U}$, and (X, T) is called a uniform space if $T = T(\mathcal{U})$ for some uniformity $\mathcal{U}$ on X. Often, (X, T) is then written as $(X, \mathcal{U})$.

b) $T(\mathcal{U})$ is Hausdorff if and only if $\cap \mathcal{U} = \Delta$.

Now define: if $(X, \mathcal{U})$ and $(Y, \mathcal{M})$ are uniform spaces ($\mathcal{U}$ and $\mathcal{M}$ uniformities on X and Y respectively), a function $f \colon X \to Y$ is $(\mathcal{U}, \mathcal{M})$ uniformly continuous if, given $A \in \mathcal{U}$, there is some $B \in \mathcal{M}$ such that $(x, y) \in A$ implies $(f(x), f(y)) \in \mathcal{M}$.

c) If ϱ is a metric on X, let, for each $\varepsilon > 0$, $B(\varepsilon) = (X \times X) \cap \{(x, y)|\varrho(x, y) < \varepsilon\}$. The set $\mathcal{U}_\varrho$ of all sets $B(\varepsilon)$, for $\varepsilon < 0$, is a uniformity on X. Further, $T(\mathcal{U}_\varrho) = T_\varrho$. If d is a metric on Y, then $f \colon X \to Y$ is (T_ϱ, T_d) uniformly continuous exactly when f is $(\mathcal{U}_\varrho, \mathcal{U}_d)$ uniformly continuous.

d) (X, T) is said to be uniformizable if there is a uniformity $\mathcal{U}$ on X such that $T(\mathcal{U}) = T$. Show that any compact Hausdorff space is uniformizable. (Choose as sets in $\mathcal{U}$ all sets containing an open nghd. of Δ in $X \times X$). Note, however, that not every compact Hausdorff space is metrizable.

e) If $\mathcal{U}$ is a uniformity on X and $\mathcal{F}$ a filter on X, then $\mathcal{F}$ is a $\mathcal{U}$-Cauchy filter if, given $A \in \mathcal{U}$, there is some $V \in \mathcal{F}$ with $V \times V \subset A$.

  i) Each $T(\mathcal{U})$-convergent filter on X is a $\mathcal{U}$-Cauchy filter. Define: $(X, T(\mathcal{U}))$ is complete if each $\mathcal{U}$-Cauchy filter on X is $T(\mathcal{U})$ convergent.
  ii) Each closed subspace of a complete space is complete.
  iii) Each complete subspace of a uniform Hausdorff space is closed.

iv) A product of non-empty uniform spaces is complete
   if and only if
   each factor is complete.
   (Here, the product uniform space is the product set $\prod_{\alpha \in \mathscr{A}} X_\alpha$, with the smallest uniformity $\mathscr{U}$ such that the projections $p_\alpha : \prod_{\beta \in \mathscr{A}} X_\beta \to X_\alpha$ are $(\mathscr{U}, \mathscr{U}_\alpha)$ uniformly continuous. Note that this is entirely analogous to the definition of the product topology.
f) Ω is a uniformizable, non-metrizable space. To show that Ω is uniformizable, note that Ω is normal, hence completely regular. Embed Ω in a cube. It now suffices to uniformize subspaces of cubes.

28) Suppose that $(\hat{X}, \hat{T})$ and $(\hat{Y}, \hat{M})$ are complete spaces, with (X, T) homeomorphic to the proper dense subspaces $(\hat{X} - \{x_0\}, T_{\hat{X}-\{x_0\}})$ and $(\hat{Y} - \{y_0\}, M_{\hat{Y}-\{y_0\}})$. Then, $(\hat{X}, \hat{T}) \cong (\hat{Y}, \hat{M})$. Thus the completion of (X, T) is unique up to homeomorphism.

CHAPTER IX

Topological groups

9.1 Basic terminology, notation and examples

IT IS OFTEN THE CASE that there is more than one kind of structure on the set in which one is working. For example, the real numbers have their usual topological structure as E^1, and an algebraic one as well, as a group under addition. Similarly, $E^1 - \{0\}$ has the relative Euclidean topology, and is a group under ordinary multiplication.

In the case of E^1, there is a great deal of interplay between the two structures. One can speak not only of real-valued continuous functions, but also of sums and products of such functions; of sums and products of sequences and limits, and so on.

A natural generalization of this situation is the topological group. Suppose a set G has a topology T, and is at the same time a group under some operation which we shall call multiplication. If the topology and the group operation are sufficiently compatible, then G is called a topological group. Experience has shown the following conditions to comprise a good definition of "sufficiently compatible".

1.1 DEFINITION $(G, \cdot, T)$ *is a topological group if and only if*

(1) $(G, \cdot)$ *is a group.*

(2) (G, T) *is a topological space.*

(3) The map $\varphi\colon G \times G \to G$ *defined by* $\varphi(x, y) = x \cdot y$ *for* $x, y \in G$, *is continuous in the product topology on* $G \times G$.

(4) The map $v\colon G \to G$ *defined by* $v(x) = x^{-1}$ *for* $x \in G$, *is continuous.*

To simplify matters, we assume for the rest of this chapter that $(G, \cdot)$ is a group and that T is a topology on G. If $x, y \in G$, we write $x \cdot y$ as xy. We also fix the maps φ and v as defined in (3) and (4) of the definition, and in addition define the map $\xi\colon G \times G \to G$ by: $\xi(x, y) = xy^{-1}$ for $x, y \in G$.

When referring to continuity of φ or ξ, the product topology on $G \times G$ determined by T is always understood. Finally, when there is no danger of confusion, we write G in place of $(G, \cdot, T)$.

The definition of a topological group can be restated more economically in terms of ξ.

1.1 THEOREM *G is a topological group if and only if ξ is continuous.*

Proof Suppose first that G is a topological group, so that φ and v are continuous.

Define $i: G \to G$ by $i(x) = x$ for each $x \in G$. Clearly i is continuous.

Map $\alpha: G \times G \to G \times G$ by $\alpha(x, y) = (x, y^{-1})$ for $x, y \in G$. Denoting the projections of $G \times G \to G$ by p_1 and p_2, we have $p_1 \circ \alpha = i$ and $p_2 \circ \alpha = v \circ p_2$, both continuous, the latter by Theorem 1.3.II and Theorem 2.5.III. By Theorem 2.7.III, α is continuous. Then, $\xi = \varphi \circ \alpha$ is continuous.

Conversely, suppose that ξ is continuous.

To show that v is continuous, let $f: G \to G \times G$ be defined by $f(x) = (e, x)$ for each $x \in G$. Then, $p_1 \circ f$ and $p_2 \circ f$ are continuous, so f is continuous. Hence, $\xi \circ f = v$ is continuous.

To show that φ is continuous, define α as above. As before, α is continuous, so $\varphi = \xi \circ \alpha$ is continuous. ∎

1.1 EXAMPLE E^1 is a topological group with respect to addition and the usual Euclidean topology.

To show that ξ is continuous at $(x, y) \in E^1 \times E^1$, let $\varepsilon > 0$. Then, $|(x - y) - (a - b)| \leq |x - a| + |y - b| < \varepsilon$ if $|x - a| < \varepsilon/2$ and $|y - b| < \varepsilon/2$. Thus, if $(a, b) \in B_{\varrho_2}((x, y), \varepsilon/2)$, we have $(x - a)^2 + (y - b)^2 < \varepsilon^2/4$, so $|x - a| < \varepsilon/2$ and $|y - b| < \varepsilon/2$, hence $|\xi(x, y) - \xi(a, b)| < \varepsilon$.

Similarly, E^n is a topological group under addition:
$$(x_1, \ldots, x_n) + (y_1, \ldots, y_n) = (x_1 + y_1, \ldots, x_n + y_n).$$ ∎

1.2 EXAMPLE Let G be any group and T the discrete topology on G. Then, G is a topological group. ∎

1.3 EXAMPLE Let (X, M) be any space, and $C(X, G)$ the set of continuous functions $f: X \to G$.

If $f, g \in C(X, G)$, define $fg: X \to G$ by $(fg)(x) = f(x) g(x)$ for each $x \in X$.

It is easy to show that $C(X, G)$ is a group under this operation.

Now, $C(X, G)$ may be thought of as a subset of $\prod_{x \in X} G_x$, where $G_x = G$ for each $x \in X$. Give $C(X, G)$ the relative product topology from $\prod_{x \in X} G_x$. The reader can check that $C(X, G)$ is then a topological group. ∎

1.4 EXAMPLE Let M_n be the set of all real $n \times n$ matrices, with $\varrho(A, B) = \max_{\substack{1 \leq i \leq n \\ 1 \leq j \leq n}} |A_{ij} - B_{ij}|$ for $A, B \in M_n$. Then, ϱ is a metric on M_n and determines a topology on M_n. With the usual addition of matrices, M_n is a topological group. ∎

1.5 EXAMPLE Any normed linear space $(V, \| \ \|)$, considered as an additive group, is a topological group with the topology induced by the norm metric. ∎

1.2 DEFINITION Let $x \in G$. Then,

(1) $t_x: G \to G$ is defined by $t_x(y) = yx$ for each $y \in G$.

(2) $_xt: G \to G$ is defined by $_xt(y) = xy$ if $y \in G$.

(3) $\mathscr{I}_x: G \to G$ is defined by: $\mathscr{I}_x(y) = xyx^{-1}$ if $y \in G$.

The functions t_x and $_xt$ are the right and left translations by x respectively. From group theory, t_e, $_et$ and $\mathscr{I}_x$ are automorphism (isomorphisms of G onto itself) for each $x \in G$. Sometimes $\mathscr{I}_x$ is called an inner automorphism of G. We shall show that left and right translations, inner automorphisms, and the inverse map v are all homeomorphisms of G onto itself.

1.2 THEOREM

(1) v is an homeomorphism: $G \to G$.

(2) If $x \in G$, then t_x and $_xt$ are homeomorphisms: $G \to G$.

(3) If $x \in G$, then $\mathscr{I}_x$ is an homeomorphism: $G \to G$.

Proof of (1) Immediately v is a bijection, and v is continuous by assumption. Since $v = \operatorname{inv} v$, then v is an homeomorphism.

Proof of (2) Let $x \in G$. Immediately, t_x is a bijection. Now note that if $y \in G$, then $t_x(y) = yx = \varphi(y, x)$. Now, φ is continuous by assumption. By Theorem 2.8.III, t_x is continuous.

It is now easy to verify that $\operatorname{inv}(t_x) = t_{x^{-1}}$ is continuous, hence t_x is an homeomorphism.

A similar argument applies to $_xt$.

Proof of (3) If $x \in G$, then $\mathscr{I}_x = t_{x^{-1}} \circ {}_x t$ is an homeomorphism by (2). ∎

This theorem is surprisingly useful in proving theorems about topological groups. As an immediate application, we shall show that every topological group is a homogeneous space. That is, given any two points of G, there is an homeomorphism of G mapping one point to the other.

1.3 THEOREM *Let $x \in G$ and $y \in G$.*
Then,
there is an homeomorphism $f: G \to G$ with $f(x) = y$.

Proof Let $f = {}_{yx^{-1}}t$. ∎

One consequence of this theorem is that not every topological space is a topological group. There are non-homogeneous spaces. In such spaces, it is impossible to impose a group structure which is compatible with the existing topological structure (see (Problem 1).

One advantage of homogeneity in topological spaces is that many theorems can be proved locally (at or in a nghd. of a point) and then automatically extended to other points in the space, as the translations shift neighborhoods from one point to another. We shall see use of this device very often in the remainder of the chapter (see Theorem 3.6, or Theorem 4.1, for example). Particularly important in this connection is the existence of certain kinds of neighborhoods (especially of the identity).

A brief review of notation from algebra first:
If $W \subset G$ and $V \subset G$, then,

$WV = \{xy | x \in W \text{ and } y \in V\}$.
$W^{-1} = \{x^{-1} | x \in W\}$.
$xW = \{xy | y \in W\}$ for each $x \in G$.
$Wx = \{yx | y \in W\}$ for each $x \in G$.
$WW = W^2$; $W^2 W = W^3$, and, by induction, W^n denotes $W^{n-1}W$

for each positive integer $n \geq 2$.

1.4 THEOREM *If N is a neighborhood of e, then,*
(1) There is a nghd. V of e such that $V^2 \subset N$.
(2) There is a nghd. V of e such that $V = V^{-1}$ and $V \subset N$.
(3) There is a nghd. V of e such that $V = V^{-1}$ and $V^2 \subset N$.
(4) If N' is a nghd. of x, then there is a nghd. V of e with $xV \subset N'$.
(5) If N' is a nghd. of x, then there is a nghd. V of e with $Vx \subset N'$.
(6) If $x \in G$, then $xVx^{-1} \subset N$ for some nghd. V of e.

Proof of (1) Since $\varphi(e, e) = e$ and φ is continuous at e, there are nghds. A and B of e with $\varphi(A \times B) \subset N$. Let $V = A \cap B$.

Proof of (2) Let $V = N \cap N^{-1}$.

Proof of (3) By (1), there is a nghd. P of e with $P^2 \subset N$. Let $V = P \cap P^{-1}$. Then, $V^2 \subset P^2 \subset N$ and $V = V^{-1}$.

Proof of (4) Let $x \in N'$. Now, $_xt$ is continuous at e, and $_xt(e) = xe = x$. Since N' is a nghd. of x, then there is a nghd. V of e with $_xt(V) = xV \subset N'$.

Proof of (5) Use t_x in place of $_xt$ in the proof of (4).

Proof of (6) Let $x \in G$. Since $\mathscr{I}_x$ is continuous at x, and $\mathscr{I}_x(e) = e$, then there is a nghd. V of e with $\mathscr{I}_x(V) \subset N$. Then, $xVx^{-1} \subset N$. ∎

Theorem 1.4 (2) is often rephrased: each nghd. of e contains a symmetric nghd. of e. We shall see next section that the symmetric neighborhoods of the identity generate the topology in a certain sense.

The following three theorems are important because they give us more information about the interrelationship between the algebraic and topological structures on G. The proofs are typical applications of Theorem 1.2.

1.5 Theorem *Let F be T-closed. Then,*

(1) F^{-1} is T-closed.

(2) If $a \in G$, then aF and Fa are T-closed.

Proof of (1) $F^{-1} = v(F)$ is closed by Theorem 1.2 (1).

Proof of (2) Immediate by Theorem 1.2 (2). ∎

1.6 Theorem *Let H be T-open. Then,*

(1) H^{-1} is T-open.

(2) If $a \in G$, then aH and Ha are open.

(3) If $A \subset G$, then AH and HA are open.

Proof (1) and (2) are immediate by Theorem 1.2. To prove (3), note that $AH = \bigcup_{a \in A} aH$ and $HA = \bigcup_{a \in A} Ha$. ∎

In view of Theorem 1.6 (3), it is natural to ask whether Theorem 1.5 (2) can be strengthened. If F and K are closed in G, is FK also closed? An example will show that this need not be the case. Consider E^1 as a topological group under addition. The set Z of integers is a closed subset, as is $\sqrt{2}Z$. But $Z + \sqrt{2}Z$ is not closed $(\overline{Z + \sqrt{2}Z} = E^1)$. However, a product of compact sets is compact.

1.7 Theorem *Let A and B be T-compact. Then, AB is T-compact.*

Proof By Tychonov's Theorem, $A \times B$ is compact in $G \times G$. Then, $\varphi(A \times B) = AB$ is compact in G by Theorem 1.7.VII. ∎

Actually, with some more effort, one can show that AB is closed if one set is closed and the other compact (see Hewitt and Ross, Vol. I). However, Theorem 1.7 will suffice for our purposes.

9.2 Bases and separation

The translates of the symmetric neighborhoods of the identity form a base for the topology on G.

2.1 Theorem *Let $\mathscr{B} = \{V | V \text{ is a nghd. of } e \text{ and } V = V^{-1}\}$. Then,*

(1) $\{xV | x \in G \text{ and } V \in \mathscr{B}\}$ is a base for T.
(2) $\{Vx | x \in G \text{ and } V \in \mathscr{B}\}$ is a base for T.

Proof By Theorem 1.6 (2), $\{xV | x \in G \text{ and } V \in \mathscr{B}\} \subset T$. Now let $x \in G$ and let U be any nghd. of x. Then, $x^{-1}U$ is a nghd. of e. By Theorem 1.4, there is some $V \in \mathscr{B}$ with $V \subset x^{-1}U$. Then, xV is a nghd. of x and $xV \subset U$. (1) now follows from Theorem 3.1.I.

The proof of (2) is similar. ∎

One indication of the effect of the merger between algebra and topology on the topological structure of G can be seen in the area of separation. A space is called T_0 if, given any two distinct points, one (we don't know which) has a neighborhood not containing the other. This is much weaker than the Hausdorff axiom, as shown by Sierpinski space, which is T_0 but not Hausdorff. However, in a topological group, the T_0 axiom is sufficient to insure not only the Hausdorff, but also the regularity (T_3) axiom.

2.2 Theorem *The following are equivalent:*
(1) Given two distinct points of G, there is a neighborhood of one of the points not containing the other point.
(2) T is regular.

Proof (2) ⇒ (1) is immediate, as a regular space is Hausdorff.
(1) ⇒ (2):

Assume (1). Let $x \in G$, $y \in G$ and $x \neq y$. We now proceed in steps.

i) There is a nghd. V of x and a nghd. U of y such that $x \notin U$ and $y \notin V$.

To prove this, produce a nghd. V of, say, x with $y \notin V$, by (1). Let $A = x^{-1}V$. Then, A is a nghd. of e. Let $B = A \cap A^{-1}$. Then, B is a nghd. of e and $B = B^{-1}$. Let $U = yB$. Then, U is a nghd. of y.

We claim that $x \notin U$.

For suppose $x \in U$. Then, $x \in yB$, so $x = yb$ for some $b \in B$. Then, $x^{-1} = b^{-1}y^{-1}$. But, $b^{-1} \in B^{-1} = B$, so $x^{-1} \in B^{-1}y^{-1} = By^{-1}$. Now, $B \subset A$. Then, $x^{-1} \in By^{-1} \subset Ay^{-1} = x^{-1}Vy^{-1}$. Then, $e = xx^{-1} \in Vy^{-1}$, so $y \in V$, a contradiction.

ii) $\{x\}$ is closed.

Let $z \in G - \{x\}$. By i), produce a nghd. U of z with $U \subset G - \{x\}$. Then, $G - \{x\}$ is open.

iii) T is Hausdorff.

Since $y \neq x$, then $y \in G - \{x\} \in T$ by ii). Then, $y^{-1}(G - \{x\})$ is a nghd. of e. For some V, V is a nghd. of e and $V = V^{-1}$ and $V^2 \subset y^{-1}(G - \{x\})$. Then, yV is a nghd. of y.

Let $W = G - \overline{(yV)}$.

Then, W is open and $W \cap (yV) = \phi$.

We claim that $x \in W$. If $x \notin W$, then $x \in \overline{(yV)}$. Since xV is a nghd. of x, then $(xV) \cap (yV) \neq \phi$. Let $t \in (xV) \cap (yV)$. Then, $t = xv_1 = yv_2$ for some $v_1, v_2 \in V$. Then,

$$y^{-1}x = v_2 v_1^{-1} \in VV^{-1} = V^2 \subset y^{-1}(G - \{x\}).$$

But then $x \in G - \{x\}$, impossible, implying that $x \in W$.

iv) If U is a nghd. of e, then there is a nghd. V of e such that $\bar{V} \subset U$.

Produce first a nghd. V of e with $V = V^{-1}$ and $V^2 \subset U$. We claim that $\bar{V} \subset U$.

For suppose that $z \in \bar{V}$. Since zV is a nghd. of z, then $(zV) \cap V \neq \phi$. Let $t \in (zV) \cap V$. Then $t = z\alpha$ for some $\alpha \in V$. Then, $z = t\alpha^{-1} \in V \subset V^2 \subset U$. Hence, $\bar{V} \subset U$.

v) If U is a nghd. of x, then there is a nghd. V of x such that $\bar{V} \subset U$.

To see this, note that $x^{-1}U$ is a nghd. of e. Produce by iv) some nghd. P of e with $\bar{P} \subset x^{-1}U$. Then, xP is a nghd. of x, and $\overline{xP} = x \cdot \bar{P} \subset U$, by Theorem 1.2 (2).

vi) T is regular.
This is now immediate by v) and Theorem 2.2.V. ∎

With a great deal more effort, one can prove that a T_0 topological group is completely regular. One proof of this is very like that of Urysohn's Lemma (see Husain, or Hewitt and Ross). Another proof (see Thron) is based on the theory of uniformities, which we have not treated. It is not true that every T_0 topological group is normal.

We shall return to separation briefly again in section 4.

9.3 Subgroups and connectedness

If H is a subgroup of G, then we can consider H as a subspace of G as well. In the relative topology T_H, H becomes a topological group (a topological subgroup of G).

We shall adhere to the usual algebraic notation:

$H < G$ if H is a subgroup of G;

$H \triangleleft G$ if H is a normal (in the algebraic sense) subgroup of G.

It might be better to use the synonym "invariant" for "normal" in the algebraic sense of group theory, but it should always be clear from context whether normal refers to the group property or the topological separation axiom.

3.1 THEOREM *Let $H < G$. Then,*
$(H, \cdot, T_H)$ is a topological group.

Proof Left as an exercise. ∎

3.2 THEOREM
(1) If $H < G$, then $\bar{H} < G$.
(2) If $H \triangleleft G$, then $\bar{H} \triangleleft G$.

Proof of (1) By continuity of φ, by Theorem 1.1.II and Theorem 2.2(1).III,
$\bar{H} \cdot \bar{H} = \varphi(\bar{H} \times \bar{H}) = \varphi(\overline{H \times H}) \subset \overline{\varphi(H \times H)} = \overline{H \cdot H} = \bar{H}$.
By continuity of v, and Theorem 1.1.II, $(\bar{H})^{-1} = v(\bar{H}) \subset \overline{v(H)} = \overline{(H^{-1})} = \bar{H}$.
Thus, $\bar{H} < G$.

Proof of (2) Suppose that $H \triangleleft G$. Then, $H < G$, so $\bar{H} < G$. To show that $\bar{H}$ is a normal subgroup, let $x \in G$, then $xHx^{-1} = \mathscr{I}_x(H) = H$. Hence, by Theorem 1.2 (3), $x\bar{H}x^{-1} = \mathscr{I}_x(\bar{H}) = \overline{\mathscr{I}_x(H)} = \overline{xHx^{-1}} = \bar{H}$. ∎

It is perhaps surprising that every open subgroup of a topological group is also closed. This means that a topological group with a nontrivial open subgroup cannot be connected. Thus, for example, the only open subgroup of E^1 is E^1 (although Z constitutes a closed, but not open, subgroup).

3.3 Theorem *Let $H < G$.*
Let H be T-open.
Then,
H is T-closed.

Proof By Theorem 1.6 (2), $G - H = \cup \{gH | g \in G \text{ and } gH \neq H\}$ is open. ∎

As we saw immediately preceding the last theorem, a closed subgroup need not be open. In the next section we shall give necessary and sufficient conditions for a normal subgroup to be open and for a normal subgroup to be closed.

Even more remarkable than Theorem 3.3 is the fact that a subgroup of G is either open, or has no interior points at all.

3.4 Theorem *Let $H < G$.*
Then,
$\text{Int}_T (H) = H$ or $\text{Int}_T (H) = \phi$.

Proof Suppose that $\text{Int}_T (H) \neq \phi$. Let $x \in H$. It suffices to show that $x \in \text{Int}_T (H)$.

For some y, $y \in \text{Int}_T (H)$. Then there is a nghd. V of y with $V \subset H$. Then, $y^{-1}V$ is a nghd. of e, and $xy^{-1}V$ is a nghd. of x. Further, $xy^{-1} \in H$, and $V \subset H$, so $xy^{-1} V \subset H$. ∎

Since a subgroup is never empty, another way of stating the theorem is the following.

1 Corollary *Let $H < G$.*
Then,
H is open $\Leftrightarrow \text{Int}_T (H) \neq \phi$.

Proof Immediate by Theorem 3.4. ∎

This makes it easy to test whether a subgroup is open or not, since only one interior point need be exhibited (usually one examines e).

We remarked above that Theorem 3.3 has obvious implications in the area of connectivity of G. We shall now show that any connected topological group is a union of powers of any symmetric neighborhood of the identity. Thus, in an algebraic sense, each symmetric neighborhood of e generates the (connected) group.

3.5 THEOREM *Let G be T-connected.*
Let U be a nghd. of e such that $U = U^{-1}$.
Then,

$$G = \bigcup_{n=1}^{\infty} U^n.$$

Proof Let $H = \bigcup_{n=1}^{\infty} U^n$. It is easy to check that $H < G$, using the fact that $U^{-1} = U$. Further, H is open. By Theorem 3.3, H is closed. Then, $H = G$ by Theorem 1.1.VI. ∎

In the event that G is not connected, it is natural to look at the components of G. To do this, it suffices to exhibit just the component of e, which in addition turns out to be a closed, normal subgroup of G.

3.6 THEOREM

(1) $\text{Comp}_T(e)$ is a closed, normal subgroup of G.
(2) If $x \in G$, then $\text{Comp}_T(x) = x \, \text{Comp}_T(e) = \text{Comp}_T(e) \, x$.

Proof of (1) By Theorem 1.8 (5).VI, $\text{Comp}_T(e)$ is closed. To show that $\text{Comp}_T(e) < G$, note first that $v(\text{Comp}_T(e)) = \text{Comp}_T(e)$ by Theorem 1.2(1) and Theorem 1.9 (1).VI.

Thus, $x \in \text{Comp}_T(e)$ implies $x^{-1} \in \text{Comp}_T(e)$.

Now, by Theorem 1.6.VI, Theorem 1.8 (1).VI and Theorem 1.4.VI, $\varphi(\text{Comp}_T(e) \times \text{Comp}_T(e))$ is T-connected. Since $e \in (\varphi(\text{Comp}_T(e) \times \text{Comp}_T(e))) \cap \text{Comp}_T(e)$, then $(\varphi(\text{Comp}_T(e) \times \text{Comp}_T(e))) \cup \text{Comp}_T(e)$ is T-connected by Theorem 1.3.VI. By Theorem 1.8 (2).VI, $\varphi(\text{Comp}_T(e) \times \text{Comp}_T(e)) \cup \text{Comp}_T(e) \subset \text{Comp}_T(e)$. Then, $\varphi(\text{Comp}_T(e) \times \text{Comp}_T(e)) \subset \text{Comp}_T(e)$, hence $\text{Comp}_T(e) < G$.

Finally, if $x \in G$, then Theorem 1.9 (1).VI and Theorem 1.2 (3) imply that $\mathscr{I}_x(\text{Comp}_T(e)) = \text{Comp}_T(e)$, so that $\text{Comp}_T(e) \triangleleft G$.

Proof of (2) Immediate by Theorem 1.2 (2) and Theorem 1.9 (1).VI. ∎

9.4 Compactness, local compactness and quotient groups

In this section we shall be primarily concerned with quotient topological groups, although the first theorem is solely concerned with sufficient conditions for G to be locally compact. This theorem is another illustration of how a local property can be examined in a topological group by looking at just one point and then translating.

4.1 Theorem *G is locally compact
if and only if
there is a nghd. V of e such that $\bar{V}$ is compact.*

Proof Suppose there is a nghd. V of e with $\bar{V}$ compact. Let $x \in G$. Then, xV is a nghd. of e, and $\overline{(xV)} = \overline{{}_xt(V)} = {}_xt(\bar{V})$ is compact by Theorem 1.2 (2), Theorem 1.7.VII and Theorem 2.3 (4).II.

The converse is immediate. ∎

We shall concentrate for the remainder of this section on quotient spaces. If $H \triangleleft G$, then G/H is a group. The next theorem says that G/H is a topological group relative to the quotient topology of Chapter III, sec. 3, induced by the natural map.

Recall that the natural map $\eta: G \to G/H$ is a surjection (and an algebraic homomorphism) defined by $\eta(x) = xH$ for each $x \in G$.

4.2 Theorem *Let $H \triangleleft G$.
Then,
G/H is a topological group relative to the quotient topology I_η on G/H.*

Proof Left as an exercise. ∎

When $H \triangleleft G$ and H is open, then G/H turns out to be a discrete space. That is, $\eta^{-1}(A)$ is open in G for each subset A of G/H when H is open. The converse is also true.

4.3 Theorem *Let $H \triangleleft G$.
Then,
G/H is discrete (as a topological space)
if and only if
H is open in G.*

Proof Left as an exercise. ∎

When H is open, we know from Theorem 3.3 that H is also closed, though not conversely. A necessary and sufficient condition for a normal subgroup to be closed is that the quotient space satisfy the T_1 separation axiom (which is given in the statement of the theorem—compare this with the T_0 axiom given in Theorem 2.2).

4.4 Theorem *Let $H \triangleleft G$.*
Then,
H is closed
if and only if
if x and y are distinct points of G/H, then there are I_η-nghds. U of x and V of y such that $x \notin V$ and $y \notin U$.

Proof Left as an exercise. ∎

It is immediate that the T_1 separation axiom implies the T_0 axiom. In view of Theorems 4.4 and 2.2, we have immediately:

4.5 Theorem *Let $H \triangleleft G$.*
Let H be T-closed.
Then,
I_η is a regular topology on G/H.

Proof Left to the reader. ∎

The last two theorems of this section give sufficient conditions for the quotient group to be compact and locally compact respectively.

4.6 Theorem *Let $H \triangleleft G$.*
Let G be T-compact.
Then,
G/H is I_η-compact.

Proof Immediate from Theorem 3.1 (2).III, the fact that η is a (T, I_η) continuous surjection, and Theorem 1.7.VII. ∎

4.7 Theorem *Let G be T-locally compact.*
Let H be a closed, normal subgroup of G.
Then,
G/H is I_η-locally compact.

Proof Suppose that V is open in G. The reader can check that $\eta^{-1}(\eta(V)) = VH$.

By Theorem 1.6 (3), VH is T-open. Then, $\eta(V) \in I_\eta$, so η is a (T, I_η) continuous, open surjection. Now, by Theorem 4.4 and Theorem 2.2, G/H is Hausdorff. Hence by Theorem 2.2.VII, G/H is I_η-locally compact. ∎

9.5 Uniform continuity on topological groups

It is essential, in the development of analysis on topological groups, to have a concept of uniform continuity in this new setting. Definition 5.1 is the natural translation of Definition 3.1.VIII, from metric space to topological group terminology. Right and left uniform continuity are defined separately because the group may not be abelian. If G is abelian, then of course the two notions coincide.

5.1 DEFINITION *Let $f: G \to E^1$.*
Let $D \subset G$.
Then,
(1) f is left uniformly continuous on D
  if and only if
  given $\varepsilon > 0$, there is some nghd. V of e such that $|f(x) - f(y)| < \varepsilon$ whenever $x, y \in D$ and $x^{-1}y \in V$.
(2) f is right uniformly continuous on D
  if and only if
  given $\varepsilon > 0$, there is some nghd. V of e such that $|f(x) - f(y)| < \varepsilon$ whenever $x, y \in D$ and $yx^{-1} \in V$.

On the real line, we can write $V = \,]{-\varepsilon},\varepsilon[$, and $x^{-1}y \in V$ becomes $-x + y \in \,]{-\varepsilon},\varepsilon[$, or $|y - x| < \varepsilon$, the group operation on E^1 being ordinary addition. In this case, $yx^{-1} \in V$ also becomes $|y - x| < \varepsilon$, and the above definition reduces to the usual calculus definition.

There is a direct topological group analogue of Theorem 3.4.VIII.

For most applications, the following terminology and statement are the most convenient.

The support of a real valued function on G is the closure of the set of points in G which do not map to 0. We shall prove that a continuous, real valued function with compact support is uniformly continuous (on both sides).

5.2 DEFINITION *Let $f: G \to E^1$.*
Then,
$\mathrm{Supp}\,(f) = \overline{G \cap \{x | f(x) \neq 0\}}.$

5.1 THEOREM *Let $f: G \to E^1$ be continuous.*
Let $\mathrm{Supp}\,(f)$ be compact in G.
Then,
f is left and right uniformly continuous on G.

Proof We shall prove that f is left uniformly continuous on G. The proof is in several steps.

i) Given $\varepsilon > 0$, there is a nghd. W of e such that $|f(x) - f(xt)| < \varepsilon$ for each $x \in \text{Supp}(f)$ and $t \in W$.

To prove this, define $\zeta = f \circ \varphi \colon G \times G \to E^1$. Then ζ is continuous, hence continuous at (x, e) if $x \in \text{Supp}(f)$. Then there is a nghd. V_x of x and a nghd. $V_x(e)$ of e such that $|\zeta(y, t) - \zeta(x, e)| = |f(yt) - f(x)| < \varepsilon/2$ for $y \in V_x$, $t \in V_x(e)$.

By continuity of f, there is some nghd. U_x of x such that $|f(y) - f(x)| < \varepsilon/2$ for $y \in U_x$.

If then $y \in U_x \cap V_x$ and $t \in V_x(e)$, we have

$$|f(y) - f(yt)| \leq |f(y) - f(x)| + |f(yt) - f(x)| < \varepsilon.$$

Now, $\{U_x \cap V_x | x \in \text{Supp}(f)\}$ is an open cover of $\text{Supp}(f)$, which is compact. Then there are points $x_1, \ldots, x_n \in \text{Supp}(f)$ such that $\text{Supp}(f) \subset \bigcup_{j=1}^{n} (U_{x_j} \cap V_{x_j})$.

Let $W = \bigcap_{j=1}^{n} V_{x_j}(e)$.

Then, W is a nghd. of e. Further, if $x \in \text{Supp}(f)$ and $t \in W$, we claim that $|f(x) - f(xt)| < \varepsilon$. For, $x \in U_{x_i} \cap V_{x_i}$ for some i, $1 \leq i \leq n$. Since $t \in W$, then $t \in V_{x_i}(e)$, hence $|f(x) - f(xt)| < \varepsilon$.

ii) Given $\varepsilon > 0$, then $\{x \mid |f(x)| \geq \varepsilon\}$ is closed and $\{x \mid |f(x)| \geq \varepsilon\} \subset \text{Int}(\text{Supp}(f))$.

Immediately $\{x \mid |f(x)| \geq \varepsilon\} = G - \{x \mid |f(x)| < \varepsilon\} = G - \{x \mid -\varepsilon < f(x) < \varepsilon\} = G - f^{-1}(]-\varepsilon, \varepsilon[)$ is closed, since $f^{-1}(]-\varepsilon, \varepsilon[)$ is open by continuity of f.

It is also immediate that $\{x \mid |f(x)| \geq \varepsilon\} \subset \text{Supp}(f)$.

Finally, let $x \in G$ and $|f(x)| \geq \varepsilon$. If $x \notin \text{Int}(\text{Supp}(f))$, then $x \in \text{Bdry}(\text{Supp}(f))$. If V is a nghd. of x, there is some $y \in V$ such that $y \notin \text{Supp}(f)$. Then, $f(y) = 0$, so $|f(x) - f(y)| \geq \varepsilon$, contradicting the continuity of f at x.

iii) Given $\varepsilon > 0$, there is a nghd. M of e such that $|f(xt)| < \varepsilon$ for each $x \in G - \text{Supp}(f)$ and $t \in M$.

To prove this, let $S = \{x \mid |f(x)| \geq \varepsilon\} \cap G$. If $x \in S$, then by ii), Theorem 3.2.1, and Theorem 1.4 (1), there is a nghd. V_x of e such that $V_x^2 \subset V_x$ and $xV_x \subset \text{Supp}(f)$. Now, $\{xV_x | x \in S\}$ is an open cover of S, hence has a finite reduction by Theorem 1.8.VII. Let $x_1, \ldots, x_m \in S$ such that $S \subset \bigcup_{i=1}^{m} x_i V_{x_i}$.

Let $M = \bigcap_{i=1}^{m} V_{x_i}^{-1}$. We then have that M is a nghd. of e.

We now claim that $|f(xt)| < \varepsilon$ if $x \in G - \text{Supp}(f)$ and $t \in M$.

Let $x \in G - \text{Supp}(f)$ and $t \in M$. Suppose that $xt \in S$. Then $xt \in x_j V_{x_j}$ for some $j = 1, \ldots, m$. Then, $x \in x_j V_{x_j} t^{-1}$; but $t \in M$, so $t \in V_{x_j}^{-1}$, hence $t^{-1} \in V_{x_j}$. Then, $x \in x_j V_{x_j} t^{-1} \subset x_j V_{x_j} V_{x_j} = x_j V_{x_j}^2 \subset \text{Supp}(f)$, a contradiction.

iv) f is left uniformly continuous on G.

Choose M and W as in i) and iii) and let $V = M \cap W$.
Then, V is a nghd. of e.
Further, if $x \in G$ and $y \in G$ and $x^{-1}y \in V$, then:
if $x \in \text{Supp}(f)$, then $|f(x) - f(x(x^{-1}y))| = |f(x) - f(y)| < \varepsilon$, by i);
if $x \notin \text{Supp}(f)$, then $|f(x) - f(y)| = |f(x(x^{-1}y))| < \varepsilon$ by iii).
This proves left uniform continuity of f on G.
Right uniform continuity is proved similarly. ∎

9.6 Application: analysis on topological groups

In this section we shall indicate briefly how the modern analyst works in the topological group setting. Specifically, we shall sketch the construction of an integral on certain topological groups which mirrors the properties of the Riemann integral on E^1, and, in fact, reduces to it when E^1 is the topological group under consideration.

The problem of "generalizing" integration to topological groups was first considered for groups of matrices. An integral was developed by Hurwitz in 1897 for the multiplicative group M_n of complex, n by n, orthogonal matrices of determinant 1. (The topology of the group was obtained by embedding M_n in E^{2n^2} in the obvious way and using the relative Euclidean topology.) Peter and Weyl solved the problem for compact Lie groups (these are matrix transformation groups which arose first in connection with problems in physics). But the fundamental breakthrough was left to Haar, who in 1933 gave a solution for compact, $2°$-countable groups. Andre Weil extended Haar's result to locally compact, Hausdorff groups. It is Weil's elegant existence proof which we shall now sketch. The details of proof of the preliminary lemmas will be omitted.

We begin by restricting our attention to the set $\mathcal{K}$ of continuous f: $G \to E^1$ having compact support and such that $f(x) \geq 0$ for each $x \in G$.

Here, G is a locally compact, Hausdorff group. We seek a function $\int_G : \mathscr{K} \to E^1$ such that:

(1) $\int_G (f + g) = \int_G f + \int_G g.$

(2) $\int_G \alpha f = \alpha \int_G f.$

for α real, non-negative, $f, g \in \mathscr{K}$, $x \in G$.

(3) $\int_G f \circ {}_x t = \int_G f.$

Note that (1) and (2) reflect the usual linearity properties of integrals; (3) is left translation invariance, and is also enjoyed by the Riemann (or Lebesgue) integral. Right translation invariance could be substituted for left translation invariance, yielding a right Haar integral on G instead of the left Haar integral we shall obtain.

If $g \in \mathscr{K}$ and $g(x) \neq 0$ for some $x \in G$, write $g \not\equiv 0$.

If $f, g \in \mathscr{K}$ and $g \not\equiv 0$, there is some positive integer n, and there are positive reals $c_1, \ldots, c_n$ and elements $t_1, \ldots, t_n$ of G such that $f(x) = \sum_{j=1}^{n} c_j f(t_j x)$ for each $x \in G$. The Haar covering function $(f : g)$ of f with g is the infimum over all possible sums $\sum_{j=1}^{n} c_j$ of finitely many positive numbers $c_1, \ldots, c_n$ such that, for some $t_1, \ldots, t_n \in G$, $f(x) = \sum_{j=1}^{n} c_j f(t_j x)$ for each $x \in G$. (See Nachbin for a nice discussion of the motivation behind this and the remaining parts of the construction leading to the Haar integral.)

Properties of the Haar covering function are:

$(f : g) \geq 0$.

If $f \not\equiv 0$, then $(f : g) > 0$.

$(f : g) = ({}_y f : g)$ for each $y \in G$ $({}_y f = f \circ {}_y t)$.

If $\alpha \geq 0$, then $(\alpha f : g) = \alpha(f : g)$.

$(f + h : g) \leq (f : g) + (h : g)$.

If $h \not\equiv 0$, then $(f : h) \leq (f : g)(g : h)$.

Now fix some $f_0 \in \mathscr{K}$, with $f_0 \not\equiv 0$. If $f, g \in \mathscr{K}$ and $g \not\equiv 0$, let $I_g(f) = (f : g)/(f_0 : g)$. This is the approximate integral of f with respect to g (relative to f_0). Its properties are:

$I_g(f) \geq 0$.

$I_g(f) > 0$ if $f \not\equiv 0$.

$I_g(_y f) = I_g(f)$ for each $y \in G$.

If $\alpha \geq 0$, then $I_g(\alpha f) = \alpha I_g(f)$.

If $f \not\equiv 0$, then $1/(f_0 : f) \leq I_g(f) \leq (f : f_0)$.

If $f(x) \leq h(x)$ for each $x \in G$, then $I_g(f) \leq I_g(h)$.

$I_g(f + h) \leq I_g(f) + I_g(h)$.

Thus, $I_g(f)$ is almost the $\int_G f$ we are looking for, except that it is subadditive instead of additive. Additivity of the final integral is the delicate part, and comes from the following:

If $f, h \in \mathcal{K}$ and $f \not\equiv 0$, and $h \not\equiv 0$ and $\varepsilon > 0$, then there is a nghd. U of e such that $I_\varphi(f) + I_\varphi(h) < I_\varphi(f + h) + \varepsilon$ for each $\varphi \in \mathcal{K}$ with $\varphi \not\equiv 0$ and Supp $(\varphi) \subset U$.

With this as background, Weil's proof proceeds as follows.

Let $A = \mathcal{K} \cap \{f | f \not\equiv 0\}$.

If V is a neighborhood of e, let

$N_V = \{I_g | g \in A \text{ and Supp } (g) \subset V\}$.

Note that:

(1) $N_V \neq \phi$ for each neighborhood V of e.

(2) If $g \in A$, then $1/(f_0 : f) \leq I_g(f) \leq (f : f_0)$ for each $f \in A$. Thus,

$$I_g \in \prod_{f \in A} [1/(f_0 : f), (f : f_0)].$$

(3) If V and W are neighborhoods of e, so is $V \cap W$, and

$$N_{V \cap W} \subset N_V \cap N_W.$$

Now let

$\mathscr{F} = \{N_V | V \text{ is a nghd. of } e\}$.

Then, by the above remarks, $\mathscr{F}$ is a filterbase on $\prod_{f \in A} [1/(f_0 : f), (f : f_0)]$.

By Tychonov's Theorem, $\prod_{f \in A} [1/(f_0 : f), (f : f_0)]$ is compact. Hence $\bigcap_{a \in \mathscr{F}} \bar{a} \neq \phi$.

Let $\lambda \in \bigcap_{a \in \mathscr{F}} \bar{a}$.

Then, λ is a function on A and, if $f \in A$, then

$$1/(f_0 : f) \leq \lambda_f \leq (f : f_0).$$

Define:

$$\int_G f = \begin{cases} \lambda_f & \text{if } f \in \mathcal{K} \text{ and } f \not\equiv 0 \\ 0 & \text{if } f \equiv 0. \end{cases}$$

Using previous lemmas about the Haar covering function and the approximate integral, one can now show that:

$$\int_G f \geq 0 \text{ if } f \in \mathscr{K}.$$

If $f \not\equiv 0$, then $\int_G f > 0$.

$$\int_G {}_x f = \int_G f \text{ for each } x \in G.$$

$$\int_G (f + g) = \int_G f + \int_G g.$$

If $f(x) \leq g(x)$ for each $x \in G$, then $\int_G f \leq \int_G g.$

$$\int_G f_0 = 1.$$

Finally, using the Daniell procedure of classical integration theory (see, for example, Royden or Loomis), $\int_G$ can be extended to a wider class of functions on G.

All the machinery needed to prove the preliminary lemmas sketched above is contained in the first seven chapters and Chapter Nine, although considerable ingenuity is required in constructing the arguments. Hewitt and Ross (Vol. I) have a very complete treatment of a variation of Weil's argument which uses nets instead of filterbases. Other references are Nachbin, Husain, and Montgomery and Zippin.

The reader who wishes to pursue the subject of harmonic analysis can read Hewitt and Ross, Loomis, Katznelson, Reiter, or Bachman.

Problems

1) Consider $[0, 1]$ as a subspace of E^1. Show that there is no group operation on $[0, 1]$ with respect to which $[0, 1]$ becomes a topological group.

2) Let $f: G \to G'$ be a homomorphism which is continuous at e. Then, f is continuous.

3) E^1 has no nontrivial, compact subgroups.

4) The center $C(G)$ of a group G is the set
$C(G) = \{x|xy = yx \text{ for each } y \in G\} \cap G$.
Show that the center of a Hausdorff group is a closed, normal subgroup.

5) Show that $\overline{\{e\}}$ is a closed, normal subgroup of the topological group G.

6) A subgroup H of a topological group G is discrete if and only if H has an isolated point.

7) Let G be a topological group and H a normal subgroup. Then, G/H is Hausdorff if and only if H is closed in G.

8) Let H be a closed normal subgroup of G. Then, G/H is 1°-(2°-)countable if G is 1°-(2°-)countable.

9) A topological group G is Hausdorff if and only if $\{e\}$ is closed if and only if $\cap\{V|V \text{ is a nghd. of } e\} = \{e\}$.

10) Let H be a normal subgroup of G. If H and G/H are compact, then G is compact.

11) Let (X, T) be a locally compact, Hausdorff space. Suppose that $(X, \cdot)$ is a group, and φ is continuous. Then, $(X, \cdot, T)$ is a topological group.

12) Suppose, if $\alpha \in \mathcal{O}$, G_α is a group. The direct product group $\prod_{\alpha \in \mathcal{O}} G_\alpha$ is the group consisting of the set $\prod_{\alpha \in \mathcal{O}} G_\alpha$ and the operation $\{x_\alpha\}_{\alpha \in \mathcal{O}} \{y_\alpha\}_{\alpha \in \mathcal{O}} = \{x_\alpha y_\alpha\}_{\alpha \in \mathcal{O}}$.

   (a) $\prod_{\alpha \in \mathcal{O}} G_\alpha$ is a topological group relative to the usual product topology.

   (b) The projections $p_\beta: \prod_{\alpha \in \mathcal{O}} G_\alpha \to G_\beta$ are continuous, open homomorphisms.

   (c) Let $H = \left(\prod_{\alpha \in \mathcal{O}} G_\alpha\right) \cap \{x|x_\alpha = e_\alpha \text{ for all but at most finitely many } \alpha\}$. Then, H is a dense, normal subgroup of $\prod_{\alpha \in \mathcal{O}} G_\alpha$.

*13) Each topological group is a uniform space (see Problem 27, Chapter VIII). Hint: Let $\mathcal{B}$ consist of all subsets of G containing an open nghd. of e. If $A \in \mathcal{B}$, let $A_r = \{(x, y)|y \in xA\} \cap (G \times G)$. Then, $\mathcal{U} = \{A_r|A \in \mathcal{B}\}$ is a uniformity on G and $T(\mathcal{U})$ is the original topology on G.

Further, show that, if G is locally compact and Hausdorff, then G is complete relative to the uniform topology $T(\mathcal{U})$ on G.

14) Let $\mathscr{I} = \{f | f$ is an isometry: $S^n \to S^n\}$. If $f, g \in \mathscr{I}$, let $d(f, g) = \sup \{\varrho_{n+1}(f(x), g(x)) | x \in S^n\}$. Then, d is a metric on $\mathscr{I}$, which is a group under composition $\circ$. Show that $\langle \mathscr{I}, \circ, T_d \rangle$ is a topological group.

15) Let H_n be the set of non-singular, n by n, real matrices, considered as a group under multiplication. Give H_n a topology as follows. Let X be the subset of E^{n^2} consisting of n^2-tuples

with
$$(x_{11}, \ldots, x_{1n}, x_{21}, \ldots, x_{2n}, \ldots, x_{n1}, \ldots, x_{nn}),$$

$$\det \begin{pmatrix} x_{11} & \cdots & x_{1n} \\ \cdots & \cdots & \cdots \\ x_{n1} & \cdots & x_{nn} \end{pmatrix} \neq 0.$$

Define a bijection $\varphi: X \to G$ in the obvious way:

$$\varphi(x_{11}, \ldots, x_{1n}, \ldots, x_{n1}, \ldots, x_{nn}) = \begin{pmatrix} x_{11} & \cdots & x_{1n} \\ \cdots & \cdots & \cdots \\ x_{n1} & \cdots & x_{nn} \end{pmatrix}.$$

Give H_n the identification topology induced by φ. Show that H_n is then a topological group.

16) Let G be the multiplicative group of 2 by 2 real, non-singular matrices. Topologize G as in Problem 15. That is, let

$$X = E^4 \cap \{(x_1, x_2, x_3, x_4) | x_1 x_4 - x_2 x_3 \neq 0\}.$$

Consider X as a subspace of E^4, map $\varphi: X \to G$ by:

$$\varphi(x_1, x_2, x_3, x_4) = \begin{pmatrix} x_1 & x_2 \\ x_3 & x_4 \end{pmatrix}$$

and let G have the identification topology I_φ.
Note that φ is an homeomorphism and an algebraic isomorphism of X onto G, where X is considered as a group under the product

$(x_1, x_2, x_3, x_4)(y_1, y_2, y_3, y_4)$
$= (x_1 y_1 + x_2 y_3, x_1 y_2 + x_2 y_4, x_3 y_1 + x_4 y_3, x_3 y_2 + x_4 y_4).$

Thus, to construct a Haar integral for G, it suffices to construct one on X.

If $f: X \to E^1$ is continuous with compact support, let

$$\int_X f = \iiiint_X \frac{f(x_1, x_2, x_3, x_4)}{(x_1 x_4 - x_2 x_3)^2} dx_1\, dx_2\, dx_3\, dx_4,$$

the integral on the right being an ordinary multiple integral over the region X of E^4.

Prove that $\int_X$ is left translation invariant. (Hint: use the calculus theorem on transformation of multiple integrals, and show that the Jacobian of the transformation $_a t$ is $(a_1 a_4 - a_2 a_3)^2$, where $a = (a_1, a_2, a_3, a_4)$.)

In this case, $\int_X$ is also a right Haar integral over X.

17) Let G be as in (16), and let H be the subgroup consisting of all real matrices $\begin{pmatrix} x & y \\ 0 & 1 \end{pmatrix}$ with $x \neq 0$. Embed H, both algebraically and topologically, in E^2 minus the y-axis by mapping $\begin{pmatrix} x & y \\ 0 & 1 \end{pmatrix} \to (x, y)$, $x \neq 0$. Thus, consider H as $M = E^2 - \{(0, y)\,|\,y \text{ is real}\}$. We can now define a left and right Haar integral over H in terms of a double integral over M.

Show that a left Haar integral is given by:

$$\int_{(L)M} f = \iint_M \frac{f(x_1, x_2)\, dx_1\, dx_2}{x_1^2}$$

and a right Haar integral by

$$\int_{(R)M} f = \iint_M \frac{f(x_1, x_2)}{|x_1|} dx_1\, dx_2$$

for $f: M \to R$ continuous with compact support.
In this case the left and right Haar integrals do not agree. This example is due to von Neumann.

CHAPTER X

The concept of homotopy

10.1 Introduction

THE FIRST SEVEN chapters of this book were devoted exclusively to set topology. Chapter Eight dealt with metric spaces, with some applications to analysis. In Chapter Nine we developed some of the theory of topological groups, with a brief introduction to the Haar integral and harmonic analysis.

We shall now begin to turn toward algebraic topology, in which one associates groups, or sequences of groups, with topological spaces in such a way that topological problems may be reformulated algebraically. The strategy in this is the hope that the algebraic problems thus encountered may have already been solved, or at least may be easier to solve than the original topological problems, and in fact this has been the case in a number of instances. The algebraic point of view has also led to the development of new and powerful techniques for handling topological problems.

In Chapter Eleven we shall introduce the homotopy groups of a topological space. However, while it is possible to do this immediately, it would also be in a sense misleading, as homotopy plays a considerable role in set topology irrespective of the homotopy groups of algebraic topology. We shall therefore begin with homotopy of maps in general.

10.2 Homotopy

For the remainder of this chapter, let (X, T) and (Y, M) be spaces. As usual, Y^X denotes the set of all (T, M) continuous functions on X to Y. Elements of Y^X are also often called maps, the term map in this context always carrying with it the connotation of continuity. Finally, let $I = [0, 1]$, considered as a subspace of E^1.

2.1 DEFINITION *Let $f, g \in Y^X$.*
Then,
f is homotopic to g
if and only if

there is a continuous $\Phi: X \times I \to Y$ such that, for each $x \in X$, $\Phi(x, 0) = f(x)$ and $\Phi(x, 1) = g(x)$.

We then call Φ a homotopy from f to g.

When f is homotopic to g, we shall write $f \simeq g$. And, if Φ is a homotopy from f to g, we write $\Phi: f \simeq g$.

Often, we let $\Phi_t: X \to Y$ be defined by $\Phi_t(x) = \Phi(x, t)$ for $x \in X$. Φ_t is continuous, as Φ is.

If you think of f and g as subsets of $X \times Y$, then $f \simeq g$ iff f can be deformed continuously into g in $X \times Y$ through a sequence of intermediary maps, Φ_t being the tth stage of the deformation. Of course, the initial stage ($t = 0$) is f, and the final stage ($t = 1$), g.

2.1 EXAMPLE In the more familiar spaces, homotopy is particularly easy to visualize geometrically.

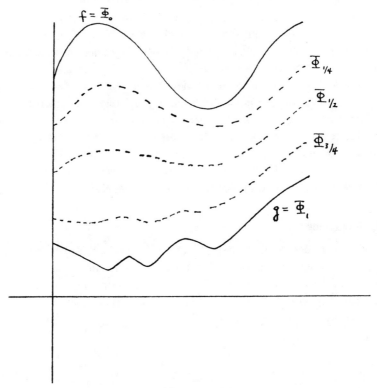

FIGURE 2.1

THE CONCEPT OF HOMOTOPY

For example, let $X = [0, 2]$, $Y = E^1$, and $f, g\colon [0, 2] \to E^1$. Typical graphs are shown in Figure 2.1. If $f \simeq g$, then we may fill in intermediary maps Φ_t which gradually change from f to g as t varies from 0 to 1.

To give an explicit example, let $f(x) = x^2$ and $g(x) = 5x$, $0 \leq x \leq 2$. Define $\Phi(x, t) = 5xt + (1 - t)x^2$ for $0 \leq x \leq 2$ and $0 \leq t \leq 1$. Then, is a homotopy from f to g, the graphs of $\Phi_{1/3}$, $\Phi_{1/2}$, and $\Phi_{3/4}$ being shown in Figure 2.2. In this case, the scheme for defining the homotopy is very simple. For each x in $[0, 2]$, the y-coordinates on the straight line segment from (x, x^2) to $(x, 5x)$ are given by $t(5x) + (1 - t)x^2$, $0 \leq t \leq 1$. Φ conveys the operation of pulling f up onto g. ■

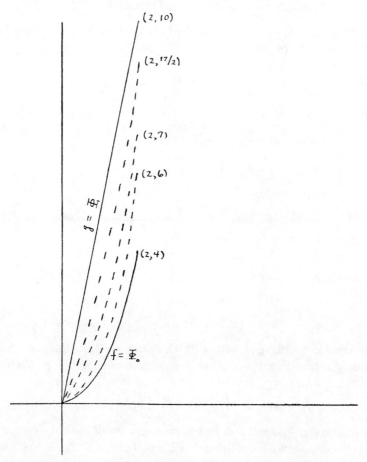

FIGURE 2.2

Homotopy arises quite naturally in considering the problem of extending continuous functions. Given maps f, $g: X \to Y$, let $\varphi(x, 0) = f(x)$ and $\varphi(x, 1) = g(x)$ for each $x \in X$. Then, $\varphi: (X \times \{0\}) \cup (X \times \{1\}) \to Y$, and φ is continuous. Further, $f \simeq g$ exactly when φ can be extended to a map Φ $X \times I \to Y$. We shall consider some extension problems and other applications of homotopy theory after determining some of its basic properties.

Our first theorem says that $\simeq$ is an equivalence relation on Y^X. Before proving this, however, we shall insert a Lemma which will be used very often throughout this and the next chapter. As will be seen, homotopy theory often requires that we define a new map by pasting together several given maps defined on subspaces. Lemma 1 will allow us to do this rather automatically in the situations which will arise.

1 LEMMA *Let $A_1, \ldots, A_n$ be closed subsets of X such that $\bigcup_{i=1}^{n} A_i = X$.*
Let $f_i: A_i \to Y$ be continuous for $i = 1, \ldots, n$.
Suppose that $f_i|A_i \cap A_j = f_j|A_i \cap A_j$ for $i, j = 1, \ldots, n$.
Let $F(x) = f_i(x)$ for each $x \in A_i$, $i = 1, \ldots, n$.
Then,
$F: X \to Y$ is continuous.

Proof Left as an exercise. ∎

2.1 THEOREM $\simeq$ *is an equivalence relation on Y^X.*

Proof We proceed in three steps.

1) Let $f \in Y^X$. Let $\Phi(x, t) = f(x)$ for $x \in X$, $0 \leq t \leq 1$. Then, $\Phi: f \simeq f$.

2) Suppose $\Phi: f \simeq g$.
Let $\Psi(x, t) = \Phi(x, 1 - t)$ for $x \in X$ and $0 \leq t \leq 1$. Then, $\Psi: g \simeq f$.

3) Suppose $\Phi: f \simeq g$ and $\Psi: g \simeq h$. Let

$$\Theta(x, t) = \begin{cases} \Phi(x, 2t) \text{ for } 0 \leq t \leq \tfrac{1}{2} \text{ and } x \in X, \\ \Psi(x, 2t - 1) \text{ for } \tfrac{1}{2} \leq t \leq 1 \text{ and } x \in X. \end{cases}$$

By Lemma 1 (letting $A_1 = X \times [0, \tfrac{1}{2}]$, $f_1(x, t) = \Phi(x, 2t)$, $A_2 = X \times [\tfrac{1}{2}, 1]$, and $f_2(x, t) = \Psi(x, 2t - 1)$), Θ is continuous: $X \times I \to Y$. Finally, for each $x \in X$,

$$\Theta(x, 0) = \Phi(x, 0) = f(x) \text{ and } \Theta(x, 1) = \Psi(x, 1) = h(x). \blacksquare$$

In the future, we shall usually claim continuity of the constructed function (Θ above) without explicit reference to Lemma 1. The student should always be clear, however, that the continuity claims are well founded.

2.2 Theorem *Let $f, g \in Y^X$.*
Let $F, G \in S^Y$, where (S, N) is a space.
Suppose that $f \simeq g$ and $F \simeq G$.
Then,
$F \circ f \simeq G \circ g$.

Proof Let $\Phi: f \simeq g$ and $\Psi: F \simeq G$.
Let
$$\Theta(x, t) = \begin{cases} F \circ \Phi(x, 2t) \text{ for } 0 \leq t \leq \tfrac{1}{2} \text{ and } x \in X, \\ \Psi(g(x), 2t - 1) \text{ for } \tfrac{1}{2} \leq t \leq 1 \text{ and } x \in X. \end{cases}$$
It is easy to check that $\Theta: F \circ f \simeq G \circ g$. ∎

2.3 Theorem *Let $f, g \in Y^X$.*
Let $f \simeq g$.
Let $A \subset X$.
Then,
$f|A \simeq g|A$.

Proof Left as an exercise. ∎

The converse of the last theorem is not true. For example, let $X = I$, $Y = E^2 - \{(0, 0)\}$, and $f(x) = (\cos(2\pi x), \sin(2\pi x))$, $g(x) = (1, 0)$, for $x \in I$. Let $A = \{0\}$. Then, $f|A = g|A$, so $f|A \simeq g|A$. At least intuitively, however, f and g are not homotopic, as the circle cannot be continuously shrunk to a point (constant map) without passing over the origin, which is punched out of the space Y. In fact, let $\Phi(t, s) = s(1, 0) + (1 - s)(\cos(2\pi t), \sin(2\pi t))$ for $0 \leq t \leq 1$, $0 \leq s \leq 1$. Then, $\Phi: I \times I \to E^2$ and $\Phi_0 = f$ and $\Phi_1 = g$. Thus $f \simeq g$ if we think of f and g as maps of I into E^2. However, Φ fails to be a homotopy from f to g if f and g are thought of as maps of I to E^2 – the origin, since $\Phi(\tfrac{1}{2}, 0) = (0, 0)$.

A partial converse to Theorem 2.3 can be achieved through the notion of a deformation.

2.2 Definition *Let $A \subset X$.*
Then,
X is deformable to A
if and only if
there is a continuous $\varphi: X \to A$ such that $\varphi \simeq 1_X$.

Here, 1_X is the identity map on X, given by $1_X(x) = x$ for each $x \in X$. A homotopy Φ of 1_X with a continuous map of X into A is sometimes called a deformation of X to A.

Geometrically, X is deformable to A if X can be continuously compressed into A. For example, the real line can be compressed to the origin, a deformation of E^1 to $\{0\}$ being given by $\Phi(x, t) = (1 - t) x$, $x \in E^1$ and $0 \leq t \leq 1$.

2.2 Example The unit ball $B_2 = \{(x, y) | x^2 + y^2 \leq 1\}$ in E^2 is deformable to the origin. Let $\Phi((x, y), t) = ((1 - t) x, (1 - t) y)$ for $0 \leq t \leq 1$, $(x, y) \in B_2$. At each t, Φ_t maps B_2 to a ball of radius $1 - t$, compressing B_2 to $(0, 0)$ as $t \to 1$. ∎

2.3 Example $E^2 - \{(0, 0)\}$ is deformable to S^1. Let

$$\Phi((x, y), t) = (1 - t)(x, y) + \frac{t(x, y)}{\sqrt{x^2 + y^2}}$$

for $0 \leq t \leq 1$, $(x, y) \in E^2 - (0, 0)\}$. ∎

2.4 Theorem *Let X be deformable to A. Let $f, g \in Y^X$ and suppose that $f|A \simeq g|A$. Then,*
$f \simeq g$.

Proof Let $\varphi: X \to A$ be continuous and let $\Phi: X \times I \to X$ be a homotopy from 1_X to φ. Now, by Theorem 2.2, $f \circ \varphi \simeq g \circ \varphi$.

Let $\Psi(x, t) = f \circ \Phi(x, t)$ for $0 \leq t \leq 1$ and $x \in X$. Then, Ψ is continuous: $X \times I \to Y$, and $\Psi(x, 0) = f(\Phi(x, 0)) = f(1_X(x)) = f(x)$, and $\Psi(x, 1) = f(\Phi(x, 1)) = f(\varphi(x))$. Thus, Ψ is a homotopy from f to $f \circ \varphi$.

Let $\Theta(x, t) = g \circ \Phi(x, t)$ for $0 \leq t \leq 1$ and $x \in X$. Then, $\Theta: g \simeq g \circ \varphi$, by an argument similar to that just used.

We then have
$$f \simeq f \circ \varphi \simeq g \circ \varphi \simeq g,$$
hence $f \simeq g$. ∎

We conclude this section with a theorem on maps into spheres. In this connection, it is customary to call points x and y of S^n antipodal if $x = -y$, that is, if x and y are at opposite ends of the intersection of S^n with a straight line through the origin.

2.5 Theorem *Let f, g be continuous: $X \to S^n$. Suppose that $f(x)$ and $g(x)$ are not antipodal for each $x \in X$. Then,*
$f \simeq g$.

Proof Note first that, for any $x \in X$ and $t \in I$, $tg(x) + (1 - t)f(x) = 0$ implies $tg(x) = (t - 1)f(x)$ implies $t|g(x)| = (1 - t)|f(x)|$ implies $t = 1 - t$ implies $t = 1/2$ implies $g(x) = -f(x)$, a contradiction. Thus, $tg(x) + (1 - t) \times f(x) \neq 0$ for $t \in I$ and $x \in X$. Let

$$\Phi(x, t) = \frac{tg(x) + (1 - t)f(x)}{|tg(x) + (1 - t)f(x)|} \quad \text{for } x \in X \text{ and } t \in I.$$

It is now routine to check that Φ is a homotopy from f to g. ∎

10.3 Nullhomotopy

It is usually surprising to the beginning student that maps which are nullhomotopic (homotopic to a constant map) are particularly interesting. In this and the next section we shall investigate this notion.

For convenience, if $y \in Y$, denote by c_y the constant map $X \to \{y\}$.

3.1 DEFINITION *Let $f \in Y^X$.*
Then,
f is nullhomotopic
if and only if
for some $y \in Y$, $f \simeq c_y$.

The statement "f is nullhomotopic" is customarily abbreviated $f \simeq 0$. Be careful with this symbol! If $f \simeq 0$ and $g \simeq 0$, it does not follow that $f \simeq g$ ($f \simeq 0$ is a symbol, and Theorem 2.1 does not apply). This is intuitively clear with a little thought. If $f \simeq c_a$ and $g \simeq c_b$, then $f \simeq g$ exactly when $c_a \simeq c_b$. But c_a cannot always be twisted to c_b. In fact, this can be done exactly when a and b can be joined by a path in Y.

3.1 THEOREM *Let $a, b \in Y$.*
Then,
$c_a \simeq c_b$
if and only if
a and b lie in the same path component of Y.

Proof Suppose first that $P \text{ Comp}(a) = P \text{ Comp}(b)$. Let $\gamma: I \to Y$ be a path from a to b in Y. Let $\Phi(x, t) = \gamma(t)$ for each $x \in X$, $t \in I$. Then, $\Phi: c_a \simeq c_b$.

Conversely, suppose $c_a \simeq c_b$. Let Φ be a homotopy from c_a to c_b. Choose any $x \in X$, and let $\gamma: I \to Y$ be defined by $\gamma(t) = \Phi(x, t)$ for $0 \leq t \leq 1$. Then, Φ is a path from a to b in Y. ∎

An immediate consequence of Theorem 3.1 is that all nullhomotppic maps to a path connected space are themselves homotopic.

Nullhomotopy comes rather easily to maps into and out of E^n. More generally, a non-empty subset A of E^n is called convex if the points on the straight line between any two elements of A are again in A. That is, $\{tx + (1 - t)y | t \in I\} \subset A$ for each $x, y \in A$. Of course, E^n is convex, as is the interior of an n-sphere. It is easy to conjure up examples of non-convex sets (e.g. S^1 in E^2). We now show that maps into and out of convex sets are nullhomotopic.

3.2 Theorem *Let A be a convex subset of E^n.*
Let $f \in A^X$, where (X, T) is any space.
Then,
$f \simeq 0$.

Proof Let $a \in A$. Let $\Phi(x, t) = ta + (1 - t)f(x)$ for $x \in X$ and $0 \leq t \leq 1$. Then, Φ is a homotopy from f to the constant map: $X \to \{a\}$. ∎

3.3 Theorem *Let A be a convex subset of E^n.*
Let $f \in X^A$, where (X, T) is any space.
Then,
$f \simeq 0$.

Proof Choose any $a_0 \in A$, and define $\Phi(a, t) = f(ta_0 + (1 - t)a)$ for $a \in A$ and $0 \leq t \leq 1$. Then, Φ is a homotopy from f to the constant map $A \to \{f(a_0)\}$. ∎

In particular, for any space (X, T), any continuous $f: X \to E^n$ is nullhomotopic, and any continuous $f: E^n \to X$ is nullhomotopic.

As we shall later be concerned with maps into spheres, it is natural to ask whether all maps into S^n are also nullhomotopic. The answer is no. For example, let $f: I \to S^1$ be given by $f(x) = (\cos(2\pi x), \sin(2\pi x))$ for $0 \leq x \leq 1$. Then, f is not nullhomotopic. The reader might try showing this directly now. It will follow easily from later work.

However, non-surjective maps into S^n are always nullhomotopic.

3.4 Theorem *Let (X, T) be any space.*
Let $f: X \to S^n$ be continuous.
Suppose that f is not surjective.
Then,
$f \simeq 0$.

Proof Shoose any $a \in S^n - f(X)$. Then, f and the constant map d: $X \to \{-a\}$ are never antipodal, hence $f \simeq d$ by Theorem 2.5. ∎

Nullhomotopy can yield surprising information about maps and spaces. For example, a famous theorem due to K. Borsuk says that a compact subset A of E^n ($n \geq 2$) separates E^n (i.e. $E^n - A$ is disconnected) exactly when there is a nonnullhomotopic map of A to S^{n-1}. The proof of this would take us rather far afield. However, we shall prove two results on nullhomotopy and extensions of maps.

Recall that B_{n+1} denotes the unit ball $\{x \mid |x| \leq 1\}$ in E^{n+1}.

3.5 THEOREM *Let f be continuous: $S^n \to X$, where (X, T) is any space. Then,*
f has a continuous extension to B_{n+1}
if and only if
$f \simeq 0$.

Proof Suppose first that there is a continuous extension F of f to B_{n+1}. Let $\Phi(s, t) = F((1 - t)s)$ for $s \in S^n$, $t \in I$. Then Φ is a homotopy of f to the constant map: $S^n \to \{F(0)\}$.

Conversely, suppose that $f \simeq 0$. Let $x_0 \in X$ and $\Phi: f \simeq d$, where $d: S^n \to \{x_0\}$. If $y \in B_{n+1}$, write $y = |y| z_y$, where $z_y \in S^n$ is chosen on the straight line from the origin through y, if $y \neq 0$, and $z_0 = 0$. Let $F(y) = \Phi(z_y, 1 - |y|)$ for each $y \in B_{n+1}$. Then, F is continuous: $B_{n+1} \to X$, and $F|S^n = f$. ∎

The second extension theorem requires two preliminary Lemmas. The first is stated more generally than necessary for now in anticipation of a later use.

1 LEMMA *Let (X, T) be normal.*
Let A be a closed subset of X.
Let f be continuous: $A \to S^n$.
Then,
there is a neighborhood U of A such that f has a continuous extension to U.

Proof Note that $f: A \to S^n \subset E^{n+1}$. Further, S^n is closed in E^{n+1}, so $f: A \to E^{n+1}$ is continuous.

Now think of E^{n+1} as $\prod_{i=1}^{n+1} E_i$, where $E_i = E^1$ for $i = 1, \ldots, n+1$. Let $p_i: \prod_{i=1}^{n+1} E_i \to E^1$ be the projection map. Then $p_i \circ f: A \to E^1$. Since X is normal, then by Tietze's Theorem $p_i \circ f$ has a continuous extension g_i:

$X \to E^1$. Let $G(x) = (g_1(x), \ldots, g_{n+1}(x))$ for $x \in X$. Then G is a continuous extension of f to a map: $X \to E^{n+1}$.

Now let $U = \{x | G(x) \neq 0\} \cap X$. Then, U is open in X and $A \subset U$. Let $F(x) = \dfrac{G(x)}{|G(x)|}$ for $x \in U$. Then, $F: U \to S^n$ is continuous, and $F|A = f$. ∎

2 LEMMA *Let $A \subset E^m$ and let U be a neighborhood of $A \times I$ in $E^m \times I$. Then, there is a neighborhood V of A such that $V \times I \subset U$.*

Proof Let $p: E^m \times I \to E^m$ be the projection map. We first show that $p(F)$ is closed in E^m if F is closed in $E^m \times I$.

Let $a \in E^m - p(F)$. Then, $(\{a\} \times I) \cap F = \phi$. Since F is closed, there is, for each $t \in I$, a basic neighborhood $U_t(a) \times V_t$ of (a, t) in $E^m \times I$ such that $(U_t(a) \times V_t) \cap F = \phi$.

Now, $\{U_t(a) \times V_t | t \in I\}$ is an open cover of the compact set $\{a\} \times I$. Hence, there are elements $t_1, \ldots, t_r$ of I such that

$$\{a\} \times I \subset \bigcup_{i=1}^{r} (U_{t_i}(a) \times V_{t_i}).$$

Now, $\bigcap_{i=1}^{r} U_{t_i}(a)$ is a neighborhood of a in E^m. But, $\left(\bigcap_{i=1}^{r} U_{t_i}(a)\right) \cap p(F) = \phi$. For suppose that $((b_1, \ldots, b_m), t_0) \in F$ such that $p((b_1, \ldots, b_m), t_0) = (b_1, \ldots, b_m) \in \bigcap_{i=1}^{r} U_{t_i}(a)$. Now, $(a, t_0) \in U_{t_j}(a) \times V_{t_j}$ for some j, $1 \leq j \leq r$. Since $(b_1, \ldots, b_m) \in U_{t_j}(a)$, then $((b_1, \ldots, b_m), t_0) \in (U_{t_j}(a) \times V_{t_j}) \cap F$, a contradiction.

Hence, $a \in \bigcap_{i=1}^{r} U_{t_i}(a) \subset E^m - p(F)$, so that $E^m - p(F)$ is open, and $p(F)$ is closed.

Now choose $V = E^m - p((E^m \times I) - U)$.

Then, V is open in E^m, as $(E^m \times I) - U$ is closed in $E^m \times I$.

Finally,

$$V \times I = p^{-1}(V) = (E^m \times I) - p^{-1}(p((E^m \times I) - U))$$

$$\subset (E^m \times I) - ((E^m \times I) - U) = U. \quad \blacksquare$$

We can now prove Borsuk's extension theorem.

3.6 THEOREM *Let A be a closed subset of E^m.*
Let $f: A \to S^n$.
Then,
f has a continuous extension $F: E^m \to S^n$
if and only if
$f \simeq 0$.

Proof Suppose first that $f \simeq 0$. Let Φ be a homotopy: $d \simeq f$, where $d: A \to \{y\}$ for some $y \in S^n$.

Let $\Psi(x, 0) = y$ for $x \in E^m$,
$\Psi(a, t) = \Phi(a, t)$ for $a \in A$, $0 \leq t \leq 1$.

Then, Ψ is continuous: $(E^m \times \{0\}) \cup (A \times I) \to S^n$.

Now, $(E^m \times \{0\}) \cup (A \times I)$ is a closed subset of E^{m+1}. By Lemma 1, there is a neighborhood U of $(E^m \times \{0\}) \cup (A \times I)$ such that Ψ has a continuous extension $\Theta: U \to S^n$.

But, U is also a neighborhood of $A \times I$. By Lemma 2, then, there is a neighborhood V of A such that $V \times I \subset U$.

By Urysohn's Lemma, there is a continuous $g: E^m \to I$ such that $g(A) = 1$ and $g(E^m - V) = 0$.

Let $v(x, t) = \Theta(x, tg(x))$ for $x \in E^m$ and $0 \leq t \leq 1$.

Then, v is continuous: $E^m \times I \to S^n$. Further, $v|(A \times I) = \Phi$.

Finally, let $F(x) = v(x, 1)$ for $x \in E^m$. Then, F is continuous: $E^m \to S^n$ and $F|A = f$.

Conversely, suppose f has a continuous extension $F: E^m \to S^n$. Then, $F \simeq 0$ by Theorem 3.3, so that $F|A = f \simeq 0$ by Theorem 2.3. ∎

We shall conclude this section with a characterization by S. Eilenberg of nullhomotopic maps into S^1. The proof uses two very simple Lemmas. In the notation, we identify the ordered pair (a, b) in E^2 with the complex number $a + ib$, and $(a, 0)$ with a, as customary.

3 LEMMA *Let $f, g: X \to S^1$, and suppose f and g are continuous.*
Suppose $f(x)$ and $g(x)$ are not antipodal for each $x \in X$.
Suppose that there is a continuous $\varphi: X \to E^1$ such that $f = e^{i\varphi}$.
Then,
there is a continuous $\psi: X \to E^1$ such that $g = e^{i\psi}$.

Proof Note that, if $x \in X$, then $g(x)/f(x) \neq -1$, and $g(x)/f(x) \in S^1$. Choose $\theta(x)$ as the length of the arc on S^1 from 1 to $g(x)/f(x)$ not passing through

-1. Then, θ is continuous: $X \to E^1$, and $g(x)/f(x) = e^{i\theta(x)}$ for each $x \in X$. Then, $g(x) = e^{i\theta(x)}e^{i\varphi(x)}$, and we may choose $\psi = \theta + \varphi$. ■

The choice of $\theta(x)$ is illustrated in Figure 3.1. Although $\theta(x)$ was described geometrically in the proof, the reader who wishes a more analytic description can use a suitably defined argument function from elementary complex analysis (say $\theta(x) = \arg(g(x)/f(x))$, where $-\pi \leq \arg(z) < \pi$).

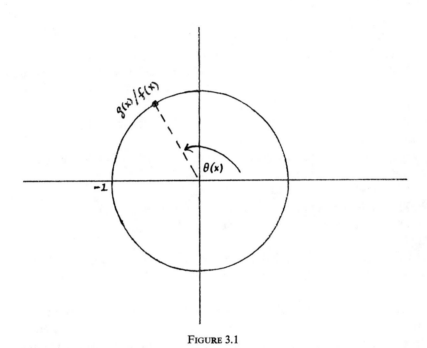

FIGURE 3.1

4 LEMMA *Let $f_0, ..., f_n$ be continuous: $X \to S^1$.*
Suppose $f_j(x)$ and $f_{j+1}(x)$ are not antipodal for each $x \in X$, if $0 \leq j \leq n-1$. Let φ be continuous: $X \to E^1$ such that $f_0 = e^{i\varphi}$.
Then,
there is a continuous $\psi: X \to E^1$ such that $f_n = e^{i\psi}$.

Proof A straightforward induction, using Lemma 3. ■

3.7 Theorem *Let (X, ϱ) be a compact metric space.*
Let f be continuous: $X \to S^1$.
Then,
$f \simeq 0$
if and only if
there is a continuous $\varphi: X \to E^1$ such that $f = e^{i\varphi}$.

Proof Suppose first that $\varphi: X \to E^1$ is continuous, and $f = e^{i\varphi}$. Let $\Phi(x, t) = e^{i(1-t)\varphi(x)}$ for $x \in X$, $0 \leq t \leq 1$. Then, Φ is a homotopy from f to the constant map: $X \to \{(1, 0)\}$.

Conversely, suppose that $f \simeq 0$.

Let $y \in S^1$ and let Φ be a homotopy from c_y to f. Note that, if $y = a + ib$, then by choosing $\alpha(x) = \arg(y)$ for each $x \in X$, we have $c_y = e^{i\alpha}$. Further, $\alpha: X \to E^1$ is continuous.

Now, (X, ϱ) is a compact metric space, as is I, so $X \times I$ is a compact metric space. Hence Φ is uniformly continuous. Choose some $\delta > 0$ such that $|\Phi(x, t) - \Phi(x', t')| < 1$ whenever $x, x' \in X$, $t, t' \in I$, and $|t - t'| < \delta$ and $\varrho(x, x') < \delta$. Next, choose points $t_0, \ldots, t_n$ in I with $0 = t_0 < t_1 \ldots < t_n = 1$ and $t_j - t_{j-1} < \delta$ for $j = 1, \ldots, n$. Then, $|\Phi_{t_{j+1}}(x) - \Phi_{t_j}(x)| < 1$ for $j = 0, \ldots, n - 1$ and $x \in X$. Then, $\Phi_{t_{j+1}}$ are not antipodal for each $x \in X$.

By Lemma 4, $\Phi_{t_n} = f = e^{i\varphi}$ for some continuous $\varphi: X \to E^1$. ∎

10.4 Brouwer's theorem for S^1

In this section we shall use homotopy to prove a classical theorem due to L. E. J. Brouwer. In the next chapter we shall give two algebraic proofs based on homotopy groups. Comparison of the different techniques involved provides an interesting contrast.

We begin by showing the equivalence of several formulations of the theorem.

4.1 Theorem *The following three statements are equivalent.*

(1) 1_{S^1} is not nullhomotopic.
(2) There is no continuous $f: B_2 \to S^1$ such that $f|S^1 = 1_{S^1}$.
(3) Each continuous $f: B_2 \to B_2$ has a fixed point.

Proof (1) ⇒ (2):

Assume (1). Suppose f is continuous: $B_2 \to S^1$ and $f|S^1 = 1_{S^1}$. Then, 1_{S^1} has a continuous extension to B_2, hence $1_{S^1} \simeq 0$ by Theorem 3.5.

(2) ⇒ (3):

Assume (2). Suppose $f: B_2 \to B_2$ is continuous. If f has no fixed point, then $f(x) \neq x$ for each $x \in B_2$. Define $g: B_2 \to S^1$ by specifying, for each $x \in B_2$, $g(x)$ to be the point of intersection of S^1 with the line from $f(x)$ through x. Then, g is continuous, and $g|S^1 = 1_{S^1}$.

(3) ⇒ (1):

Assume (3). Suppose that 1_{S^1} is nullhomotopic. Then, 1_{S^1} extends to a continuous map $f: B_2 \to S^1$ by Theorem 3.5. Let $g(x) = -f(x)$ for each $x \in B_2$. Then, g has no fixed point. For, if $y \in B_2$ and $g(y) = y$, then $f(y) = -y \in S^1$, so $y \in S^1$ also. But then $-y = f(y) = 1_{S^1}(y) = y$, which is impossible. ∎

Usually Theorem 4.1 (1) is called Brouwer's Theorem, and (3), Brouwer's Fixed Point Theorem. In connection with (2), if $A \subset X$, a continuous $r: X \to A$ such that $r|A = 1_A$ is called a retraction, and, if such r exists, A is a retract of X. (See Problem 4, Ch. III, and Prob. 8, Ch. V.) Then (2) is often phrased: S^1 is not a retract of B_2.

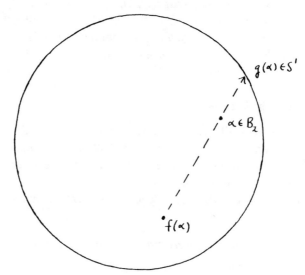

FIGURE 4.1 Brouwer's Theorem: If we write $f(x, y) = (u(x, y), v(x, y))$, for $(x, y) \in B_2$, then $g(x, y)$ is given analytically by:

$$g(x, y) = \left(\frac{D(y - v) + (x - u)\sqrt{d^2 - D^2}}{d^2},\ \frac{-D(x - u) + (y - v)\sqrt{d^2 - D^2}}{d^2} \right)$$

where $d^2 = (x - u)^2 + (y - v)^2$ and $D = \begin{vmatrix} u & v \\ x & y \end{vmatrix}$.

We have, of course, not proved any of the three statements of Theorem 4.1, simply that they are equivalent. We shall prove the first one.

One proof is quite easy. Note that S^1 is a compact metric space. Hence by Eilenberg's Theorem (3.7) 1_{S^1} is nullhomotopic only if $1_{S^1} = e^{i\varphi}$ for some continuous $\varphi: S^1 \to E^1$. That is, we must have $e^{i\varphi(x)} = x$ for each $x \in S^1$. Elementary complex analysis tells us that there is no such function φ. (It may seem at first that $\varphi(x) = \arg(x)$ would work. However, S^1 is compact and connected, so $\varphi(S^1)$ must be a closed interval. This means $\varphi(S^1)$ would be $[0, 2\pi]$. But then $\varphi(1, 0) = 0 = 2\pi$, which is impossible if φ is a function.)

We shall base our proof on the notion of the degree of a map from S^1 to S^1. One reason for doing this is that the construction of the degree function generalizes to maps of S^n to S^n, while Eilenberg's Theorem fails for $n \geq 2$.

There are several ways of defining the degree function. We shall take the triangulation approach, both for its geometric appeal and, again, because the reader may wish to pursue the n-dimensional case, which is a direct generalization of triangulation of S^1.

The triangulation procedure may be outlined as follows.

1) A simplex on S^1 is a set $\{p, q\}$ of distinct, non-antipodal points on S^1. The points p and q are the vertices of the simplex. The cell $[p, q]$ consists of all points of S^1 on the short arc between p and q, including p and q. More carefully, if $p = e^{i\theta_1}$ and $q = e^{i\theta_2}$, then $[p,q] = \{e^{i\tau} | \theta_1 \leq \tau \leq \theta_2 \text{ or } \theta_2 \leq \tau \leq \theta_1\}$. Here, θ_1 and θ_2 are in $[0, 2\pi[$.

If a simplex is denoted, say s, without explicitly listing its vertices, then $[s]$ denotes the cell of s.

2) If $\{p, q\}$ is a simplex on S^1, we form an ordered simplex by choosing an ordering of the vertices p and q. The two ordered simplexes with vertices p and q are (p, q) and (q, p).

3) Let $p = (x_1, x_2)$ and $q = (y_1, y_2)$. The sign of the ordered simplex (p, q) is the sign of the quantity $x_1 y_2 - x_2 y_1$. We write sign $(p, q) > 0$ if $x_1 y_2 - x_2 y_1 > 0$; and sign $(p, q) < 0$ if $x_1 y_2 - x_2 y_1 < 0$. Note that sign $(p, q) = -\text{sign}(q, p)$. Also,

$$x_1 y_2 - x_2 y_1 = \det \begin{pmatrix} x_1 & x_2 & 1 \\ y_1 & y_2 & 1 \\ 0 & 0 & 1 \end{pmatrix} =$$

± 2. Area of the triangle with vertices $(0, 0)$, (x_1, x_2), (y_1, y_2). This is non-zero if p and q are not antipodal.

250   FUNDAMENTAL CONCEPTS OF TOPOLOGY

Observe that, geometrically, sign $(p, q) > 0$ means that p comes before q if you walk around S^1 counterclockwise from $(1, 0)$; similarly, sign $(p, q) < 0$ means q is encountered first (Fig. 4.2).

4) Let $\{p, q\}$ and $\{p, r\}$ be simplexes on S^1, and let L be the straight line through $(0, 0)$ and p. Then, sign $(p, q) =$ sign (q, r) if and only if q and r lie on the same side of L (that is, the line segment from q to r does not intersect L). The simple geometric proof of this is left to the reader (see Figure 4.3).

5) A triangulation is a finite set T of simplexes on S^1 with the following two conditions:

i) $\bigcup\limits_{s \in T} [s] = S^1$ (the cells cover S^1).

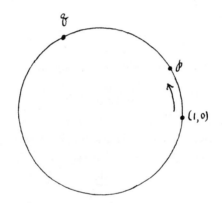

FIGURE 4.2

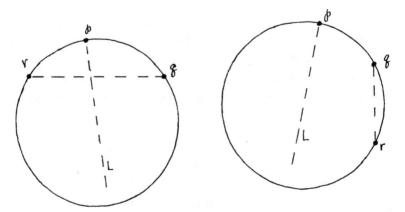

FIGURE 4.3

ii) If s and s' are distinct simplexes in T, then $[s] \cap [s'] = \phi$, or $[s] \cap [s']$ is a vertex of s and of s'.

Figure 4.4 shows a triangulation and a non-triangulation of S^1.

We now outline the procedure for defining the degree of a map from S^1 to S^1.

1) Begin with a triangulation T of S^1, and a vertex map v which maps the set V of vertices of simplexes in T to S^1. Assume, if $\{p, q\} \in T$, then

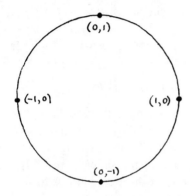

$\{(1,0),(0,1)\}, \{(0,1),(-1,0)\},$

$\{(-1,0),(0,-1)\}, \{(0,-1),(1,0)\}$

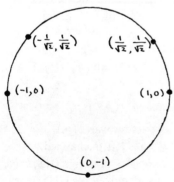

$\{(1,0),(\frac{1}{\sqrt{2}},\frac{1}{\sqrt{2}})\}, \{(\frac{1}{\sqrt{2}},\frac{1}{\sqrt{2}}),(-\frac{1}{\sqrt{2}},\frac{1}{\sqrt{2}})\},$

$\{(-\frac{1}{\sqrt{2}},\frac{1}{\sqrt{2}}),(-1,0)\}, \{(-1,0),(0,-1)\}, \{(0,-1),(\frac{1}{\sqrt{2}},\frac{1}{\sqrt{2}})\}$

FIGURE 4.4

$v(p)$ and $v(q)$ are not antipodal (although $\{v(p), v(q)\}$ need not be a simplex, as possibly $v(p) = v(q)$).

2) Choose any point $P \in S^1$, with $P \notin v(V)$. Order each simplex in T to form an ordered simplex with positive sign. If (p, q) is such an ordered simplex in T, and if $v(p) \neq v(q)$, order the simplex $\{v(p), v(q)\}$ as $(v(p), v(q))$, which may have positive or negative sign. We call $(v(p), v(q))$ an image simplex.

Define

$\varrho(v, T, P)$ = number of image simplexes s with $P \in \text{Int}\,[s]$ and sign $(s) > 0$,

$\eta(v, T, P)$ = number of image simplexes s with $P \in \text{Int}\,[s]$ and sign $(s) < 0$.

We then let
$$\mathscr{D}(v, T, P) = \varrho(v, T, P) - \eta(v, T, P).$$

3) We now claim that $\mathscr{D}(v, T, P)$ is unchanged if we replace P by any other point on $S^1 - v(V)$. We prove this as follows. Imagine that P is set in motion about S^1. It is obvious that $\mathscr{D}(v, T, P)$ can change only when P crosses over some $v(x)$ from (or to) one image simplex, possibly to (or from) another (or possibly several others). It therefore suffices to show that $\mathscr{D}(v, T, P)$ does not change when P moves across some $v(x)$.

We consider a number of cases and subcases.

Case I) Each $\{p, q\} \in T$ maps to a simplex $\{v(p), v(q)\}$ on S^1.

Suppose P passes over $v(p_i)$, say from $\{v(p_i), v(p_j)\}$ to $\{v(p_i), v(p_k)\}$.

Draw a line L from $(0, 0)$ to $v(p_i)$.

Case I–i) $v(p_j)$ and $v(p_k)$ are on the same side of L (Fig. 4.5).

Case I–i$_1$) sign $(p_j, p_i) > 0$ and sign $(p_i, p_k) > 0$ (Fig. 4.6).

In this case, we order the image simplexes $\{v(p_i), v(p_j)\}$ and $\{v(p_i), v(p_k)\}$ as $(v(p_j), v(p_i))$ and $(v(p_i), v(p_k))$. Now, by 4) above, $(v(p_i), v(p_j))$ and $(v(p_i), v(p_k))$ have the same sign. Thus:

if sign $(v(p_j), v(p_i)) > 0$, then sign $(v(p_i), v(p_j)) < 0$,

so sign $(v(p_i), v(p_k)) < 0$. Hence as P passes over $v(p_i)$, $\varrho(v, T, P)$ and $\eta(v, T, P)$ each increase by 1, and $\mathscr{D}(v, T, P)$ is left unchanged.

If sign $(v(p_j), v(p_i)) < 0$, then sign $(v(p_i), v(p_j)) > 0$, so sign $(v(p_i), v(p_k)) > 0$, and again ϱ and η each increase by 1 as P moves over $v(p_i)$, leaving $\mathscr{D}$ unchanged.

Case I–i$_2$) sign $(p_j, p_i) < 0$ and sign $(p_i, p_k) < 0$.

This case is similar to case I–i$_1$, and is left to the reader.

Case I–ii) $v(p_j)$ and $v(p_k)$ are on opposite sides of L (Fig. 4.7).

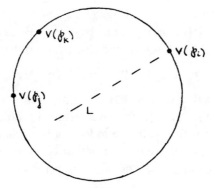

FIGURE 4.5

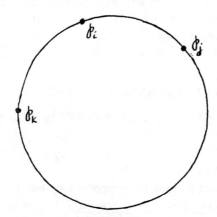

FIGURE 4.6

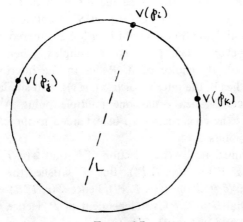

FIGURE 4.7

Case I–ii$_1$) sign $(p_j, p_i) > 0$ and sign $(p_i, p_k) > 0$.

We then order $\{v(p_i), v(p_j)\}$ and $\{v(p_i), v(p_k)\}$ as $(v(p_j), v(p_i))$ and $(v(p_i), v(p_k))$. Here, $(v(p_i), v(p_j))$ and $(v(p_i), v(p_k))$ have opposite signs. Thus:

if sign $(v(p_i), v(p_j)) > 0$, then sign $(v(p_j), v(p_i)) < 0$,

and sign $(v(p_i), v(p_k)) < 0$ (Fig. 4.7). In this event, as P moves from $\{v(p_i), v(p_j)\}$ to $\{v(p_i), v(p_k)\}$, P moves from an ordered simplex with negative sign to one with negative sign, leaving $\mathscr{D}$ unchanged.

If sign $(v(p_i), v(p_j)) < 0$, we reach a similar conclusion.

Case I–ii$_2$) sign $(p_j, p_i) < 0$ and sign $(p_i, p_k) < 0$. This is handled in similar fasion.

Thus in case I, $\mathscr{D}(v, T, P) = \mathscr{D}(v, T, P')$ for any $P, P' \in S^1 - v(V)$.

Case II) There is at least one $\{p, q\} \in T$ such that $\{v(p), v(q)\}$ is not a simplex (that is, $v(p) = v(q)$).

Reduce case II to case I as follows. Produce a function $\bar{v}: V \to S^1$ such that:

1) $\{\bar{v}(x), \bar{v}(y)\}$ is a simplex whenever $\{x, y\} \in T$.

2) If $\{x, y\} \in T$, and $\{v(x), v(y)\}$ is a simplex on S^1, then sign $(\bar{v}(x), \bar{v}(y))$ = sign $(v(x), v(y))$.

3) P lies in the interior of $[\bar{v}(x), \bar{v}(y)]$ exactly when P lies in the interior of $[v(x), v(y)]$, and similarly for P'.

$\bar{v}$ can be produced as follows. Suppose first that $V = \{p_1, ..., p_r\}$, and $v(p_1) = \cdots = v(p_m) = y \in S^1$ for some $1 \leq m \leq r$, while $v(p_m)$, $v(p_{m+1}), ..., v(p_r)$ are all distinct if $m < r$. Let $\bar{v}(p_i) = v(p_i)$ for $i = m + 1, ..., r$ if $m < r$. Let δ be the infimum of arc lengths between successive points of $v(V) \cup \{P\} \cup \{P'\}$ on S^1. Replace y by distinct points $v(p_1), ..., v(p_m)$, all lying on the arc of length $\delta/2$ centered at y, and ordered so that, whenever $\{v(x), v(y)\}$ is a simplex, then sign $(v(x), v(y))$ = sign $(\bar{v}(x), \bar{v}(y))$. By choice of δ, P lies in the interior of $[\bar{v}(x), \bar{v}(y)]$ exactly when P lies in the interior of $[v(x), v(y)]$, and similarly for P'.

This constructs $\bar{v}$ when v has one multiple point. A straightforward induction allows the construction to be extended to the case of arbitrarily many multiple points.

Now, it is immediate by the selection of $\bar{v}$, that $\mathscr{D}(\bar{v}, T, P) = \mathscr{D}(v, T, P)$ and that $\mathscr{D}(\bar{v}, T, P') = \mathscr{D}(v, T, P')$. But, $\bar{v}$ satisfies the requirements of fcase I, so that $\mathscr{D}(\bar{v}, T, P) = \mathscr{D}(\bar{v}, T, P')$, hence $\mathscr{D}(v, T, P) = \mathscr{D}(v, T, P')$.

This shows that $\mathscr{D}(v, T, P)$ is independent of P. Hence we shall drop P rom the symbol and just write $\mathscr{D}(v, T)$.

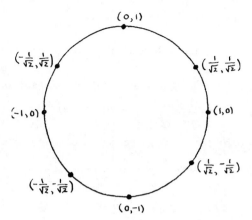

Figure 4.8a

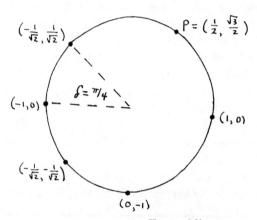

Figure 4.8b

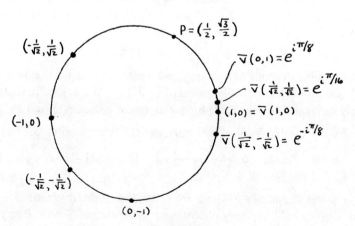

Figure 4.8c

4.1 EXAMPLE For clarity, we pause here to illustrate the construction of $\tilde{v}$ from v with an example.

Consider the triangulation of Figure 4.8a (simplexes consist of successive vertices). Say $v(1, 0) = v\left(\frac{1}{\sqrt{2}}, \frac{1}{\sqrt{2}}\right) = v(0, 1) = v\left(-\frac{1}{\sqrt{2}}, \frac{1}{\sqrt{2}}\right) = (1, 0)$, while $v(a, b) = (a, b)$ for the other vertices. Here, we may choose $\delta = \pi/4$. Define, say, $\tilde{v}(1, 0) = (1, 0), \tilde{v}\left(\frac{1}{\sqrt{2}}, \frac{1}{\sqrt{2}}\right) = e^{\pi i/16} = (\cos(\pi/16), \sin(\pi/16)), \tilde{v}(0, 1) = e^{\pi i/8}$, and $\tilde{v}\left(\frac{1}{\sqrt{2}}, -\frac{1}{\sqrt{2}}\right) = e^{-i\pi/8}$. ∎

4) We shall now show that $\mathscr{D}(v, T)$ is independent of T.

We begin with the following observation. Let T be a triangulation of S^1, and introduce a new vertex to form a new triangulation T'. More specifically, say $\{x, y\} \in T$, and we let $z \in \text{Int }[x, y]$ and form T' as the set of all the simplexes of T, except $\{x, y\}$, together with new simplexes $\{x, z\}$ and $\{z, y\}$. Then, $\mathscr{D}(v, T) = \mathscr{D}(v, T')$. For, we can choose some P interior to a simplex $\{a, b\} \neq \{x, y\}$ in T, and then $\mathscr{D}(v, T) = \mathscr{D}(v, T, P) = \mathscr{D}(v, T', P) = \mathscr{D}(v, T')$.

By induction, we can insert n new vertices to form a new triangulation from T without changing $\mathscr{D}(v, T)$.

Now suppose that T and T' are any two triangulations of S^1. Use the vertices of T of T' and to form a new triangulation T''. Then, $\mathscr{D}(v, T) = \mathscr{D}(v, T'') = \mathscr{D}(v, T')$.

5) Now suppose f is continuous: $S^1 \to S^1$. Then, f is uniformly continuous, so there is a triangulation T of S^1 such that $f(x)$ and $f(y)$ are not antipodal whenever $\{x, y\} \in T$. If V is the vertex set of T, we let

$$\deg(f) = \mathscr{D}(f|V, T).$$

This number depends only upon f, independence of T having been shown in 4). Of course, in actually calculating $\deg(f)$ for a specific map, one usually calculates $\mathscr{D}(f|V, T, P)$ for a convenient choice of P and T.

4.2 EXAMPLE $\deg(1_{S^1}) = 1$. Simply use the triangulation of Figure 4.4 and choose P, say, as $\left(\frac{1}{\sqrt{2}}, \frac{1}{\sqrt{2}}\right)$. Then, $\varrho(1_{S^1}|V, T, P) = 1$ and $\eta(1_{S^1}|V, T, P) = 0$. ∎

4.3 EXAMPLE Let $y \in S^1$ and let d be the constant map: $S^1 \to \{y\}$. Then, $\deg(d) = 0$. We may choose any triangulation T with P any point of $S^1 - \{y\}$. Then, $\varrho(d|V, T, P) = 0 = \eta(d|V, T, P)$. ∎

We now prove the fundamental theorem relating homotopy and degree.

4.2 Theorem *Let f, g be continuous: $S^1 \to S^1$.*
Let $f \simeq g$.
Then,
$\deg(f) = \deg(g)$.

Proof Let $\Phi: f \simeq g$. Since $S^1 \times I$ is a compact metric space, then Φ is uniformly continuous, so there is an $\eta > 0$ such that $|\Phi(x, t) - \Phi(y, r)| < 1$ if $x, y \in S^1$, $t, r \in I$, $\varrho_2(x, y) < \eta$ and $|t - r| < \eta$.

Choose a triangulation T of S^1 such that, if $\{x, y\} \in T$, then the length of arc of $[x, y]$ is less than η.

Choose any $P \in S^1$ with $P \neq f(v)$ for each vertex v of T. Let $t_0 \in I$, and choose ε sufficiently small that any variation of the vertices $\Phi_{t_0}(p)$, p a vertex of T, to new positions $\alpha(p)$ by an arc length less than ε leaves $\mathscr{D}(\Phi_{t_0}|V, T, P) = \mathscr{D}(\alpha, T, P)$.

Again by uniform continuity, there is some $\delta > 0$ such that $|\Phi(x, t) - \Phi(y, r)| < \varepsilon$ if $\varrho_2(x, y) < \delta$ and $|t - r| < \delta$. In particular, $|\Phi_t(x) - \Phi_{t_0}(x)| < \varepsilon$ if $|t - t_0| < \delta$ for any $x \in S^1$.

Then, $\mathscr{D}(\Phi_t|V, T, P) = \mathscr{D}(\Phi_{t_0}|V, T, P)$, so $\deg(\Phi_t) = \deg(\Phi_{t_0})$ for $|t - t_0| < \delta$.

Now choose a partition $0 = r_0 < r_1 < \cdots < r_{n-1} < r_n = 1$ of I such that $r_j - r_{j-1} < \delta$ for $j = 1, \ldots, n$. Then, $\deg(\Phi_{r_j}) = \deg(\Phi_{r_{j-1}})$ for $j = 1, \ldots, n$. It is now easy to check that $\deg(f) = \deg(\Phi_0) = \deg(\Phi_1) = \deg(g)$. ∎

We now prove Brouwer's Theorem for S^1.

4.3 Theorem (*Brouwer*) 1_{S^1} *is not nullhomotopic.*

Proof Suppose, for some $y \in S^1$, $1_{S^1} \simeq d: S^1 \to \{y\}$. Then $1 = \deg(1_{S^1}) = \deg(d) = 0$, a contradiction. ∎

10.5 Applications

We shall conclude this chapter with two applications of the theorems and techniques developed this far.

The first illustrates the use of triangulation techniques in an extension problem. Basically, the idea is to extend the given function over the pieces of the triangulation, then paste the pieces together.

5.1 Theorem *Let A be a compact subset of E^1.*
Let f be continuous: $A \to S^1$.
Then,
f has a continuous extension $F: E^1 \to S^1$.

Proof Let $h: S^1 - \{(0, 1)\} \to E^1$ be the stereographic projection homeomorphism.

Then, $f \circ h: (\text{inv } h)(A) \to S^1$.

We now proceed in six steps, the details being left to the reader.

1) $(\text{inv } h)(A)$ is a closed subset of S^1.

2) There is a neighborhood U of $(\text{inv } h)(A)$ in S^1 such that $f \circ h$ can be extended to a continuous $g: U \to S^1$.

3) $\inf \{\varrho_2(x, y) | x \in (\text{inv } h)(A) \text{ and } y \in S^1 - U\} > 0$.

4) Let $\varepsilon = \frac{1}{2} \inf \{\varrho_2(x, y) | x \in (\text{inv } h)(A) \text{ and } y \in S^1 - U\}$.

Let T be a triangulation of S^1 with the length of $[a, b] < \varepsilon$ for each $\{a, b\} \in T$. Let $C = \cup \{[a, b] | \{a, b\} \in T \text{ and } [a, b] \cap (\text{inv } h(A)) \ne \phi\}$.

Then, $(\text{inv } h)(A) \subset C \subset U$, so $f \circ h$ has a continuous extension $g: C \to S^1$.

5) If $\{a, b\} \in T$ and $[a, b] \cap (\text{inv } h)(A) = \phi$, define $G(a)$ and $G(b)$ arbitrarily on S^1, so that $G(a)$ and $G(b)$ are not antipodal. Define G to map $[a, b]$ onto $[G(a), G(b)]$.

Finally, let $G(x) = g(x)$ for $x \in C$. Then, $G: S^1 \to S^1$ is continuous, and $G|C = g$.

6) Let $F = G \circ \text{inv } h$.

Then, $F: E^1 \to S^1$ is continuous, and $F|A = f$. ∎

Finally, we shall apply the degree function to a derivation of a result on continuous vector fields and normals.

Suppose $v_1, v_2: B_2 \to E^1$. Then, corresponding to $(x, y) \in B_2$, we can associate a vector $\vec{v}(x, y) = (v_1(x, y), v_2(x, y)) = v_1(x, y) i + v_2(x, y) j$ in the plane. If you represent vectors as arrows, think of $\vec{v}$ as placing an arrow from (x, y) for each (x, y) in B_2.

If v_1 and v_2 are continuous, we call $\vec{v}$ a continuous vector field on B_2; if $\vec{v}(x, y)$ is never the zero vector, we call $\vec{v}$ a non-vanishing vector field.

We say that $\vec{v}$ has an inner normal on S^1 if $\vec{v}(x, y)$ is an inner normal to S^1 for some $(x, y) \in S^1$ (e.g. say $v(1, 0) = -i$). Similarly, $\vec{v}$ has an outer normal on S^1 if $\vec{v}(x, y)$ is an outer normal to S^1 for some $(x, y) \in S^1$ (e.g. say $v(1, 0) = i$).

5.2 THEOREM *Each continuous, non-vanishing vector field on B_2 has an inner normal on S^1 and an outer normal on S^1.*

Proof Let $\vec{v} = (v_1, v_2)$ be a continuous, non-vanishing vector field on B_2.

Since $\vec{v}(x, y) \neq (0, 0)$ for each $(x, y) \in B_2$, we can define a continuous map $V: S^1 \to S^1$ by setting, for each $(x, y) \in S^1$,

$$V(x, y) = \left(\frac{v_1(x, y)}{\sqrt{v_1(x, y)^2 + v_2(x, y)^2}}, \frac{v_2(x, y)}{\sqrt{v_1(x, y)^2 + v_2(x, y)^2}} \right).$$

Now, V has a continuous extension $\left(\dfrac{\vec{v}}{|\vec{v}|}\right)$ to B_2, hence is nullhomotopic by Theorem 3.5.

Then, $\deg(V) = 0$.

Now, let $\mathcal{O}$ denote the continuous vector field of unit outer normals on S^1. That is, $\mathcal{O}(x, y) = x\vec{i} + y\vec{j}$ for $(x, y) \in S^1$. Let $\mathcal{I}$ be the continuous vector field of unit inner normals on S^1: $\mathcal{I}(x, y) = -x\vec{i} - y\vec{j}$ for $(x, y) \in S^1$.

If $V(x, y)$ and $\mathcal{O}(x, y)$ are never antipodal, then $V \simeq \mathcal{O}$, by Theorem 2.5. But then $0 = \deg(V) = \deg(\mathcal{O}) = 1$, a contradiction. Hence $\mathcal{O}(x, y)$ and $V(x, y)$ are antipodal for some $(x_0, y_0) \in S^1$, and then $V(x_0, y_0) = -\mathcal{O}(x_0, y_0)$ is an inner normal to S^1.

Similarly, if V and $\mathcal{I}$ are never antipodal, then $V \simeq \mathcal{I}$. But $\deg \mathcal{I} = 1$ also (use the triangulation and choice of P of Figure 4.9, where $\varrho(\mathcal{I}, T, P) = 1$ and $\eta(\mathcal{I}, T, P) = 0$). Thus, for some $(x_1, y_1) \in S^1$, $V(x_1, y_1) = -(x_1, y_1)$ is an outer normal to S^1. ∎

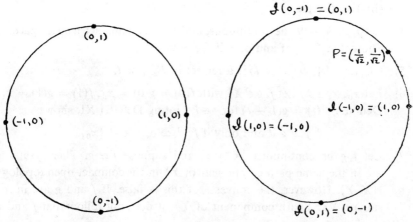

FIGURE 4.9

To illustrate how such a geometric result might be used in an analytic situation, suppose $f, g: B_2 \to E^1$ are continuous, and $f(x, y)^2 + g(x, y)^2 \neq 0$ for $(x, y) \in B_2$. We may then think of $V(x, y) = f(x, y) i + g(x, y) j$ as a continuous, non-vanishing vector field on B_2. Then V has an inner and an outer normal on S^1. This means that:

for some $(x_0, y_0) \in S^1$, and some $\alpha > 0$,
$$f(x_0, y_0) = \alpha x_0 \text{ and } g(x_0, y_0) = \alpha y_0;$$

and,

for some $(x_1, y_1) \in S^1$, and some $\beta < 0$,
$$f(x_1, y_1) = \beta x_1 \text{ and } g(x_1, y_1) = \beta y_1.$$

Just as an exercise, the reader should try an analytic proof of this as a basis for comparison.

Problems

1) Let $f(x) = 4 - x^2$ and $g(x) = -18 + x$, for $-1 \leq x \leq 1$. Consider f, g as functions on $[-1, 1]$ to $E^1 - \{0\}$. Is $f \simeq g$?

*2) Let $X = \left(\bigcup_{n \in Z^+} (\{1/n\} \times I) \right) \cup (I \times \{0\}) \cup (\{0\} \times I)$.
   (a) $1_X \simeq c_{(0,1)}: X \to \{(0, 1)\}$.
   (b) There is no homotopy $\Phi: 1_X \simeq c_{(0,1)}$ with $\Phi((0, 1), t) = (0, 1)$ for each $t \in I$.

3) Discuss homotopy of $f, g: X \to Y$ when
   (a) X is discrete.
   (b) Y is discrete.

4) Let $f, g: X \to S^1$ be continuous, and (X, ϱ) a compact metric space. Show that $f \simeq g$ if and only if $f/g \simeq 0$.

*5) $(B_{n+1} \times \{0\}) \cup (S^n \times I)$ is a retract of $B_{n+1} \times I$.

6) Let $x_0, x_1 \in X$. Let $f, g \in X^I$ with $f(0) = g(0) = x_0, f(1) = g(1) = x_1$. Denote $g^{-1}(t) = g(1 - t)$ for $t \in I$. Noting Def. 1.1.XI, show:
$$f \simeq g \text{ if and only if } fg^{-1} \simeq c_{x_0}: X \to \{x_0\}.$$

*7) Let f, g be continuous: $X \to Y$, and suppose $f \simeq g$. Then, f and g are in the same path component of Y^X in the compact-open topology $c(X, Y)$. However, the converse of this is false. If f and g are in the same $c(X, Y)$ path component of Y^X, it does not follow that f and g are homotopic.

*8) Generalize the triangulation procedure and the definition of the degree function to S^2, then to S^n for arbitrary n. Use this to prove Brouwer's Theorem for S^n: 1_{S^n} is not nullhomotopic.

Prove also that this is equivalent to the n-dimensional analogues of the other two parts of Theorem 4.1.

*9) Give examples of homotopic and non-homotopic $f, g\colon X \to Y$ in the following cases:

(a) $X = [0, 1] \times [0, 1]$, with $T_<$ the topology of Example 3.7.I, and $Y = E^1$.

(b) X the space of Example 1.7.I and Y the ordinal space Ω.

(c) X the reals with the cofinal topology and Y the reals with the topology consisting of ϕ and all complements of countable subsets.

CHAPTER XI

Homotopy groups

11.1 The fundamental group of a space

IN THE PREVIOUS chapter we sketched some of the concepts and methods of homotopy theory in set topology. We shall now use homotopy to associate algebraic groups with topological spaces in what will turn out to be a fairly natural way.

The objects of the first homotopy group (or fundamental group) of a space will be equivalence classes (under a homotopy relation) of loops at fixed base points. We shall be more explicit about this soon. For the time being, we define an operation which will later develop into a group product.

1.1 DEFINITION Let $f, g \in X^I$.
Suppose $f(1) = g(0)$.
Then, we define $fg: I \to X$ by:

$$(fg)(t) = \begin{cases} f(2t) \text{ for } 0 \leq t \leq \tfrac{1}{2} \\ g(2t - 1) \text{ for } \tfrac{1}{2} \leq t \leq 1. \end{cases}$$

1.1 THEOREM Let $f, g \in X^I$ and $f(1) = g(0)$.
Then,
$fg \in X^I$.

Proof Left to the reader. ■

Thus, we can form a product fg of paths f and g in X when the initial point of g coincides with the terminal point of f. Geometrically, fg is the path from $f(0)$ to $g(1)$ obtained by stringing f and g together (Figure 1.1). Note, however, that gf may not be defined.

Suppose now that $f \simeq g$, where f and g are paths in X. Then there is a continuous $\Phi: I \times I \to X$ with $\Phi(t, 0) = f(t)$ and $\Phi(t, 1) = g(t)$, $0 \leq t \leq 1$. Geometrically, we may think of Φ as mapping the square $I \times I$ into X, with the maps of the bottom and top of the bottom and top of the rectangle

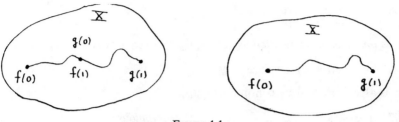

FIGURE 1.1

behaving like the maps f and g respectively. We are going to impose another condition on the homotopy Φ and one on the paths f and g themselves.

1.2 DEFINITION *Let $f, g \in X^I$.*
Suppose that $f(0) = g(0)$ and $f(1) = g(1)$.
Then,
$f \cong g$
if and only if
there is a homotopy $\Phi: f \simeq g$ such that $\Phi(0, s) = f(0)$ and $\Phi(1, s) = f(1)$ for $0 \leq s \leq 1$.
When such a Φ exists, we write $\Phi: f \cong g$.

Thus, on the bottom and top of $I \times I$, Φ behaves like f and g respectively, but in addition f and g coincide at their initial and terminal points, and on each of the vertical sides Φ behaves like a constant map. When $f \cong g$, then f can be deformed to g, with the end points of the stages of the deformation anchored at the end points of f (and g) throughout (Figure 1.2).

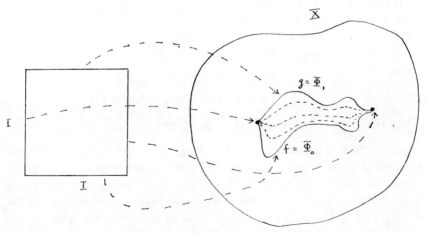

FIGURE 1.2

1.2 THEOREM *Let $x_0, x_1 \in X$.*
Then,
$\cong$ *is an equivalence relation on $X^I \cap \{f | f(0) = x_0 \text{ and } f(1) = x_1\}$.*

Proof Left as an exercise. ∎

$\cong$ is also preserved by continuous maps.

1.3 THEOREM *Let $\varphi: X \to Y$ be (T, M) continuous.*
Let $f, g \in X^I$ with $f(0) = g(0)$ and $f(1) = g(1)$.
Suppose $f \cong g$.
Then,
$\varphi \circ f \cong \varphi \circ g$.

Proof Note first that $\varphi(f(0)) = \varphi(g(0))$ and $\varphi(f(1)) = \varphi(g(1))$. Now let $\Phi: f \cong g$. The reader can check that $\varphi \circ \Phi: \varphi \circ f \cong \varphi \circ g$. ∎

We now examine the above product of paths in the light of our new homotopy relation $\cong$.

1.4 THEOREM *Let $f, g, f', g' \in X^I$.*
Let $f(0) = g(0), f(1) = g(1) = f'(0) = g'(0)$.
Let $f \cong g$ and $f' \cong g'$.
Then,
$ff' \cong gg'$.

Proof Note that ff' and gg' are both defined, as $f(1) = f'(0)$ and $g(1) = g'(0)$. Further, $(ff')(0) = f(0) = g(0) = (gg')(0)$, and, since $f' \cong g'$, $(ff')(1) = f'(1) = g'(1) = (gg')(1)$.

We must now produce a homotopy $\Phi: ff' \cong gg'$.
Since $f \cong g$, there is a homotopy $\Theta: f \cong g$.
Since $f' \cong g'$, there is a homotopy $\Theta': f' \cong g'$.
Let
$$\Phi(t, s) = \begin{cases} \Theta(2t, s) \text{ for } 0 \leq t \leq \tfrac{1}{2}, 0 \leq s \leq 1, \\ \Theta'(2t - 1, s) \text{ for } \tfrac{1}{2} \leq t \leq 1, 0 \leq s \leq 1. \end{cases}$$

The reader can check that $\Phi: ff' \cong gg'$. ∎

In constructing homotopies, such as Φ in the last proof, it is often useful to think in terms of geometric operations on $I \times I$. For example (Fig. 1.3) shrink the t-side of the Θ-box from $[0, 1]$ to $[0, \tfrac{1}{2}]$; and that of the Θ'-box from $[0, 1]$ to $[\tfrac{1}{2}, 1]$. Then paste the two rectangles together to form the Φ copy of $I \times I$. We may then think of Φ as deforming ff' to gg' in two stages: f to f', then g to g'.

Multiplication as defined above is not associative, as $(fg)h$ and $f(gh)$ will in general be different. Specifically:

$$((fg)h)(t) = \begin{cases} f(4t), 0 \leq t \leq \tfrac{1}{4}, \\ g(4t-1), \tfrac{1}{4} \leq t \leq \tfrac{1}{2}, \\ h(2t-1), \tfrac{1}{2} \leq t \leq 1, \end{cases}$$

and,

$$f(gh)(t) = \begin{cases} f(2t), 0 \leq t \leq \tfrac{1}{2}, \\ g(4t-2), \tfrac{1}{2} \leq t \leq \tfrac{3}{4}, \\ h(4t-3), \tfrac{3}{4} \leq t \leq 1. \end{cases}$$

However, we do have a form of the associative law with $\cong$ replacing $=$.

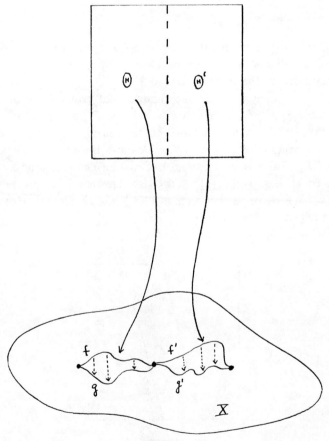

FIGURE 1.3

1.5 Theorem *Let $f, g, h \in X^I$.
Let $f(1) = g(0)$ and $g(1) = h(0)$.
Then,*
$(fg)h \cong f(gh).$

Proof Let

$$\Phi(t, s) = \begin{cases} f\left(\dfrac{4t}{1+s}\right) \text{ for } 0 \leq t \leq \dfrac{s+1}{4}, 0 \leq s \leq 1, \\ g(4t - 1 - s) \text{ for } \dfrac{s+1}{4} \leq t \leq \dfrac{s+2}{4}, 0 \leq s \leq 1, \\ h\left(1 - \dfrac{4(1-t)}{2-s}\right), \dfrac{s+2}{4} \leq t \leq 1, 0 \leq s \leq 1. \end{cases}$$

Then, $\Phi: (fg)h \cong f(gh)$. ∎

Again, thinking geometrically, we split $I \times I$ into three sections (Figure 1.4) and define Φ separately on each section, being careful that the definitions agree on the common boundaries.

We are now ready to form the fundamental group of X at a point. Choose some $x_0 \in X$ (x_0 is the base point). Let P_{x_0} be the set of all loops at x_0 (paths in X from x_0 to x_0). Note that there is no difficulty in multiplying any two elements of P_{x_0}. Let $\pi(X, x_0)$ denote the set of $\cong$ equivalence classes of P_{x_0}. In view of Theorem 1.4, we can define a product of such equivalence classes by $[f][g] = [fg]$. By Theorem 1.5, this product is associative. We claim further that $\pi(X, x_0)$ has an identity element and that each $[f]$ has an inverse.

In summary:

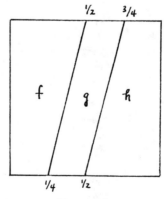

FIGURE 1.4

1.3 Definition *Let $x_0 \in X$.*
Then,
(1) $P_{x_0} = X^I \cap \{f | f(0) = f(1) = x_0\}$.
(2) $\pi(X, x_0) = P_{x_0}/\cong$.
(3) If $f \in P_{x_0}$, then $[f] = P_{x_0} \cap \{g | f \cong g\}$.

1.6 Theorem
(1) The product of elements of $\pi(X, x_0)$ given by $[f][g] = [fg]$ is well defined and associative.
(2) Let $e_{x_0}: I \to \{x_0\}$.
Then, $[f][e_{x_0}] = [e_{x_0}][f] = [f]$ for each $[f] \in \pi(X, x_0)$.
(3) If $f \in P_{x_0}$, let $f^{-1}: I \to X$ be defined by $f^{-1}(t) = f(1-t), 0 \le t \le 1$.
Then, $f^{-1} \in P_{x_0}$ and $[f][f^{-1}] = [f^{-1}][f] = [e_{x_0}]$ for each $[f] \in \pi(X, x_0)$.

Proof
(1) Is immediate by Theorems 1.4 and 1.5.
(2) We shall show that $fe_{x_0} \cong f$.
Note that
$$(fe_{x_0})(t) = \begin{cases} f(2t), & 0 \le t \le \tfrac{1}{2}, \\ x_0, & \tfrac{1}{2} \le t \le 1. \end{cases}$$
Let
$$\Phi(t, s) = \begin{cases} f\left(\dfrac{2t}{1+s}\right), & \text{for } 0 \le t \le \dfrac{s+1}{2}, 0 \le s \le 1, \\ x_0, & \dfrac{s+1}{2} \le t \le 1, 0 \le s \le 1. \end{cases}$$

The reader can check that $\Phi: fe_{x_0} \cong f$. Similarly, $e_{x_0}f \cong f$.
(3) It is obvious that $f^{-1} \in P_{x_0}$.
Let
$$\Phi(t, s) = \begin{cases} x_0 & \text{for } 0 \le t \le \dfrac{s}{2}, 0 \le s \le 1, \\ f(2t - s), & \dfrac{s}{2} \le t \le \dfrac{1}{2}, 0 \le s \le 1, \\ f(2 - 2t - s), & \dfrac{1}{2} \le t \le \dfrac{2-s}{2}, 0 \le s \le 1, \\ x_0, & \dfrac{2-s}{2} \le t \le 1, 0 \le s \le 1. \end{cases}$$

Then, $\Phi: ff^{-1} \cong e_{x_0}$.
Similarly, $f^{-1}f \cong e_{x_0}$. ∎

$\pi(X, x_0)$, which is a group by the last theorem, is called the fundamental, or first homotopy, group of X at x_0. Following custom with groups, and in view of (3) of the theorem, we shall denote $[f^{-1}]$ by $[f]^{-1}$.

There are many obvious questions to ask about $\pi(X, x_0)$. For example, if x_0 and x_1 are points in X, are $\pi(X, x_0)$ and $\pi(X, x_1)$ related? For the time being, however, it seems worthwhile to compute some examples of fundamental groups.

1.1 Example $\pi(E^1, 0)$ is the trivial group.

This will fall rather easily out of some later, more general results, but it is not difficult to prove directly. Let f be a loop in E^1 at 0. Then $\Phi(t, s) = (1 - s)f(t)$, $0 \leq t \leq 1$, $0 \leq s \leq 1$, defines a homotopy from f to $e_0: E^1 \to \{0\}$. Hence $\pi(E^1, 0) = \{[e_0]\}$. ∎

We shall use this result in the next section in computing the fundamental group of the circle.

1.2 Example If C is any convex subset of E^n and $x_0 \in C$, then $\pi(C, x_0)$ is trivial.

A proof of this may be modeled after the argument of Example 1.1, or you may use results from section 6. ∎

Of course, there are nontrivial homotopy groups. We shall compute one in the next section.

11.2 $\pi(S^1, (1, 0))$

In this section we shall prove that $\pi(S^1, (1, 0))$ is infinite cyclic. There are, of course, several ways to do this. One argument is to let $g(t) = (\cos(2\pi t), \sin(2\pi t))$, $0 \leq t \leq 1$, and then to show that each homotopy class of loops at $(1, 0)$ in S^1 is represented by a power of g, where g^n represents the path wrapping around S^1 n times in a counterclockwise sense for n positive, and g^{-n}, n times in a clockwise sense. Full details of this approach appear in Massey.

We shall give an elegant proof based on a method of A. W. Tucker. Beginning with an exponential function, we shall explicitly write an isomorphism of $\pi(S^1, (1, 0))$ onto the additive group Z of integers.

Let $\text{Exp}(x) = e^{2\pi i x} (= \cos(x) + i \sin(x))$ for x real. Here, $i^2 = -1$ as usual, and, as before, we shall write $a + ib$ and (a, b) interchangeably, and $(a, 0)$ as a.

HOMOTOPY GROUPS

By elementary properties of complex numbers and of sine and cosine, note that
$$\text{Exp}(x_1) = \text{Exp}(x_2) \Leftrightarrow x_1 - x_2 \in Z.$$

Further, $\text{Exp}\,]-\tfrac{1}{2},\tfrac{1}{2}[$ is an homeomorphism of $]-\tfrac{1}{2},\tfrac{1}{2}[$ onto $S^1 - \{(-1,0)\}$. For convenience, denote $\text{inv}(\text{Exp}\,|]-\tfrac{1}{2},\tfrac{1}{2}[)$ by Log. Then, Log: $S^1 - \{(-1,0)\} \to]-\tfrac{1}{2},\tfrac{1}{2}[$ is an homeomorphism, and has the usual multiplicative properties of logarithms.

We shall eventually need three Lemmas. The first says that we can lift a path in S^1 to a path in E^1 — that is, there is a path g such that the following diagram commutes.

$$\begin{array}{ccc} & & E^1 \\ & \nearrow{\scriptstyle g} & \downarrow{\scriptstyle \text{Exp}} \\ I & \xrightarrow{f} & S^1 \end{array}$$

1 LEMMA *Let f be a path to S^1 with initial point $(1, 0)$. Then, there is a unique path g in E^1 with initial point 0 such that $f(x) = \text{Exp}(g(x))$ for each $x \in I$.*

Proof We first prove existence. Since $f: I \to S^1$ is continuous, then f is uniformly continuous. Choose some $\delta > 0$ such that $\varrho_2(f(x), f(y)) < 1$ if $|x - y| < \delta$, for each $x, y \in I$. Now choose a positive integer N such that $1/N < \delta$. For each integer j, $0 \leq j < N$, and each $x \in I$, note that

$$\left| \frac{(j+1)x}{N} - \frac{jx}{N} \right| = \frac{x}{N} \leq \frac{1}{N} < \delta.$$

Then,
$$\varrho_2\left(f\left(\frac{(j+1)x}{N}\right), f\left(\frac{jx}{N}\right)\right) < 1,$$
so that
$$\frac{f((j+1)x/N)}{f(jx/N)} \in S^1 - \{(-1,0)\}.$$

Now define $g: I \to E^1$ by letting, for each $x \in I$,
$$g(x) = \sum_{j=0}^{N-1} \text{Log}\left(\frac{f((j+1)x/N)}{f(jx/N)}\right).$$

Then, g is continuous, and, for each $x \in I$,
$$g(x) = \text{Log}\left(\prod_{j=0}^{N-1}\left(\frac{f((j+1)x/N)}{f(jx/N)}\right)\right) = \text{Log}\left(\frac{f(x)}{f(0)}\right) = \text{Log}(f(x)).$$

Hence, $f(x) = \text{Exp}(g(x))$.

Finally, $g(0) = \text{Log}(1) = 0$, as $\text{Exp}(0) = 1$.

We now prove uniqueness.

Suppose that $G: I \to E^1$ is continuous, $G(0) = 0$, and $f = \text{Exp } G$. Then, for each $x \in I$,

$$\text{Exp}(g(x) - G(x)) = \frac{f(x)}{f(x)} = 1.$$

Then, $g(x) - G(x) \in Z$ for each $x \in I$. But, $g - G$ is continuous, so $g - G$ must be a constant function.

Then, for each $x \in I$,

$$g(x) - G(x) = g(0) - G(0) = 0. \quad \blacksquare$$

Lemma 2 is simply Lemma 1 with $I \times I$ in place of I. The proof is virtually the same (replace $|x - y|$, for $x, y \in I$, by $\varrho_2(x, y)$ for $x, y \in I \times I$, and choose N so that $\sqrt{2}/N < \delta$ instead of just $1/N < \delta$), and so will be omitted.

2 LEMMA *Let F be continuous: $I \times I \to S^1$, and suppose that $F(0,0) = (1,0)$. Then,*
there is a unique continuous $G: I \times I \to E^1$ such that $G(0,0) = 0$ and $F = \text{Exp} \circ G$.

Proof Left to the reader. $\blacksquare$

We can now define the function we want to use from $\pi(S^1, (1, 0))$ to Z. Let f be a loop at $(1, 0)$ in S^1. Then, $f: I \to S^1$ and $f(0) = f(1) = 1$. By Lemma 1, there is a unique $f^*: I \to E^1$ with $f^*(0) = 0$ and $f(x) = e^{2\pi i f^*(x)}$ for each $x \in I$. We define $\alpha_f = f^*(1)$.

Note that $e^{2\pi i f^*(0)} = f(0) = (1, 0) = f(1) = e^{2\pi i f^*(1)}$. Hence $f^*(0) - f^*(1) \in Z$. Since $f^*(0) = 0$, then $f^*(1) \in Z$. Thus α_f is an integer for each loop f at $(1, 0)$ in S^1.

Lemma 3 enables us to use α to define a well-defined function β on the homotopy classes of $\pi(S^1, (1, 0))$ by the rule $\beta([f]) = \alpha_f$.

3 LEMMA *Let f and g be loops at $(1, 0)$ in S^1.*
Let $f \cong g$.
Then,
$\alpha_f = \alpha_g.$

Proof Let $\Phi: f \cong g$. By Lemma 2, there is a unique, continuous $G: I \times I \to E^1$ with $G(0, 0) = 0$ and $\Phi(x, y) = e^{2\pi i G(x,y)}$ for $(x, y) \in I \times I$.

Now, if $s \in I$, then $\Phi(0, s) = e^{2\pi i G(0,s)} = 1$.

Then, $G(0, s) \in Z$ for each $s \in I$. Since the function $s \to G(0, s)$ of $I \to E^1$ is continuous, then there is some integer a with $G(0, s) = a$ for each $s \in I$.

Similarly, $\Phi(1, s) = e^{2\pi i G(1,s)} = 1$, so for some integer b, $G(1, s) = b$ for each $s \in I$.

Now, $G(0, 0) = 0$, so $a = 0$.

Define $q, r \colon I \to E^1$ by $q(t) = G(t, 0)$, $r(t) = G(t, 1)$, $0 \le t \le 1$. Then, $q(0) = G(0, 0) = 0$, and, for each $t \in I$,

$$e^{2\pi i q(t)} = e^{2\pi i G(t,0)} = \Phi(t, 0) = f(t).$$

Then, $\alpha_f = q(1) = G(1, 0) = b$.

Similarly,
$$r(0) = G(0, 1) = a = 0, \text{ and, for each } t \in I,$$
$$e^{2\pi i r(t)} = e^{2\pi i G(t,1)} = \Phi(t, 1) = g(t).$$

Then, $\alpha_g = r(1) = G(1, 1) = b$.

Thus, $\alpha_f = \alpha_g$. ∎

2.1 THEOREM $\pi(S^1, (1, 0))$ *is isomorphic to Z.*

Proof If $[f] \in \pi(S^1, (1, 0))$, let $\beta([f]) = \alpha_f (= f^*(1))$. In view of Lemma 3, β is a well-defined function from $\pi(S^1, (1, 0))$ to Z.

We now proceed in three steps.

1) β is an homomorphism.

Let $[f], [g] \in \pi(S^1, (1, 0))$. Then,

$$\beta([f] [g]) = \beta([fg]) = (fg)^* (1)$$
and
$$\beta([f]) + \beta([g]) = f^*(1) + g^*(1).$$

Thus, we must show that $(fg)^* (1) = f^*(1) + g^*(1)$.

Now, $f = e^{2\pi i f^*}$ and $g = e^{2\pi i g^*}$ and $fg = e^{2\pi i (fg)^*}$. For $0 \le t \le \frac{1}{2}$,

$$e^{2\pi i (fg)^*(t)} = (fg)(t) = f(2t) = e^{2\pi i f^*(2t)}.$$

Then, $(fg)^* (t) - f^*(2t)$ is an integer for each t in $[0, \frac{1}{2}]$. Since the function $t \to (fg)^* (t) - f^*(2t)$, $0 \le t \le \frac{1}{2}$, is continuous, then there is an integer a such that
$$(fg)^* (t) - f^*(2t) = a \quad \text{for} \quad 0 \le t \le \tfrac{1}{2}. \tag{1}$$

For $\frac{1}{2} \le t \le 1$,

$$e^{2\pi i (fg)^*(t)} = (fg)(t) = g(2t - 1) = e^{2\pi i g^*(2t-1)}.$$

Reasoning as before, there is an integer b with
$$(fg)^*(t) - g^*(2t - 1) = b \quad \text{for } \tfrac{1}{2} \leq t \leq 1. \tag{2}$$
By (1),
$$(fg)^*(0) - f^*(0) = 0 - 0 = a.$$
Also by (1),
$$(fg)^*(\tfrac{1}{2}) - f^*(1) = 0.$$
Thus,
$$(fg)^*(\tfrac{1}{2}) = \alpha_f.$$
By (2),
$$(fg)^*(\tfrac{1}{2}) - g^*(0) = b.$$
Then,
$$b = (fg)^*(\tfrac{1}{2}) = \alpha_f.$$
Also by (2),
$$(fg)^*(1) - g^*(1) = \alpha_{fg} - \alpha_g = b = \alpha_f.$$
Thus,
$$\alpha_{fg} = \alpha_f + \alpha_g.$$

2) β is a surjection.

Let n be an integer. Let $f(t) = e^{2\pi i n t}$ for $0 \leq t \leq 1$. Then, $[f] \in \pi(S^1, (1,0))$. Further, by uniqueness in Lemma 1, $f^*(t) = tn$ for $0 \leq t \leq 1$. Thus, $\beta([f]) = f^*(1) = n$.

3) β is an injection.

It suffices to show that $\ker(\beta) = \{[e_{(1,0)}]\}$, where $e_{(1,0)}: S^1 \to \{(1,0)\}$. Suppose $[f] \in \pi(S^1, (1,0))$ and $\beta([f]) = 0$. Then, $f^*(1) = 0$. Now, $f^*: I \to E^1$, and $f^*(0) = 0$, so that $[f^*] \in \pi(E^1, 0) = \{[e_0]\}$, where $e_0: E^1 \to \{0\}$. Thus, $[f^*] = [e_0]$. But then $f^* \cong e_0$, so that $f = \text{Exp} \circ f^* \cong \text{Exp} \circ e_0 = e_{(1,0)}$. Thus, $[f] = [e_{(1,0)}]$, and the proof is complete. ∎

2.1 EXAMPLE As an illustration of the techniques developed in this section, we shall prove the Fundamental Theorem of Algebra. This theorem says that each equation $p(z) = 0$, with p a non-constant, complex-coefficient polynomial, has a complex root. Surprisingly, this "fundamental theorem" has no known proof employing only elementary algebra (that is, only algebra at the level needed to state the theorem).

Begin with a polynomial p of degree n, say
$$p(z) = a_0 + a_1 z + \cdots + a_{n-1} z^{n-1} + a_n z^n,$$
where the a_j's are complex and $n \geq 0$. We lose no generality by assuming the coefficient of z^n to be 1. If $q(z) = b_0 + b_1 z + \cdots + b_n z^n$ and $b_n \neq 0$, then $q(z)/b_n$ and $q(z)$ have exactly the same roots.

The proof is by contradiction. Suppose that $p(z) \neq 0$ for each complex number z, and that $n \geq 1$.

For each $r \geq 0$, we can then let $\gamma_r(t) = \dfrac{|p(r)|}{p(r)} \dfrac{p(re^{2\pi i t})}{|p(re^{2\pi i t})|}$, $0 \leq t \leq 1$.

Then, γ_r is a loop at $(1, 0)$ in S^1. Further, $\gamma_0 = e_{(1,0)}: I \to \{(1,0)\}$.

We claim that $\gamma_r \cong \gamma_0$ for each $r \geq 0$. For, if $r > 0$, let

$$\Phi(t, s) = \gamma_{sr}(t), \quad 0 \leq t \leq 1, \quad 0 \leq s \leq 1.$$

Then, $\Phi: \gamma_0 \cong \gamma_r$.

By Lemma 3, $\alpha_{\gamma_0} = \alpha_{\gamma_r}$. But, $\alpha_{\gamma_0} = 0$, so we must have $\alpha_{\gamma_r} = 0$ for each $r \geq 0$.

We shall now produce a number $\varrho > 0$ such that $\alpha_{\gamma_\varrho} = n$.

Take ϱ to be any number such that $\varrho \geq n + 1$ and also $\varrho \geq (n + 1)|a_j|$ for $j = 0, 1, \ldots, n - 1$.

Let $q(t) = \varrho e^{2\pi i t}$, for $0 \leq t \leq 1$, and let

$$\Psi(t, s) = \frac{|s\varrho^n + (1 - s)p(\varrho)|}{(s\varrho^n + (1 - s)p(\varrho))} \frac{(s(q(t))^n + (1 - s)p(q(t)))}{|s(q(t))^n + (1 - s)p(q(t))|}$$

for $0 \leq t \leq 1$, $0 \leq s \leq 1$. The reader should check that the choice of ϱ insures that the denominator is never zero, hence $\Psi: I \times I \to S^1$. Further, Ψ is continuous, $\Psi(t, 0) = \gamma_\varrho(t)$, $\Psi(t, 1) = e^{2\pi i n t}$, $\Psi(0, s) = (1, 0)$, and $\Psi(1, s) = (1, 0)$. Now, let $r(t) = e^{2\pi i n t}$, $0 \leq t \leq 1$. Then, r is a loop at $(1, 0)$ in S^1, and $\alpha_r = n$. Further, $\Psi: \gamma_\varrho \cong r$, so that $\alpha_{\gamma_\varrho} = n$, and we have a contradiction. Hence, $n = 0$ (so p is a constant polynomial) or $p(z) = 0$ for some complex number z, and the theorem is proved. ∎

11.3 Dependence upon the base point

Since $\pi(X, x_0)$ consists of equivalence classes of loops at x_0, we might naturally expect that $\pi(X, x_1)$ might be a different group if x_1 is a different point of X. We shall now show, however, that $\pi(X, x_0)$ and $\pi(X, x_1)$ are isomorphic if x_0 and x_1 belong to the same path component of X.

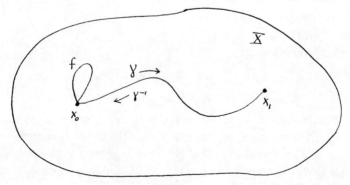

FIGURE 3.1

It is easy to construct a suitable isomorphism. If γ is a path from x_0 to x_1 in X, then with any loop f at x_0 we can associate a loop $\gamma^{-1}f\gamma$ at x_1, with $\gamma^{-1}(t) = \gamma(1 - t)$ for $0 \leq t \leq 1$. Think of $\gamma^{-1}f\gamma$ as the path beginning at x_1, tracing over γ^{-1} to x_0, then around f, and back along γ in the opposite direction to x_1 again (Fig. 3.1). This suggests we map $[f] \to [\gamma^{-1}f\gamma]$. All that remains is to show that this is an isomorphism.

3.1 THEOREM *Let x_0 and x_1 lie in the same path component of X. Then,*
$\pi(X, x_0)$ *and* $\pi(X, x_1)$ *are isomorphic.*

Proof Let γ be a path from x_0 to x_1 in X. If $[f] \in \pi(X, x_0)$, it is easy to check that $[\gamma^{-1}f\gamma] \in \pi(X, x_1)$. Further, if $f \cong g$, then $[\gamma^{-1}f\gamma] = [\gamma^{-1}g\gamma]$, by (two applications of) Theorem 1.4. Hence we may define a function τ: $\pi(X, x_0) \to \pi(X, x_1)$ by putting $\tau([f]) = [\gamma^{-1}f\gamma]$ for each $[f] \in \pi(X, x_0)$.

We now proceed in three steps to show that τ is an isomorphism.

1) τ is an homomorphism.

Let $[f], [g] \in \pi(X, x_0)$. Then, $\tau(([f])([g])) = \tau([fg]) = [\gamma^{-1}fg\gamma]$ and $\tau([f])\tau([g]) = [\gamma^{-1}f\gamma][\gamma^{-1}g\gamma] = [\gamma^{-1}f\gamma\gamma^{-1}g\gamma]$. It suffices to show that $[\gamma^{-1}fg\gamma] = [\gamma^{-1}f\gamma\gamma^{-1}g\gamma]$.

The reader can check that $\gamma\gamma^{-1} \cong e_{x_0}$, and that $[\gamma^{-1}f][e_{x_0}] = [\gamma^{-1}f]$. Finally, it is easy to verify that:

$$[\gamma^{-1}f\gamma\gamma^{-1}g\gamma] = [\gamma^{-1}f][\gamma\gamma^{-1}][g\gamma] = [\gamma^{-1}f][e_{x_0}][g\gamma] = [\gamma^{-1}f][g\gamma]$$
$$= [\gamma^{-1}fg\gamma].$$

2) τ is surjective.

Let $[f] \in \pi(X, x_1)$. Then, $[\gamma f \gamma^{-1}] \in \pi(X, x_0)$, and the reader can check that $\tau([\gamma f \gamma^{-1}]) = [f]$.

3) τ is injective.

Suppose that $[f], [g] \in \pi(X, x_0)$, and that $\tau([f]) = \tau([g])$. Then, $[\gamma^{-1}f\gamma] = [\gamma^{-1}g\gamma]$.

It is easy to check that:

and
$$f\gamma \cong \gamma\gamma^{-1}f\gamma \cong \gamma\gamma^{-1}g\gamma \cong g\gamma,$$
$$f \cong f\gamma\gamma^{-1} \cong g\gamma\gamma^{-1} \cong g, \text{ so } f \cong g. \blacksquare$$

Note that there is more to this proof than appears at first glance. That $[\gamma\gamma^{-1}] = [e_{x_0}]$ seems obvious, but $[\gamma]$ and $[\gamma^{-1}]$ are not elements of either $\pi(X, x_0)$ or $\pi(X, x_1)$, so the fact that $[\gamma\gamma^{-1}] = [e_{x_0}]$ must be shown (although the proof is very like that of Theorem 1.6 (3)). Similarly, it must be verified that $[\gamma^{-1}f][e_{x_0}] = [\gamma^{-1}f]$.

1 COROLLARY *Let (X, T) be path-connected.
Let $x_0, x_1 \in X$.
Then,
$\pi(X, x_0)$ is isomorphic to $\pi(X, x_1)$.*

Proof Immediate by Theorem 3.1. ∎

If (X, T) is path connected, then, the base point is of no importance (except perhaps as a device used in actually calculating the group). In this event, $\pi(X, x)$ is usually just written $\pi(X)$. Thus, for example, from now on we will write $\pi(S^1)$, $\pi(E^n)$, and so on.

11.4 The fundamental group of a product space

It seems natural to hope that $\pi(X \times Y, (x_0, y_0))$ should be obtainable from the individual groups $\pi(X, x_0)$ and $\pi(Y, y_0)$. It is, by taking a direct product (or sum, depending upon your notation) of groups.

4.1 THEOREM $\pi(X \times Y, (x_0, y_0))$ *is isomorphic to* $\pi(X, x_0) \oplus \pi(Y, y_0)$.

Proof Let $p_1 \colon X \times Y \to X$ and $p_2 \colon X \times Y \to Y$ be the projections.
Define a map $\varphi \colon \pi(X \times Y, (x_0, y_0)) \to \pi(X, x_0) \oplus \pi(Y, y_0)$ by setting, for each $[f] \in \pi(X \times Y, (x_0, y_0))$,

$$\varphi([f]) = ([p_1 \circ f], [p_2 \circ f]).$$

We leave it for the reader to check that φ is an isomorphism. ∎

Of course, a simple induction extends this to arbitrary finite products.

4.2 THEOREM $\pi(X_1 \times \cdots \times X_n, (x_1, \ldots, x_n))$ *is isomorphic to the direct sum* $\pi(X_1, x_1) \oplus \cdots \oplus \pi(X_n, x_n)$.

Proof Left as an exercise. ∎

4.1 EXAMPLE The homotopy group of a torus is the additive group of Gaussian integers, i.e. the group of complex numbers $a + ib$ with a and b integers. For, a torus is $S^1 \times S^1$, so that

$$\pi(S^1 \times S^1) = \pi(S^1) \oplus \pi(S^1) = Z \oplus Z. \quad \blacksquare$$

4.2 EXAMPLE The fundamental group of a cylinder is the group of integers. Write the cylinder as $S^1 \times I$. Now, $\pi(I) = \{0\}$, and $\pi(S^1) = Z$, so that $\pi(S^1 \times I) = Z \oplus \{0\} = Z$. (Of course, it is understood that we are writing = in place of a symbol for isomorphism here, as is customary in algebraic topology.) ∎

4.3 EXAMPLE $\pi(E^n)$ is trivial, since $\pi(E^1)$ is trivial (Example 1.1). ∎

11.5 Maps and fundamental groups

Maps between spaces induce homomorphisms of the fundamental groups in an obvious but extremely useful way.

Suppose $\varphi: X \to Y$ is continuous, and $x_0 \in X$. If f is a loop at x_0 in X, then $\varphi \circ f$ is a loop at $\varphi(x_0)$ in Y. Further, if $f \cong g$, then $\varphi \circ f \cong \varphi \circ g$. This means that we can define a function $\hat{\varphi}$ from $\pi(X, x_0)$ to $\pi(Y, \varphi(x_0))$ by putting $\hat{\varphi}([f]) = [\varphi \circ f]$ for each $[f] \in \pi(X, x_0)$. $\hat{\varphi}$ is said to be induced by φ, and turns out to have useful algebraic properties.

5.1 Theorem *Let φ be continuous: $X \to Y$.*
Let $x_0 \in X$.
Then,
$\hat{\varphi}: \pi(X, x_0) \to \pi(Y, \varphi(x_0))$ is an homomorphism.

Proof It is immediate that $\hat{\varphi}$ is a function.
Now let $[f], [g] \in \pi(X, x_0)$. It is routine to check that $\varphi \circ (fg) = (\varphi \circ f)(\varphi \circ g)$, hence $\hat{\varphi}([f][g]) = \hat{\varphi}([f]) \hat{\varphi}([g])$. ∎

It is not true that $\hat{\varphi}$ is surjective if φ is. For example, $\varphi(x) = e^{2\pi i x}$ for x real defines a continuous surjection of E^1 to S^1, but $\hat{\varphi}: \pi(E^1) = \{0\} \to \pi(S^1) = Z$ cannot be a surjection.

We shall now show that the operations of $\wedge$ and $\circ$ (composition) commute.

5.2 Theorem *Let $\varphi: X \to Y$ be (T, M) continuous.*
Let $\xi: Y \to K$ be (M, N) continuous.
Let $x_0 \in X$.
Then,
$\widehat{(\xi \circ \varphi)} = \hat{\xi} \circ \hat{\varphi}: \pi(X, x_0) \to \pi(K, \xi \circ \varphi(x_0))$.

Proof A routine calculation, which is left to the reader. ∎
The identity map induces the identity homomorphism.

5.3 Theorem *$\hat{1}_X: \pi(X, x_0) \to \pi(X, x_0)$ is the identity homomorphism.*

Proof Left to the reader. ∎

An immediate consequence of these results is that homeomorphic spaces have isomorphic homotopy groups (calculated at the appropriate base points).

5.4 Theorem *Let $h: X \to Y$ be a (T, M) homeomorphism.*
Let $x_0 \in X$.
Then,
$\hat{h}: \pi(X, x_0) \to \pi(Y, h(x_0))$ is an isomorphism.

Proof $\hat{1}_Y = \widehat{(h \circ \text{inv } h)} = \hat{h} \circ \widehat{(\text{inv } h)} =$ identity homomorphism: $\pi(Y, h(x_0)) \to \pi(Y, h(x_0))$, and $\hat{1}_X = \widehat{((\text{inv } h) \circ h)} = \widehat{(\text{inv } h)} \circ \hat{h} =$ identity homomorphism: $\pi(X, x_0) \to \pi(X, x_0)$.

Then, $\hat{h}: \pi(X, x_0) \to \pi(X, x_0)$ is an isomorphism. ∎

The converse of this theorem is false. For example, E^1 and $\{0\}$ have isomorphic homotopy groups (the trivial group), but are certainly not homeomorphic. Thus one cannot in general tell whether spaces are homeomorphic by looking at the homotopy groups. However, it is often possible to test that spaces are not homeomorphic.

5.1 Example $[0, 1]$ and S^1 are not homeomorphic. For, $\pi(S^1) = Z$ and $\pi([0, 1]) = \{0\}$. ∎

5.2 Example We shall use our theorems on induced homomorphisms to give an algebraic proof of Brouwer's Theorem.

Suppose that S^1 is a retract of B_2. That is, suppose there is a map $r: B_2 \to S^1$ such that $r|S^1 = 1_{S^1}$.

Let In be the inclusion map: $S^1 \to B_2$ defined by In $(x) = x$ for each $x \in S^1$.

Then,

$$\pi(S^1) \xrightarrow{\widehat{\text{In}}} \pi(B_2) \xrightarrow{\hat{r}} \pi(S^1)$$
$$\underset{\widehat{r \circ \text{In}}}{\longrightarrow}$$

We then have a group diagram:

$$Z \xrightarrow{\alpha} \{0\} \xrightarrow{\beta} Z$$
$$\underset{\beta \circ \alpha = 1_Z}{\longrightarrow}$$

which is algebraically impossible, as then $\beta(\alpha(Z)) = \beta(\{0\}) = \{0\} = 1_Z(Z) = Z$. ∎

Finally, we shall prove perhaps the most fundamental result on induced homomorphisms: homotopic maps induce the same homomorphism.

5.5 THEOREM *Let $\varphi, \xi \colon X \to Y$ be continuous.
Let $x_0 \in X$, and suppose that $\varphi(x_0) = \xi(x_0)$.
Let $\varphi \simeq \xi$.
Then,*
$$\hat{\varphi} = \hat{\xi} \colon \pi(X, x_0) \to \pi(Y, \varphi(x_0)).$$

Proof Routine. ∎

5.3 EXAMPLE The last theorem yields a shorter algebraic proof of Brouwer's Theorem. Suppose $1_{S^1} \simeq 0$, say $1_{S^1} \simeq d \colon S^1 \to \{y\}$ for some $y \in S^1$. Then, $\hat{1}_{S^1} = \hat{d}$. But, $\hat{1}_{S^1} \colon Z \to Z$ is the identity homomorphism, while $\hat{d} \colon Z \to \{0\}$ is the trivial homomorphism. ∎

Of course, when the spaces involved in these theorems are path connected, then the base points may be omitted from the statements.

11.6 Homotopy type

6.1 DEFINITION *(X, T) and (Y, M) are of the same homotopy type
if and only if
there are maps $f \in Y^X$ and $g \in X^Y$ such that $g \circ f \simeq 1_X$ and $f \circ g \simeq 1_Y$.*

Often spaces are said to be homotopically equivalent if they are of the same homotopy type. The map f of the definition is then called a homotopy equivalence of X to Y, and g a homotopy equivalence of Y to X.

6.2 DEFINITION *$(X, T) \equiv (Y, M)$
if and only if
(X, T) and (Y, M) are of the same homotopy type.*

We show first that homotopy equivalences induce isomorphisms between the homotopy groups.

6.1 THEOREM *Let $f \colon X \to Y$ be an homotopy equivalence.
Let $x_0 \in X$.
Then,
$\hat{f} \colon \pi(X, x_0) \to \pi(Y, f(x_0))$ is an isomorphism.*

Proof Let $g \in X^Y$ such that $g \circ f \simeq 1_X$ and $f \circ g \simeq 1_Y$.
Then,
$$\widehat{g \circ f} = \hat{g} \circ \hat{f} = \hat{1}_X = \text{identity homomorphism} \colon \pi(X, x_0) \to \pi(X, x_0),$$

and
$$\widehat{f \circ g} = \hat{f} \circ \hat{g} = \hat{1}_Y = \text{identity homomorphism: } \pi(Y, f(x_0)) \to \pi(Y, f(x_0)).$$
Then, $\hat{f}$ is an isomorphism. ∎

1 COROLLARY *Let $(X, T) \equiv (Y, M)$.*
Let (X, T) and (Y, M) be path connected.
Then,
$\pi(X)$ is isomorphic to $\pi(Y)$.

Proof Immediate by Theorem 6.1. ∎

6.2 THEOREM *Let $h: X \to Y$ be a (T, M) homeomorphism.*
Then,
h is an homotopy equivalence.

Proof Left as an exercise. ∎

2 COROLLARY *If $(X, T) \cong (Y, M)$, then $(X, T) \equiv (Y, M)$.*

Proof Immediate by Theorem 6.2. ∎

6.3 THEOREM *$\equiv$ is an equivalence relation on the class of topological spaces.*

Proof Left to the reader. ∎

It is of course not true that homotopically equivalent spaces are in general homeomorphic. We shall illustrate this and the notion of homotopy equivalence by introducing the concept of contractibility.

A space is said to be contractible if it can be shrunk smoothly to a point. More carefully:

6.3 DEFINITION *(X, T) is contractible*
if and only if
$1_X \simeq 0$.

It is easy to check that E^n (and any convex subset of E^n) is contractible, while, by Brouwer's Theorem, S^1 is not. We shall now see that each contractible space is homotopically equivalent to a one-point space, and has a trivial homotopy group at each point. However, we first show that each contractible space is path connected, so that actually a base point need not be chosen.

6.4 THEOREM *Let (X, T) be contractible.*
Then,
(X, T) is path connected.

Proof Let x_0 and y_0 be distinct points of X. We claim first that $c_{x_0} \simeq c_{y_0}$. To see this, reason as follows.
For some $z_0 \in X$, $1_X \simeq c_{z_0}$.
Now,
$$c_{x_0} = 1_X \circ c_{x_0} \simeq c_{z_0} \circ c_{x_0} = c_{z_0}$$
and, similarly,
$$c_{y_0} \simeq c_{z_0}.$$
Then,
$$c_{x_0} \simeq c_{y_0}.$$
Let
$$\Phi\colon c_{x_0} \simeq c_{y_0}.$$

Let $h\colon I \to \{x_0\} \times I$ be the homeomorphism defined by $h(t) = (x_0, t)$ for $0 \leq t \leq 1$.
Then, $\Phi \circ h\colon I \to X$, and $\Phi \circ h(0) = \Phi(x_0, 0) = x_0$, and $\Phi \circ h(1) = \Phi(x_0, 1) = y_0$. Thus, $\Phi \circ h$ is a path from x_0 to y_0 in X. ∎

6.5 THEOREM *Let (X, T) be contractible.*
Then,
for any a, $(X, T) \equiv \{a\}$.

Proof For some $x_0 \in X$, $1_X \simeq c_{x_0}$.
Now,
$$(c_{x_0}|\{x_0\}) \circ c_{x_0} = c_{x_0} \simeq 1_X$$
and
$$c_{x_0} \circ (c_{x_0}|\{x_0\}) = 1_{\{x_0\}}.$$
Thus, $(X, T) \equiv \{x_0\}$.
Since $\{x_0\} \cong \{a\}$, then $\{x_0\} \equiv \{a\}$, so $(X, T) \equiv \{a\}$. ∎

3 COROLLARY *Any two contractible spaces are of the same homotopy type.*

Proof Immediate by Theorem 6.5 and Theorem 6.3. ∎

6.6 THEOREM *Let (X, T) be contractible.*
Then,
$\pi(X)$ is trivial.

Proof Note first that (X, T) is path connected by Theorem 6.4.
Now, $(X, T) \equiv \{a\}$ for any a by Theorem 6.5.
By Corollary 1, $\pi(X)$ is isomorphic to $\pi(\{a\})$, which is trivial. ∎

11.7 Higher homotopy groups

We have used the name "first homotopy group" for $\pi(X, x_0)$. The reason for this is that there are also higher homotopy groups, $\pi_n(X, x_0)$ for $n = 2, 3, \ldots$ When these groups are under consideration as well, then $\pi(X, x_0)$ is denoted quite naturally $\pi_1(X, x_0)$.

We shall mention briefly some basic facts about $\pi_n(X, x_0)$. It is a good exercise to prove the assertions we shall make, the arguments being quite similar to those used for $\pi_1(X, x_0)$.

Let I^n be the n-cube $I \times \cdots \times I = \{(t_1, \ldots, t_n) | 0 \leq t_j \leq 1 \text{ for } j = 1, \ldots, n\}$. The boundary $\dot{I}^n$ of I^n consists of all points $(t_1, \ldots, t_n)$ in I^n with some $t_j = 0$ or 1.

Choose a base point $x_0 \in X$. If $f, g \in X^{I^n}$, and $f(1, t_2, \ldots, t_n) = g(0, t_2, \ldots, t_n)$ for each $(t_1, \ldots, t_n) \in I^n$, define

$$(fg)(t_1, \ldots, t_n) = \begin{cases} f(2t_1, t_2, \ldots, t_n) \text{ for } (t_1, \ldots, t_n) \in I^n, 0 \leq t_1 \leq \tfrac{1}{2}, \\ g(2t_1 - 1, t_2, \ldots, t_n) \text{ for } (t_1, \ldots, t_n) \in I^n, \tfrac{1}{2} \leq t_1 \leq 1. \end{cases}$$

Then, $fg \in X^{I^n}$.

Now let $\mathscr{P}_{x_0} = X^{I^n} \cap \{f | f(\dot{I}^n) = f(x_0)\}$.

If $f, g \in \mathscr{P}_{x_0}$, denote $f \cong g$ if there is a homotopy $\Phi: I^n \times I \to X$ with $\Phi((t_1, \ldots, t_n), 0) = f(t_1, \ldots, t_n)$, and $\Phi((t_1, \ldots, t_n), 1) = g(t_1, \ldots, t_n)$ for $(t_1, \ldots, t_n) \in I^n$, and $\Phi((t_1, \ldots, t_n), s) = x_0$ for $0 \leq s \leq 1$, $(t_1, \ldots, t_n) \in \dot{I}^n$.

Then, $\cong$ is an equivalence relation on $\mathscr{P}_{x_0}$, and we define $\pi_n(X, x_0) = \mathscr{P}_{x_0}/\cong$. This is a group under the multiplication: $[f][g] = [fg]$. Further, $\pi_n(X, x_0)$ is abelian if $n \geq 2$ ($\pi_1(X, x_0)$ need not be abelian).

If x_0 and x_1 are in the same path component of X, then $\pi_n(X, x_0)$ is isomorphic to $\pi_n(X, x_1)$. Otherwise, the two groups may be totally unrelated.

As with the fundamental group, a continuous function $\varphi: X \to Y$ induces a homomorphism $\hat{\varphi}: \pi_n(X, x_0) \to \pi_n(Y, \varphi(x_0))$. Also, $\widehat{\varphi \circ \xi} = \hat{\varphi} \circ \hat{\xi}$ and $\hat{1}_X =$ identity homomorphism. Finally, if X and Y are homotopically equivalent, then their higher homotopy groups are isomorphic.

As the reader has no doubt surmised by now, the theory of the higher homotopy groups develops along lines similar to those we have already seen for the fundamental group. In general, $\pi_n(X, x_0)$ is extremely difficult to calculate. For example, even $\pi_m(S^n)$ is not known in general, although $\pi_n(S^n)$ is infinite cyclic, and $\pi_m(S^n)$ is trivial for $m < n$.

Problems

1) Write out a proof that each convex subset C of E^n is contractible. Hence, $\pi(C) = \{0\}$.

2) Calculate $\pi(E^2 - \{(0, 0)\})$.

3) Let $X = \{1, 2, 3, 4, 5\}$, with (a) discrete (b) indiscrete topology. Calculate the homotopy group of X at each point of x for each case.
   Is X contractible?

4) (X, T) is contractible
   if and only if
   for any space (Y, M), and each pair of maps $f, g \colon Y \to X$, we have $f \simeq g$.

5) (X, T) is contractible
   if and only if
   (X, T) is deformable to $\{x\}$ for some $x \in X$.

6) Suppose that $(X, T) \equiv (Y, M)$.
   (a) If (X, T) is path connected, so is (Y, M).
   (b) (X, T) may be compact, and (Y, M) not compact.

7) Let (X, T) be path connected.
   Then,
   $\pi(X)$ is trivial
   if and only if
   each map $f \colon S^1 \to X$ has a continuous extension $F \colon B_2 \to X$.

*8) Is Theorem 4.2 true for infinite products?

9) Assuming that $\pi(S^n)$ is trivial, for $n \geq 2$, show that S^1 and S^n are not homeomorphic for $n \geq 2$.

10) Define: A is a deformation retract of X if and only if there is a retraction $r \colon X \to A$ and a homotopy $\Phi \colon 1_X \simeq r$ such that $\Phi(a, t) = a$ for $a \in A$ and $t \in I$.
    Prove: If A is a deformation retract of X, then the inclusion map In: $A \to X$ induces an isomorphism between $\pi(A, a)$ and $\pi(X, a)$ for any $a \in A$.

11) Let r be a retraction of X to A, where X and A are path connected. Let In: $A \to X$ be the inclusion map. Then, $\pi(X)$ is isomorphic to $(\ker \hat{r}) \oplus (\widehat{\text{In}}\, \pi(A))$.

*12) If G is a topological group and $x \in G$, then $\pi(G, x)$ is abelian.

*13) If G is a topological group and $\pi(G) = \{0\}$, and H is a discrete normal subgroup, then $\pi(G/H, H)$ is isomorphic to H. (Note that S^1 may be thought of as E^1/Z; also $\pi(E^1) = \{0\}$, Z is a discrete normal subgroup and $\pi(S^1) = Z$. Hence mirror the proof that $\pi(S^1) = Z$.)

*14) Suppose $(\bar{X}, \bar{T})$ and (X, T) are path connected and locally path connected, and $p: \bar{X} \to X$. Then, $(\bar{X}, p)$ is a covering space of X if:

1) p is a continuous surjection.

2) If $x \in X$, then there is a nghd. U_x of x such that $p^{-1}(U_x)$ is a disjoint union $\bigcup_{\alpha \in \mathcal{O}} V_\alpha(x)$ of open sets, and $p|V_\alpha(x): V_\alpha(x) \to U_x$ is a homeomorphism for each $\alpha \in \mathcal{O}$. (U_x is called an elementary neighborhood of x).

(a) Show that (E^1, p) is a covering space of S^1, where $p(x) = e^{2\pi i x}$.

(b) p is an open map.

(c) If $\bar{x}_0 \in \bar{X}$, and $p(\bar{x}_0) = x_0 \in X$, and f is a path in X with initial point x_0, then there is a unique path g in $\bar{X}$ with initial point $\bar{x}_0$ and $p \circ g = f$ (that is, f can be lifted to the covering space).

(d) If g and g' are paths in $\bar{X}$ with initial point $\bar{x}_0$, then $pg \simeq pg'$ if and only if $g \simeq g'$.

(e) The induced homomorphism $\hat{p}: \pi(\bar{X}, x) \to \pi(X, p(x))$ is an injection for each $x \in \bar{X}$.

(f) $p^{-1}(x)$ and $p^{-1}(y)$ have the same cardinality for each $x, y \in X$. (The cardinality of $p^{-1}(x)$ for any $x \in X$ is called the number of sheets of the covering space.)

15) Show that $E^{n+1} - \{(0, \ldots, 0)\}$ is of the same homotopy type as S^n for $n \geq 2$.

16) Let $A \subset E^n$ and let $x \in A$. A is said to be starlike at x if $\{ty + (1 - t)x | 0 \leq t \leq 1\} \subset A$ for each $y \in A$.

Show that A is contractible if A is starlike at x. Thus, for example, compute the first homotopy group of the region shown in Figure 1.

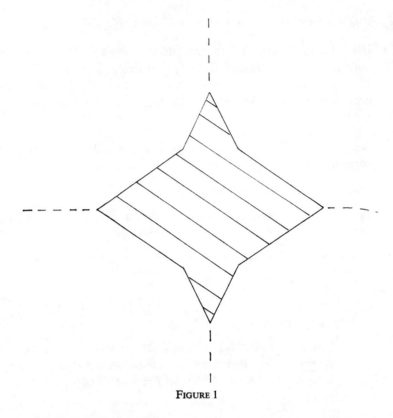

FIGURE 1

*17) Calculate the fundamental group of a Möbius band $\mathcal{M}$ by showing that $\mathcal{M}$ is of the same homotopy type as S^1.

APPENDIX A

Logical preliminaries

IN THIS APPENDIX we shall list some of the rules of logic used in manipulating propositions and implications. Nothing will be proved — a purely intuitive, working understanding is all that is needed.

If A and B are statements, or propositions, we write

$$A \Rightarrow B,$$

read "A implies B" if the truth of A implies that of B. A is said to be sufficient for B, and B is necessary for A. If $A \Rightarrow B$ and $B \Rightarrow A$, then we write $A \Leftrightarrow B$, and say that A and B are (logically) equivalent.

The negation of A is written: not A.

Note that

$$(A \Rightarrow B) \Leftrightarrow ((\text{not } B) \Rightarrow (\text{not } A)).$$

That is, "A implies B" is equivalent to "B is false implies A is false". Parentheses are inserted wherever needed to avoid ambiguity.

EXAMPLE (f is differentiable at 2) $\Rightarrow$ (f is continuous at 2).
This is equivalent to:
(f is not continuous at 2) $\Rightarrow$ (f is not differentiable at 2). ∎

It is important to be able to manipulate combinations of "and", "or", "implies" and "not" correctly. The more commonly used rules are:
(A and B) $\Leftrightarrow$ (B and A).
(A and B) $\Rightarrow$ A.
((A and B) and C) $\Leftrightarrow$ (A and (B and C)).
((A and B) and (A and C)) $\Leftrightarrow$ (A and (B and C)).
(($A \Rightarrow B$) and ($B \Rightarrow C$)) $\Rightarrow$ ($A \Rightarrow C$).
(A or B) $\Leftrightarrow$ (B or A).
$A \Rightarrow (A$ or B).
((A or B) or C) $\Leftrightarrow$ (A or (B or C)).
((A or B) or (A or C)) $\Leftrightarrow$ (A or (B or C)).
not (A and B) $\Leftrightarrow$ ((not A) or (not B)).
not (A or B) $\Leftrightarrow$ ((not A) and (not B)).

$((A$ and $B)$ or $C) \Leftrightarrow ((A$ or $C)$ and $(B$ or $C))$.
$((A$ or $B)$ and $C) \Leftrightarrow ((A$ and $C)$ or $(B$ and $C))$.
(not (for each x, $P[x]$)) $\Leftrightarrow$ (for some x, not $P[x]$).

Keep in mind that "or" is used mathematically in a non-exclusive sense. That is, A or B does not exclude the possibility of A and B.

Finally, we mention the "empty implication". If A is false, then the *implication* $(A \Rightarrow B)$ is valid for any B. Note that this says nothing about the truth of B, simply that A implies B is a valid logical implication.

APPENDIX B

Set theory

THIS APPENDIX IS a brief outline of the set theory we shall be using.

B-1 Basic vocabulary

If $P[x]$ is a statement for each object x, then the set of objects x which make $P[x]$ true is denoted
$$\{x|P[x]\}.$$

Note that the symbol x plays no special role here. That is, $\{y|P[y]\}$ denotes the same set of objects, namely those which make $P[\ ]$ a true statement when substituted into the blank space.

As examples,
$\{x|x \text{ is a real number}\}$ denotes the set of real numbers,
$\{\xi|2 \leq \xi \leq 3\}$ denotes the usual closed interval [2, 3] of calculus, and
$\{t|t \text{ is a camel}\}$ is the set of all camels.

Objects which satisfy the rule for membership in a set are called elements or members of that set. The statement "x is an element of A" is usually written: $x \in A$. If x is not an element of A, write $x \notin A$.

The basic set operations are union, intersection, difference and Cartesian product. The last will be treated later. The first three are defined:

union: $A \cup B = \{x|x \in A \text{ or } x \in B\}$.
intersection: $A \cap B = \{x|x \in A \text{ and } x \in B\}$.
difference: $A - B = \{x|x \in A \text{ and } x \notin B\}$.

For example,
$\{x|x \text{ is rational}\} \cup \{x|x \text{ is irrational}\} = \{x|x \text{ is a real number}\}$.
$\{y|2 \leq y \leq 3\} \cup \{t|5/2 \leq t \leq 5\} = \{m|2 \leq m \leq 5\}$.
$\{x|2 \leq x \leq 3\} \cap \{t|5/2 \leq t \leq 5\} = \{\alpha|5/2 \leq \alpha \leq 3\}$.
$\{x|2 \leq x \leq 5\} - \{x|5/2 \leq x < 3\} = \{x|2 \leq x < 5/2\} \cup \{x|3 \leq x \leq 5\}$.

An empty set, denoted ϕ, is defined by
$$\phi = \{x|x \neq x\}.$$

Thus, ϕ has no elements. The need for such a set is obvious if, for example, you try to take the intersection of $\{x|2 < x < 5\}$ with $\{x|7 < x < 9\}$.

We say that A is a subset of B, written $A \subset B$, if each object in A is also in B. That is,

$$A \subset B \text{ if and only if } (x \in A \Rightarrow x \in B).$$

This does not exclude the possibility that $A = B$. In fact, we say that $A = B$ if $A \subset B$ and $B \subset A$. In this event, A and B have exactly the same elements. Note that A is always a subset of A.

For any set A, $\phi \subset A$. This is true because ϕ has no elements which are not in A. Alternatively, $\phi \subset A$ is true by the empty implication: $(x \in \phi$ implies $x \in A)$ is valid because, for any x, $x \in \phi$ is false. This means that there is only one empty set. If ϕ' is also a set with no elements, then $\phi \subset \phi'$ and $\phi' \subset \phi$, so $\phi = \phi'$. Thus we speak of the empty set ϕ.

B-2 Basic set identities

The following is a list of some basic set identities.

(1) $A \cup B = B \cup A.$ ⎫
(2) $A \cap B = B \cap A.$ ⎬ commutative laws
(3) $A \cup (B \cup C) = (A \cup B) \cup C.$ ⎫
(4) $A \cap (B \cap C) = (A \cap B) \cap C.$ ⎬ associative laws
(5) $A \cup (B \cap C) = (A \cup B) \cap (A \cup C).$
(6) $A \cap (B \cup C) = (A \cap B) \cup (A \cap C).$
(7) $A \cap B \subset A \subset A \cup B.$
(8) $A - B \subset A.$
(9) $A \cap (B - C) = (A \cap B) - C.$
(10) $(A - B) \cap (A - C) = A - (B \cup C).$
(11) $(A - B) \cup (A - C) = A - (B \cap C).$
(12) $A - (B - C) = (A - B) \cup (A \cap C).$
(13) $A \cap B = A$ if and only if $A \subset B.$
(14) $A - B = A$ if and only if $A \cap B = \phi.$
(15) $A \cup \phi = A - \phi = A.$
(16) $A \cap \phi = \phi.$

Generally, proving that two sets are equal is a two step operation: each set must be shown to be a subset of the other, a process which usually involves two separate arguments. By way of illustration, we shall prove (9) and (11).

Proof of (9) Part 1: $A \cap (B - C) \subset (A \cap B) - C$.

Let $x \in A \cap (B - C)$. Then, $(x \in A$ and $x \in B - C)$. Then, $((x \in A)$ and $(x \in B$ and $x \notin C))$. Then, $((x \in A$ and $x \in B)$ and $x \notin C)$. Then, $x \in A \cap B$ and $x \notin C$, so $x \in (A \cap B) - C$, proving Part 1.

Part 2: $(A \cap B) - C \subset A \cap (B - C)$.

Let $y \in (A \cap B) - C$. Then, $y \in A \cap B$ and $y \notin C$. Then, $((y \in A$ and $y \in B)$ and $y \notin C)$. Then, $(y \in A$ and $(y \in B$ and $y \notin C))$. Then, $y \in A$ and $y \in B - C$, so that $y \in A \cap (B - C)$. ∎

We do not have any hard and fast rules about the insertion of parentheses—just put them wherever their use will prevent ambiguity.

Note the method of proof used. In proving Part 1, we showed that the implication

$$x \in A \cap (B - C) \Rightarrow x \in (A \cap B) - C$$

is a valid one. Of course, the symbol x is of no particular importance—we used y in Part 2. Also, the possibility that $A \cap (B - C) = \phi$ is automatically taken into account, as then the desired implication is true by the empty implication.

Proof of (11) Part 1: $(A - B) \cup (A - C) \subset A - (B \cap C)$.

Let $t \in (A - B) \cup (A - C)$. Then, $t \in A - B$ or $t \in A - C$. Then, $((t \in A$ and $t \notin B)$ or $(t \in A$ and $t \notin C))$. Then, $((t \in A)$ and $(t \notin B$ or $t \notin C))$. Then, $((t \in A)$ and not $(t \in B$ and $t \in C))$. Then, $((t \in A)$ and not $(t \in B \cap C))$. Then, $t \in A$ and $t \notin B \cap C$. Then, $t \in A - (B \cap C)$.

Part 2: $A - (B \cap C) \subset (A - B) \cup (A - C)$.

Let $z \in A - (B \cap C)$. Then, $z \in A$ and $z \notin B \cap C$. Then, $z \in A$ and not $(z \in B$ and $z \in C)$. Then, $((z \in A)$ and $(z \notin B$ or $z \notin C))$. Then, $((z \in A$ and $z \notin B)$ or $(z \in A$ and $z \notin C))$. Then, $z \in A - B$ or $z \in A - C$. Then, $z \in (A - B) \cup (A - C)$. ∎

Of course, after you acquire a little experience, it is no longer necessary to be so laborious about each proof. But it is a good idea to write down a few such proofs until they become fairly automatic. In the text, we shall use set identities as a matter of easy conversation, without pausing for proof.

Be careful to avoid bad notation. For example, the symbol $A - B \cup C$ is meaningless (it might stand for $(A - B) \cup C$ or $A - (B \cup C)$, which are in general quite different).

As special notation, when A consists of a single element, say φ, write A as just $\{\varphi\}$, rather than the more cumbersome $\{x | x = \varphi\}$. Similarly, if A consists of elements $\varphi_1, \ldots, \varphi_n$, just write A as $\{\varphi_1, \ldots, \varphi_n\}$. $\{\varphi\}$ is often called a singleton.

B-3 General unions and intersections

Often it is necessary to take a union or intersection of an arbitrary number of sets. More specifically, suppose that A_α is a set for each $\alpha \in \mathcal{O}$. We say that the sets A_α are indexed by $\mathcal{O}$, and $\mathcal{O}$ is the index set. Then,

$$\bigcup_{\alpha \in \mathcal{O}} A_\alpha = \{x | \text{for some } \alpha, \alpha \in \mathcal{O} \text{ and } x \in A_\alpha\}.$$

$$\bigcap_{\alpha \in \mathcal{O}} A_\alpha = \{x | x \in A_\alpha \text{ for each } \alpha \in \mathcal{O}\}.$$

When $\mathcal{O} = \phi$, then $\bigcup_{\alpha \in \mathcal{O}} A_\alpha = \phi$. But $\bigcap_{\alpha \in \mathcal{O}} A_\alpha$ presents logical difficulties because of the empty implication, so we use the symbol $\bigcap_{\alpha \in \mathcal{O}} A_\alpha$ only when $\mathcal{O} \neq \phi$. Of course, the sets A_α may themselves be empty. $\bigcup_{\alpha \in \mathcal{O}} A_\alpha = \phi$ when $A_\alpha = \phi$ for each $\alpha \in \mathcal{O}$, and $\bigcap_{\alpha \in \mathcal{O}} A_\alpha = \phi$ when $A_\alpha = \phi$ for some $\alpha \in \mathcal{O}$ (one is enough).

There are many variations on the intersection and union notation, all obvious from context. For example, if B is a set of subsets of X (that is, each element of B is a subset of X), then we can take the union or intersection without indexing. Here,

and
$$\cup B = \{x | \text{for some } b, b \in B \text{ and } x \in b\},$$
$$\cap B = \{x | x \in b \text{ for each } b \in B\} \quad (\text{if } B \neq \phi).$$

These sets might also be written $\bigcup_{b \in B} b$ and $\bigcap_{b \in B} b$.

It is easy to check that:

and
$$\bigcap_{\alpha \in \mathcal{O}} A_\alpha \subset \bigcap_{\alpha \in \mathcal{B}} A_\alpha$$
$$\bigcup_{\alpha \in \mathcal{B}} A_\alpha \subset \bigcup_{\alpha \in \mathcal{O}} A_\alpha$$

if $\mathcal{B} \subset \mathcal{O}$.

DeMorgan's Laws state that, if $\mathcal{O} \neq \phi$, and $A_\alpha \subset X$ for each $\alpha \in \mathcal{O}$, then
$$X - \bigcup_{\alpha \in \mathcal{O}} A_\alpha = \bigcap_{\alpha \in \mathcal{O}} (X - A_\alpha)$$
and
$$X - \bigcap_{\alpha \in \mathcal{O}} A_\alpha = \bigcup_{\alpha \in \mathcal{O}} (X - A_\alpha).$$

As an illustration, we shall prove the first one.

SET THEORY

Part 1 $X - \bigcup_{\alpha \in \mathcal{O}} A_\alpha \subset \bigcap_{\alpha \in \mathcal{O}} (X - A_\alpha)$.

Let $x \in X - \bigcup_{\alpha \in \mathcal{O}} A_\alpha$. Then, $x \in X$ and $x \notin \bigcup_{\alpha \in \mathcal{O}} A_\alpha$. If $\beta \in \mathcal{O}$, then $x \notin A_\beta$, so $x \in X - A_\beta$. Hence, $x \in \bigcap_{\alpha \in \mathcal{O}} (X - A_\alpha)$.

Part 2 $\bigcap_{\alpha \in \mathcal{O}} (X - A_\alpha) \subset X - \bigcup_{\alpha \in \mathcal{O}} A_\alpha$.

Let $y \in \bigcap_{\alpha \in \mathcal{O}} (X - A_\alpha)$. Since $\mathcal{O} \neq \phi$, there is some $\gamma \in \mathcal{O}$. Then, $y \in X - A_\gamma$, so $y \in X$. To show that $y \notin \bigcup_{\alpha \in \mathcal{O}} A_\alpha$, suppose instead that $y \in \bigcup_{\alpha \in \mathcal{O}} A_\alpha$. Then, for some β, $\beta \in \mathcal{O}$ and $y \in A_\beta$. But then $y \notin X - A_\beta$, contradicting $y \in \bigcap_{\alpha \in \mathcal{O}} (X - A_\alpha)$. Thus, $y \notin \bigcup_{\alpha \in \mathcal{O}} A_\alpha$, and $y \in X - \bigcup_{\alpha \in \mathcal{O}} A_\alpha$. ∎

Customarily, one writes $\bigcup_{n=1}^{\infty} A_n$ (or $\bigcup_{j=1}^{\infty} A_j$, or $\bigcup_{k=1}^{\infty} A_k$, etc.) for $\bigcup_{n \in \mathbb{Z}^+} A_n$. Here, $\mathbb{Z}^+$ is the set of positive integers. Similarly, one has $\bigcap_{n=1}^{\infty} A_n = \bigcap_{n \in \mathbb{Z}^+} A_n$. As obvious variations, we can have $\bigcup_{n=s}^{\infty} A_n$, $\bigcap_{n=1}^{\infty} A_{2n}$, and so on.

For practice, the reader might verify the following:

(1) $\bigcup_{n=1}^{\infty} \{x \mid -1/n \leq x \leq 1/n\} = \{x \mid -1 \leq x \leq 1\}$.

(2) $\bigcap_{n=1}^{\infty} \{x \mid -1/n \leq x \leq 1/n\} = \{0\}$.

(3) For any set A, $A = \bigcup_{x \in A} \{x\}$.

(4) $\bigcap_{\alpha \in \{1\} \cup \{2\}} A_\alpha = A_1 \cap A_2$ and $\bigcup_{\alpha \in \{1\} \cup \{2\}} A_\alpha = A_1 \cup A_2$.

(5) $\left(\bigcup_{\alpha \in \mathcal{O}} A_\alpha\right) \cup \left(\bigcup_{\alpha \in \mathcal{O}} B_\alpha\right) = \bigcup_{\alpha \in \mathcal{O}} (A_\alpha \cup B_\alpha)$.

(6) $\left(\bigcap_{\alpha \in \mathcal{O}} A_\alpha\right) \cap \left(\bigcap_{\alpha \in \mathcal{O}} B_\alpha\right) = \bigcap_{\alpha \in \mathcal{O}} (A_\alpha \cap B_\alpha)$.

(7) $\bigcup_{\alpha \in \mathcal{O}} (A_\alpha \cap B_\alpha) \subset \left(\bigcup_{\alpha \in \mathcal{O}} A_\alpha\right) \cap \left(\bigcup_{\alpha \in \mathcal{O}} B_\alpha\right)$, and equality may not hold.

(8) $\left(\bigcap_{\alpha \in \mathcal{O}} A_\alpha\right) \cup \left(\bigcap_{\alpha \in \mathcal{O}} B_\alpha\right) \subset \bigcap_{\alpha \in \mathcal{O}} (A_\alpha \cup B_\alpha)$, and equality may fail.

(9) $x \in A \Rightarrow x \subset \cup A$.

(10) $x \in A \Rightarrow \cap A \subset x$.

19*

B-4 Cartesian products and functions

We assume that the reader has an intuitive notion of ordered pair (rigorously, (a, b) can be defined as $\{\{a\}\} \cup \{\{a, b\}\}$, although we need not be this formal).

The Cartesian product of A and B is the set

$$A \times B = \{(x, y) | x \in A \text{ and } y \in B\}.$$

If either A or B is empty, then so is $A \times B$. Note that, in general, $A \times B \neq B \times A$.

A function f on A to B is a rule which associates with each x in A a unique object, denoted $f(x)$ or f_x, in B. We write $f: A \to B$. A neater way of looking at functions is as a set of ordered pairs: f is a function on A to B if $f \subset A \times B$ and, whenever $(x, y) \in f$, and $(x, z) \in f$, then $y = z$. We then write $y = f(x)$ (or f_x) whenever $(x, y) \in f$. The set A is called the domain of f. In actual practice one usually describes a function by giving the rule, not the set of ordered pairs, although this can be done as well. Thus, if $f = \{(x, x^2) | 0 \leq x \leq 1\}$, then $f: [0, 1] \to [0, 1]$ is a function described equally well by the rule: $f(x) = x^2$ for $0 \leq x \leq 1$.

Note that, if $f: A \to B$, and $B \subset C$, then $f: A \to C$ as well.

The following notation and terminology is now quite standard in topology. Suppose $f: A \to B$. Then,

(1) for any $C \subset A, f(C) = \{f(x) | x \in C\}$.

(2) f is a surjection (or onto function) if $f(A) = B$.

(3) f is an injection (or one-to-one function) if $f(x) = f(y)$ implies $x = y$, whenever x and y are in A.

(4) f is a bijection if f is both a surjection and an injection.

EXAMPLES $\{(x, \sin(x)) | 0 \leq x \leq \pi\}$ is a surjection (but not an injection) from $[0, \pi]$ onto $[0, 1]$.

$\{(x, \sin(x)) | 0 \leq x \leq \pi/2\}$ is an injection (but not a surjection) from $[0, \pi/2]$ to $[0, 2]$.

$\{(x, \sin(x)) | 0 \leq x \leq \pi/2\}$ is a bijection from $[0, \pi/2]$ to $[0, 1]$. ∎

If $f: A \to B$ and $C \subset A$, then $f|C: C \to B$ is defined by letting $(f|C)(x) = f(x)$ for each $x \in C$. The function $f|C$ is called the restriction of f to C.

If f is a bijection: $A \to B$, then, given $b \in B$, there is a unique $a \in A$ such that $f(a) = b$. This rule of association defines a function (bijection) from B to A, called the inverse of f, and denoted inv f. For example if $f = \{(x, x^2) | 0 \leq x \leq 2\}: [0, 2] \to [0, 4]$, then inv $f = \{(x, \sqrt{x}) | 0 \leq x \leq 4\}: [0, 4] \to [0, 2]$.

If $f: A \to B$ and $g: B \to C$, the composite function $g \circ f: A \to C$ is defined by the rule $(g \circ f)(x) = g(f(x))$ for each $x \in A$. Note that $f \circ g$ may not be defined.

Compositions, restrictions and inverses obey the following rules:

(1) $(f \circ g) \circ h = f \circ (g \circ h)$. This is the associative law for compositions. We are assuming that $f: C \to D$, $g: B \to C$, and $h: A \to B$, for some sets A, B, C and D.

(2) If $f: A \to B$ is a bijection, then

$$f \circ \operatorname{inv} f = 1_B \text{ and } (\operatorname{inv} f) \circ f = 1_A$$

where 1_B is the identity function on B given by $1_B(b) = b$ for each $b \in B$, and 1_A is the similarly defined identity function on A.

(3) If $f: A \to A$, then $f \circ 1_A = 1_A \circ f = f$.

(4) If $f: A \to B$, and $g: B \to C$, and $D \subset A$, then $(g \circ f)|D = g \circ (f|D)$.

(5) If $f: A \to B$ and $g: B \to A$, and $f \circ g = 1_B$ and $g \circ f = 1_A$, then f and g are bijections, and $g = \operatorname{inv} f$ and $f = \operatorname{inv} g$.

(6) If $f: A \to B$ and $g: B \to C$ are injections, then so is $g \circ f$, and $\operatorname{inv}(g \circ f) = (\operatorname{inv} f) \circ (\operatorname{inv} g)$. Further, $g \circ f$ is surjective if g and f are.

(7) Suppose that $A_\alpha \subset X$ for each $\alpha \in \mathcal{O}$ and that $\bigcup_{\alpha \in \mathcal{O}} A_\alpha = X$. Let $f_\alpha: A_\alpha \to Y$ for each $\alpha \in \mathcal{O}$, and $f_\alpha|(A_\alpha \cap A_\beta) = f_\beta|(A_\alpha \cap A_\beta)$ whenever α, $\beta \in \mathcal{O}$. Then, there is a unique $F: X \to Y$ such that $F|A_\alpha = f_\alpha$ for each $\alpha \in \mathcal{O}$. F is called an extension of the maps f_α to X.

Now suppose that $f: A \to B$. If $C \subset A$, we previously defined $f(C)$ as $\{f(x) | x \in C\}$. We now define, for $D \subset B$, $f^{-1}(D) = A \cap \{x | f(x) \in D\}$. It may happen that $f^{-1}(D) = \phi$. For example, let $f: [0, 2] \to [-1, 9]$ be given by $f(x) = x^2$ for $0 \leq x \leq 2$. Then, $f^{-1}([-1, 0]) = \{0\}$ and $f^{-1}([-1, -\frac{1}{2}]) = \phi$.

The reader can check that:

(1) $f(M \cup N) = f(M) \cup f(N)$.
(2) $f(M \cap N) \subset f(M) \cap f(N)$, but equality need not hold.
(3) $f^{-1}(M \cup N) = f^{-1}(M) \cup f^{-1}(N)$.
(4) $f^{-1}(M \cap N) = f^{-1}(M) \cap f^{-1}(N)$.
(5) $f(A) - f(B) \subset f(A - B)$, and equality need not hold.
(6) $f^{-1}(A - B) = f^{-1}(A) - f^{-1}(B)$.

It is obvious that (1), (3) and (4) extend to unions and intersections involving arbitrarily many sets.

We shall now extend the notion of Cartesian product, defined above for two sets, to an operation on arbitrarily many sets.

If A_α is a set for each $\alpha \in \mathcal{O}$, then the Cartesian product of the set $\{A_\alpha | \alpha \in \mathcal{O}\}$ indexed by $\mathcal{O}$ is:

$$\prod_{\alpha \in \mathcal{O}} A_\alpha = \left\{ f \,|\, f\colon \mathcal{O} \to \bigcup_{\alpha \in \mathcal{O}} A_\alpha \text{ and } f(\alpha) \in A_\alpha \text{ for each } \alpha \in \mathcal{O} \right\}.$$

This definition probably looks pretty strange if you have never seen it before, but it is a natural one if you begin a step by step generalization from $A \times B$ as defined above. Suppose first we attempt to generalize to a Cartesian product of n sets, $A_1, \ldots, A_n$. This is easy to do if you extend the notion of ordered pair to ordered n-tuple:

$$A_1 \times \cdots \times A_n = \prod_{i=1}^{n} A_i = \{(a_1, \ldots, a_n) | a_i \in A_i \text{ for } i = 1, \ldots, n\}.$$

Given an infinite collection of sets, for example sets A_i for $i \in Z^+$ (the set of positive integers), we encounter the first difficulty. There is no "infinite-tuple". However, if f is a function on Z^+, and $f_i \in A_i$, we can think of f as a sequence $(f_1, f_2, \ldots, f_{100}, \ldots)$, which serves as a generalization of an n-tuple. Then,

$$\prod_{i \in Z^+} A_i = \prod_{i=1}^{\infty} A_i = \left\{ f \,|\, f\colon Z^+ \to \bigcup_{i=1}^{\infty} A_i \text{ and } f_i \in A_i \text{ for each } i \in Z^+ \right\}.$$

This generalizes directly to the above definition when an arbitrary index set is used. Note that, if we think of $(a_1, \ldots, a_n)$ as a function $f\colon \{1, \ldots, n\} \to A_1 \cup \cdots \cup A_n$, where $f(i) = a_i$, then

$$\prod_{i=1}^{n} A_i = \{(a_1, \ldots, a_n) | a_i \in A_i \text{ for } i = 1, \ldots, n\}$$

$$= \prod_{i \in \{1, \ldots, n\}} A_i = \left\{ f \,|\, f\colon \{1, \ldots, n\} \to \bigcup_{i=1}^{n} A_i \text{ and } f(i) \in A_i \text{ for } i \in \{1, \ldots, n\} \right\},$$

so that the general definition reduces to the usual one in the more familiar, finite case.

If $\beta \in A$, the projection map $p_\beta\colon \prod_{\alpha \in A} X_\alpha \to X_\beta$ is defined by $p_\beta(f) = f_\beta$ for each $f \in \prod_{\alpha \in A} X_\alpha$.

B-5 The axiom of choice

If A_α is a set for each $\alpha \in \mathcal{O}$, then $\prod_{\alpha \in \mathcal{O}} A_\alpha$ is certainly empty if *any* A_β is empty. If, however, $\mathcal{O} \neq \phi$, and $A_\beta \neq \phi$ for each $\beta \in \mathcal{O}$, we would like to be sure that $\prod_{\alpha \in \mathcal{O}} A_\alpha$ is not empty. This intuitively plausible fact is one of the deeper axioms of set theory.

THE AXIOM OF CHOICE *Suppose that $\mathcal{O} \neq \phi$, and that $A_\alpha \neq \phi$ for each $\alpha \in \mathcal{O}$. Then, $\prod_{\alpha \in \mathcal{O}} A_\alpha \neq \phi$.*

This is equivalent to saying that it is possible to choose an element a_α from each set A_α, $\alpha \in \mathcal{O}$, as then $f = \{(\alpha, a_\alpha) | \alpha \in \mathcal{O}\}$ is an element of $\prod_{\alpha \in \mathcal{O}} A_\alpha$. Such an f is sometimes called a choice function for the sets A_α. Of course, the Axiom of Choice is only really needed when Cartesian products involving infinitely many sets are involved.

The Axiom of Choice is not accepted by every mathematician. Some prefer a set theory whose axioms do not include it or any equivalent statement. However, it has been proved that a set theory consistent without the Axiom of Choice remains consistent with the Axiom of Choice, so we shall make free use of it and several equivalent formulations.

B-6 Zorn's lemma

There are many statements which look very different from the Axiom of Choice, but are in fact logically equivalent to it. One we shall find convenient to use is Zorn's Lemma, which itself has several equivalent formulations.

Suppose A is a set. A relation on A is a subset of $A \times A$.

A relation R on A is called a partial order on A if:
(1) If $(a, b) \in R$ and $(b, c) \in R$, then $(a, c) \in R$. (transitivity)
(2) If $(a, b) \in R$ and $(b, a) \in R$, then $a = b$. (antisymmetry)
(3) $(a, a) \in R$ for each $a \in A$. (reflexivity)

If $(a, b) \in R$, we say that a and b are related by R, and write aRb. Note that not every pair of elements of A need be related by R. That is, there may be elements x and y of A with neither xRy nor yRx. Usually one thinks of a as "preceding" b, in the sense of the partial order, if aRb, and often the more suggestive notation $a \prec b$ is used instead of aRb.

EXAMPLES (1) Define aRb to mean $a \leq b$ if a, b are real numbers. This is a partial order on the reals (actually a full order, since, given any two reals x and y, either $x \leq y$ or $y \leq x$, so that any two objects are related).

(2) Let C be a set. Partially order the set $\mathscr{P}(C)$ of subsets of C by inclusion: ARB if $A \subset B$. If C consists of more than one element, this partial order is not a full order, as two objects may not be related. ∎

Now suppose that $\prec$ is a partial order on A. If $\phi \neq B \subset A$, and $\alpha \in A$, then α is an upper bound of B if $b \prec \alpha$ for each $b \in B$.

Note that a set may have many upper bounds, many of which may be unrelated by $\prec$.

EXAMPLE Partially order the set $\mathscr{P}(Z)$ of all subsets of the set Z of integers by inclusion. If n is a negative integer, let $\alpha_n = \{x | x \leq n\} \cap Z$, and let $\mathscr{B} = \{\alpha_n | n$ is a negative integer$\}$. Then any subset of Z containing all the negative integers is an upper bound of the subset $\mathscr{B}$ of $\mathscr{P}(Z)$. Two such upper bounds are $(\{x | x < 0\} \cap Z) \cup \{3\}$ and $(\{x | x \leq 0\} \cap Z)$, and these sets are unrelated. ∎

An element α of A is a maximal element (with respect to the partial order $\prec$ on A) if, whenever $b \in A$ and $\alpha \prec b$, then $\alpha = b$.

EXAMPLE Let $A = Z^+ - \{1\}$. Partially order A by: $a \prec b$ if b divides a (that is, if there is some integer c with $a = bc$). Then each prime is a maximal element of A, while, for example, 9 is not maximal, as $9 \prec 3$, but $3 \neq 9$. ∎

If A is partially ordered by $\prec$ and $\phi \neq C \subset A$, then C is a chain if each two objects in C are related by $\prec$. That is, if $a, b \in C$, then $a \prec b$ or $b \prec a$. In the last example, the set of positive powers of 2 is a chain, as 2^m divides 2^n or 2^n divides 2^m is always the case.

ZORN'S LEMMA (first version) *Let A be a partially ordered set. Suppose, if C is a chain in A, then C has an upper bound in A. Then, A has a maximal element.*

Actually, we do not need the full definition of partial ordering in Zorn's Lemma. An equivalent formulation is the following:

A is preordered by R if R is a relation on A and:

(1) aRa for each $a \in A$.

(2) If aRb and bRc, then aRc.

Thus, a partial order on A is an antisymmetric preorder on A.

If R is a preorder on A, then we define:

(1) For $\phi \neq B \subset A$, an element α of A is an upper bound of B if $bR\alpha$ for each $b \in B$. (This is the same as for partial orders.)

(2) An element α of A is a maximal element if, whenever $a \in A$ and $\alpha R a$, then also $a R \alpha$. (This reduces to the above notion when R is a partial order.)

ZORN'S LEMMA (second version) *If A is preordered by R, and each chain in A has an upper bound in A, then A has a maximal element.*

It can be shown in set theory that these two versions of Zorn's Lemma, and the Axiom of Choice, are logically equivalent. We shall use one or the other at our convenience according to the problem at hand.

EXAMPLE As an illustration of a Zorn's Lemma argument, we shall construct a Hamel basis for the set R of real numbers.

A subset H of R is a Hamel basis if:

(1) If $x_1, \ldots, x_n \in H$ and $r_1, \ldots, r_n$ are rational numbers, then $\sum_{i=1}^{n} r_i x_i = 0$ implies $r_1 = r_2 = \cdots = r_n = 0$.

(2) If $x \in R$, then there are elements $y_1, \ldots, y_m$ in H and rational numbers $r_1, \ldots, r_m$ such that $x = \sum_{i=1}^{m} r_i y_i$.

(Note that what we are really doing is producing a base for the infinite dimensional vector space of the reals over the rationals.)

Begin by letting $\mathscr{P}$ be the set of all subsets A of R with the property that, if $a_1, \ldots, a_m \in A$, and $r_1, \ldots, r_m$ are rational numbers, and $\sum_{i=1}^{m} r_i a_i = 0$, then each $r_i = 0$.

Partially order $\mathscr{P}$ by inclusion: $A \prec B$ if $A \subset B$, whenever $A, B \in \mathscr{P}$. (Note that $\mathscr{P}$ is not empty, as, for example, $\{3\} \in \mathscr{P}$.)

If C is a chain in $\mathscr{P}$, it is easy to check that $\cup C$ is an upper bound for C in $\mathscr{P}$. By Zorn's Lemma, $\mathscr{P}$ has a maximal element, say H. The reader can check that H is a Hamel basis for R.

Hamel bases are particularly useful in functional analysis. (For example, see Taylor, Introduction to Functional Analysis.)

B-7 Induction

We shall have occasion to use both mathematical induction and construction by induction.

PRINCIPLE OF MATHEMATICAL INDUCTION (first version) *Suppose P_n is a proposition for each non-negative integer n.*
Suppose P_0 is true.
Suppose, if P_n is true, then P_{n+1} is true.

Then,
P_k *is true for each non-negative integer k.*

An equivalent formulation is the following.

PRINCIPLE OF MATHEMATICAL INDUCTION (second version) *Let A be a subset of the set ω of non-negative integers.*
Let $0 \in A$.
Suppose, if $n \in A$, then $n + 1 \in A$.
Then,
$A = \omega$.

A third equivalent formulation, often called complete induction, is sometimes more convenient.

PRINCIPLE OF MATHEMATICAL INDUCTION (third version) *Suppose P_n is a proposition for each non-negative integer n.*
Suppose P_0 is true.
Suppose, if n is a non-negative integer and P_j is true for $0 \leq j \leq n$, then P_{n+1} is true.
Then,
P_k *is true for each non-negative integer k.*

Most students see induction in calculus or before, so we shall not elaborate on it here. Construction by induction is, however, usually less familiar.

To see the reason for having such a device, consider the apparently simple problem of finding a function $f: Z^+ \to \{x|x \text{ is real}\}$ such that $f(n) = 3f(n-1)$ for $n \geq 2$ and $f(1) = 5$. Thus, we need $f(1) = 5, f(2) = 15, f(3) = 45, f(4) = 135$, and so on.

It is the "and so on" which presents a difficulty. If we try to write f as a set of ordered pairs, we are led to:

$$f = \{(1, 5)\} \cup \{(x, 3f(x-1))|x \text{ is a positive integer and } x \geq 2\}.$$

But this is not a definition of f, as f appears on both sides of the equation. That such an f exists is a consequence of the theorem on construction by induction.

CONSTRUCTION BY INDUCTION *Let $A \neq \phi$.*
Suppose $h: A \to A$, and let $\alpha \in A$.
Then,
there is a unique function $\varphi: Z^+ \to A$ such that $\varphi(1) = \alpha$ and $h(\varphi(n)) = \varphi(n+1)$ for each $n \in Z^+$.

This statement looks rather more complicated than it really is. It insures the existence of a function φ on the set of positive integers to a given nonempty set A with the two properties:

(1) $\varphi(1) = \alpha$, α being a given "starting point" for φ;

(2) $\varphi(n + 1)$ depends upon $\varphi(n)$ in the sense that there is some function h such that, given $\varphi(n)$, then $\varphi(n + 1)$ is obtained by evaluating h at $\varphi(n)$.

As an example, we shall place the desired function f, on Z^+ into the reals, above, on more solid ground.

The problem is to show the existence of a function $f: Z^+ \to \{x | x \text{ is real}\}$ with $f(1) = 5$ and, thereafter, $f(n) = 3f(n - 1)$ for $n \geq 2$. Choose A as the set of real numbers, α as 5, and define $h: A \to A$ by $h(t) = 3t$, for $t \in A$. Note that h is perfectly well defined: $h = \{(t, 3t) | t \text{ is a real number}\}$. By the theorem on construction by induction, there is a function (in fact, exactly one) $f: Z^+ \to A$ such that $f(1) = 5$ and $f(n + 1) = h(f(n))$ for $n \in Z^+$.

Then, $f(n + 1) = h(f(n)) = 3f(n)$, just as we wanted.

Throughout the book, when we have occasion to define a function inductively, we shall simply write down the desired function without pausing to fill in the details. However, in each such instance, the construction can be justified by the above theorem.

Incidentally, the theorem on construction by induction is not difficult to prove. Let $\mathscr{M}$ be the set of all subsets S of $Z^+ \times A$ such that:

(1) $(1, \alpha) \in S$.

(2) Whenever $(n, a) \in S$, then $(n + 1, h(a)) \in S$.

The reader can show, using elementary properties of the positive integers, that a suitable choice for f is $\bigcap_{m \in \mathscr{M}} m$.

B-8 Ordinals

In set theory, the natural numbers are built up by starting with the empty set, forming singletons, and taking unions. Thus,

$$0 = \phi,$$
$$1 = \{\phi\},$$
$$2 = 1 \cup \{1\} = \{\phi\} \cup \{\{\phi\}\} = \{\phi, \{\phi\}\} = \{0, 1\},$$
$$3 = 2 \cup \{2\} = \{\phi, \{\phi\}, \{\phi, \{\phi\}\}\} = \{0, 1, 2\},$$

and so on. In general, having defined n, then $n + 1$ is $n \cup \{n\}$. The resulting set of natural numbers is customarily denoted ω.

Note that we can continue:
$$\omega \cup \{\omega\}$$
$$\{\omega \cup \{\omega\}, \{\omega \cup \{\omega\}\}\}$$

and so on, although such sets are not to be thought of as natural numbers.

Note also that each of these sets has the following two properties:

(1) Each nonempty subset A has its intersection as an element $\left(\bigcap_{a \in A} a \in A\right)$. For example, $\bigcap_{a \in 3} a = 2 \cap 1 \cap \phi = \phi \in 3$. Or, as a second example, $\{2, 3\} \subset 5$, and $\bigcap_{a \in \{2,3\}} a = 2 \cap 3 = 2 \in 5$.

(2) Each element m of n is the intersection of n with the set of proper subsets of m. For example, $1 \in 2$, and
$$2 \cap \{\alpha | \alpha \not\subseteq 1\} = (1 \cup \{1\}) \cap (\{\phi\}) = (1 \cup \{1\}) \cap 1$$
$$= (1 \cap 1) \cup (1 \cap \{1\}) = 1 \cup \phi = 1.$$

In general, any set which has these two properties is called an ordinal. Thus,

A is an ordinal if:

(1) $\phi \neq y \subset A$ implies $\bigcap_{t \in y} t \in y$.

(2) $y \in A$ implies $y = A \cap \{\alpha | \alpha \not\subseteq y\}$.

For example, the natural numbers are all ordinals, ω is an ordinal, $\omega \cup \{\omega\}$ is an ordinal, and so on.

In general, if α is an element of the class of all ordinals, then so is its successor, $\alpha \cup \{\alpha\}$ (α is the immediate predecessor of $\alpha \cup \{\alpha\}$). Not every ordinal has an immediate predecessor. This is the case with ω: there is no ordinal β with $\omega = \beta \cup \{\beta\}$. Such an ordinal (except ϕ) is called a limit ordinal. It has been found convenient to designate ϕ a non-limit ordinal, even though it has no predecessor.

Some basic properties of ordinals are:

(1) If A is an ordinal, and $x, y \in A$, then x and y are ordinals, and ($x \in y$ if and only if $x \not\subseteq y$).

(2) The class of all ordinals is partially ordered by inclusion. The ordering is total in the sense that, given two ordinals x and y, then either $x \subset y$ or $y \subset x$. Customarily, we write $x \leq y$ if $x \subset y$, and $x < y$ if $x \not\subseteq y$ (or, equivalently, $x \in y$).

(3) For any non-empty ordinal x, $\bigcap_{t \in x} t \in x$.

(4) If A is a set of ordinals, then $\bigcup_{a \in A} a$ is an ordinal.

(5) Each non-empty set of ordinals has a first element. That is, if A is a non-empty set of ordinals, then there is some $a \in A$ with $a \leq x$ ($a \subset x$) for each $x \in A$. In particular, 0 (or ϕ) is the first ordinal. Note, however, that A need not have a last element.

(6) Let A be a non-empty set of ordinals. If b is an ordinal, and $a \leq b$ for each $a \in A$, then we call b an upper bound of A, and say that A is bounded above. Since $\{x|x$ is an ordinal and $a \leq x$ for each $a \in A\}$ is not empty, then this set has a least element, say λ, called the least upper bound of A (lub A or sup A). λ is characterized by:

(i) $a \leq \lambda$ for each $a \in A$.
(ii) If b is an ordinal and $a \leq b$ for each $a \in A$, then $\lambda \leq b$.

An ordinal x is said to be countable if there is an injection $f: x \to \omega$. If there is a bijection $f: x \to \omega$, then x is countably infinite. In particular, ω is countably infinite. In fact, ω is the first infinite countable ordinal.

More generally, any set A is countable if there is an injection $f: A \to \omega$. Each subset of a countable set is countable, and a countable union of countable sets is countable. Further, any infinite set has a countable subset (in this sense, countability is the "smallest" order of infinity). Cantor's famous diagonalization process (see Rudin, Gamow, or Randolph) shows that [0, 1], hence the set of real numbers, is not countable.

There is an ordinal Ω which is not countable, and is a subset of any other uncountable ordinal. Usually Ω is called the first uncountable ordinal. In set theory, Ω is shown to be exactly the set of all countable ordinals. Similarly, the first countably infinite ordinal, ω, is the set of all finite ordinals.

The Principle of Transfinite Induction does for the ordinals what mathematical induction does for the positive integers. Since the positive integers are ordinals, transfinite induction may be thought of as a generalization of mathematical induction.

PRINCIPLE OF TRANSFINITE INDUCTION *Let M be a non-empty set of ordinals, such that $x \cup \{x\} \in M$ whenever $x \in M$.*

Let λ be the first element of M.
Let P_α be a proposition for each $\alpha \in M$.
Suppose that P_λ is true.
Suppose, if $\alpha \in M$ and P_β is true for each $\beta \in M$ with $\beta \leq \alpha$, then $P_{\alpha \cup \{\alpha\}}$ is true.
Then,
P_α is true for each $\alpha \in M$.

APPENDIX C

Equivalence relations

THIS APPENDIX SKETCHES the essential facts about equivalence relations.

An equivalence relation on a set A is a reflexive, symmetric and transitive relation on A. That is, R is an equivalence relation on A if:

(1) $R \subset A \times A$.
(2) $(a, a) \in R$ for each $a \in A$.
(3) $(a, b) \in R$ implies $(b, a) \in R$.
(4) $(a, b) \in R$ and $(b, c) \in R$ implies $(a, c) \in R$.

(2) is called the reflexive property, (3) is symmetry, and (4), transitivity. If $(a, b) \in R$, we sometimes write aRb, as with partial orders. Customarily, equivalence relations are denoted by symbols like $\sim$ and $\approx$.

If $\sim$ is an equivalence relation on A, and $x \in A$, we let $[x]_\sim = \{y | x \sim y\} \cap A$. Thus, $[x]_\sim$ consists of all objects in A equivalent (under $\sim$) to x, and is called the equivalence class of x.

If $x, y \in A$, it is easy to check that either $[x]_\sim = [y]_\sim$ (when $x \sim y$) or $[x]_\sim \cap [y]_\sim = \phi$ (when x is not equivalent to y). Also, $A = \bigcup_{x \in A} [x]_\sim$.

Thus, $\sim$ partitions A into disjoint equivalence classes. The set of equivalence classes $\{[x]_\sim | x \in A\}$ is denoted $A/\sim$, read "A mod $\sim$".

EXAMPLE Let $A = Z$ and write $x \sim y$ if $x - y$ is a multiple of 4. The equivalence classes are:

$$[0]_\sim = \{4n | n \in Z\},$$
$$[1]_\sim = \{4n + 1 | n \in Z\},$$
$$[2]_\sim = \{4n + 2 | n \in Z\},$$

and
$$[3]_\sim = \{4n + 3 | n \in Z\}.$$
$$Z/\sim = \{[0]_\sim, [1]_\sim, [2]_\sim, [3]_\sim\}. \blacksquare$$

If $\sim$ is an equivalence relation on A, the natural map $\eta: A \to A/\sim$ is defined by $\eta(a) = [a]_\sim$ for each $a \in A$. η is a surjection, usually not an injection.

APPENDIX D

Groups

IN THIS APPENDIX we sketch the terminology and notation from group theory needed in Chapters IX and XI.

A product on a set A is a function on $A \times A$ to A. For example, ordinary addition and multiplication of real numbers may be thought of as product operations. Generally, a product is denoted by a dot $\cdot$, but, while $\cdot (x, y)$ is consistent with function notation, usually we replace this with the more suggestive $x \cdot y$, or just xy when $\cdot$ is understood.

A product on A is associative if $(xy)z = x(yz)$ for each $x, y, z \in A$.

$\langle G, \cdot \rangle$ is a group if:

(1) $\cdot$ is an associative product on G.
(2) There is an element $e \in G$ such that $ex = xe = x$ for each $x \in G$.
(3) If $x \in G$, there is some $y \in G$ with $xy = yx = e$.

It is easy to show that a group can have only one element behaving like e in (2). Usually e is called the identity element of the group. Similarly, y in (3) is uniquely determined by x, and is usually written x^{-1}. When the product is understood, $\langle G, \cdot \rangle$ is shortened to G, and we say simply that G is a group.

EXAMPLES

(1) The reals form a group under ordinary addition.

(2) The non-zero reals form a group under ordinary multiplication.

(3) The set of bijections of any non-empty set A form a group under composition of functions. The identity element is 1_A, and the inverse of f is inv f. If A is finite, this group is called the group of permutations on A. ■

The groups in (1) and (2) are abelian. That is, $xy = yx$ whenever $x, y \in G$. The group of (3) is not abelian in general.

If G is a group and $x \in G$, a product of x with itself n times is written x^n, and $(x^n)^{-1}$ as x^{-n}.

G is cyclic if G is generated by a single element. That is, for some $g \in G$, each object in G is a power of g. The integers under addition form a cyclic group, generated by 1.

A subset H of G is a subgroup if H is a group with respect to the product defined on G. For example, the additive group of integers is a subgroup of the additive group of rationals, while the multiplicative group of non-zero rationals is not a subgroup of the additive group of rationals (though it is a subset).

If H is a non-empty subset of a group G, then H is a subgroup if and only if, for each $x, y \in H$, the product $xy^{-1} \in H$.

If H is a subgroup of G, we write $H < G$. It is easy to verify that every subgroup of a cyclic group is cyclic.

If A is any non-empty subset of G, and $x \in G$, then $xA = \{xa | a \in A\}$ and $Ax = \{ax | a \in A\}$. If $H < G$, xH and Hx are the left and right cosets of x by H in G, respectively. If $Hx = xH$ for each $x \in G$, then H is a normal subgroup of G, and we write $H \triangleleft G$. Note that $H \triangleleft G$ if and only if $xHx^{-1} = H$ for each $x \in G$. Of course, if G is abelian, then every subgroup is normal.

Now suppose that $H \triangleleft G$. If $x, y \in G$, define $x \sim y$ to mean that $xy^{-1} \in H$. This is an equivalence relation on G, hence partitions G into equivalence classes. The equivalence class $[x]_\sim$ of x is exactly the coset xH. Let $G/H = \{xH | x \in G\}$. Then G/H is a group under the product $(xH)(yH) = (xy)H$. The identity element of this group is H, and $(xH)^{-1} = x^{-1}H$. G/H is the quotient group of G by H.

If $\langle G, \cdot \rangle$ and $\langle K, * \rangle$ are groups, a function $f: G \to K$ is called an homomorphism if $f(xy) = f(x) * f(y)$ for each $x, y \in G$. For example, if $H \triangleleft G$, the natural map $\eta: G \to G/H$ defined by $\eta(x) = xH$ for each $x \in G$, is an homomorphism. If f is also a bijection, then the homomorphism is called an isomorphism, and $\langle G, \cdot \rangle$ and $\langle K, * \rangle$ are said to be isomorphic.

As an example, let G be a cyclic group of order n (having n elements). Then G is isomorphic to the quotient group $Z/[n]$, where Z is the additive group of integers and $[n]$ is the subgroup consisting of all integer multiples of n. If G is infinite cyclic, then G is isomorphic to Z.

If $f: G \to G^*$ is an homomorphism, then $\ker f = G \cap \{x | f(x) = e^*\}$, where e^* is the identity element of G^*. $\ker f$ (read kernel of f) is a normal subgroup of G, and $f(G)$ is a subgroup of G^*. The first isomorphism theorem of algebra says that $G/\ker f$ is isomorphic to $f(G)$ if $f: G \to G^*$ is an homomorphism.

If $f: G \to K$ and $g: K \to H$ are homomorphisms, then so is $g \circ f$. In the case that $H = G$, and $f \circ g = 1_K$ and $g \circ f = 1_G$, then f and g are isomorphisms, and $g = \text{inv} f$ and $f = \text{inv} g$.

The direct sum $G_1 \oplus G_2$ of two groups G_1 and G_2 is a group with set $G_1 \times G_2$ and group operation $(g_1, h_1)(g_2, h_2) = (g_1 g_2, h_1 h_2)$. This can be extended in the obvious way to direct sums of any finite number of groups. Given an infinite collection $\{G_i | i \in Z^+\}$ of groups, then $\sum_{i=1}^{\infty} G_i$ consists of all elements g of $\prod_{i=1}^{\infty} G_i$ with $g_i = e_i$ (identity element of G_i) for all but possibly finitely many values of i. Then $\sum_{i=1}^{\infty} G_i$ is a group under the operation $(gh)_i = g_i h_i$ for $i \in Z^+$.

Bibliography

The following symbols are intended as an aid in using the references:
- ST   set topology
- AT   algebraic topology
- TG   topological groups
- T   other branches of topology
- st   set theory
- a   analysis
- al   algebra
- h   history and general background.

Abian, A., *The Theory of Sets and Transfinite Arithmetic*. W. B. Saunders, Philadelphia, 1965. (st)

Aleksandrov, P. S., *Combinatorial Topology, Vol. I*, (trans. by Horace Komm). Graylock Press, Rochester, N.Y., 1956. (ST, T)

Bachman, G., *Elements of Abstract Harmonic Analysis*. Academic Press, N.Y., 1964. (a, ST, TG)

Bell, E. T., *Men of Mathematics*. Simon and Schuster, N.Y., 1937. (h)

Bers, L., *Topology*. Courant Institute Lecture Notes, New York University, N.Y., 1956. (ST, AT)

Bing, R. H., "Some Aspects of the Topology of 3-Manifolds Related to the Poincare Conjecture". Appears in *Lectures on Modern Mathematics, Vol. II*, ed. T. L. Saaty. Wiley, N.Y., 1964. (T)

Boyer, C. B., *A History of Mathematics*. Wiley, N.Y., 1968. (h)

Bourbaki, N., *Elements of Mathematics, General Topology*, Parts 1 and 2. Addison-Wesley, Reading, Mass., Translated from *Elements de Mathematique, Topologie Generale*, Hermann, Paris, 1966. (ST, TG)

Cairns, S. S., *Introductory Topology*. Ronald, N.Y., 1961. (ST, AT)

Chinn, W. G., and Steenrod, N. E., *First Concepts of Topology*. Random House, N.Y., 1966. (ST)

Chevalley, C., *Theory of Lie Groups, Vol. I*. Princeton University Press, Princeton, 1946. (TG, T, a)

Cohen, L. W., and Ehrlich, G., *The Structure of the Real Number System*. Van Nostrand, Princeton, 1963. (st)

Cooke, G. E., and Finney, R. L., *Homology of Cell Complexes*. Princeton University Press, Princeton, 1967. (AT)

Cullen, H., *Introduction to General Topology*. Heath, Boston, 1968. (ST, AT)

Dieudonne, J., *Foundations of Modern Analysis*. Academic Press, N.Y., 1960. (a, ST)

Dugundji, J., *Topology*. Allyn and Bacon, Boston, 1966. (ST, st)

Edwards, R. E., *Functional Analysis, Theory and Applications*. Holt, Rinehart and Winston, N.Y., 1965. (a)

Edwards, R. E., *Fourier Series, A Modern Introduction*, 2 Vols. Holt, Rinehart and Winston, N.Y., 1967. (a)

Eilenberg, S., "Algebraic Topology". Appears in *Lectures on Modern Mathematics, Vol. I*, ed. T. L. Saaty. Wiley, N.Y., 1963. (AT)

Eilenberg, S., and Steenrod, N. E., *Foundations of Algebraic Topology*. Princeton University Press, Princeton, 1952. (AT)

Fraleigh, J. B., *A First Course in Abstract Algebra*. Addison-Wesley, Reading, Mass., 1967. (al)

Fréchet, M., and Fan, K., *Initiation to Combinatorial Topology* (trans. and notes by Howard Eves). Prindle, Weber and Schmidt, Boston, 1967. (T)

Gaal, S., *Point Set Topology*. Academic Press, N.Y., 1964. (ST)

Gamow, G., *One, Two, Three ... Infinity*. Bantam Books, N.Y., 1967. (h)

Gemignani, M. C., *Elementary Topology*. Addison-Wesley, Reading, Mass., 1967. (ST, AT)

Godement, R., *Topologie Algebrique et Theorie des Faisceaux*. Hermann, Paris, 1964. (AT)

Goffman, C., "Preliminaries to Functional Analysis." Appears in *Studies in Modern Analysis*, R. C. Buck, ed., Vol. I of MAA Studies in Mathematics, published by the Mathematical Association of America, distributed by Prentice-Hall, Englewood Cliffs, N.J., 1962. (a)

Greenberg, M., *Lectures on Algebraic Topology*. W. A. Benjamin, Boston, 1967. (AT)

Halmos, P., *Naive Set Theory*. Van Nostrand, Princeton, 1963. (st)

Herstein, I. N., *Topics in Algebra*. Blaisdell, Waltham, Mass., 1964. (al)

Hewitt, E., and Ross, K. A., *Abstract Harmonic Analysis, Vol. I*. Springer-Verlag, N.Y., 1963. (TG, a)

Hilton, P. J., *An Introduction to Homotopy Theory*. Cambridge University Press, London, England, 1964. (AT)

Hilton, P. J., and Wylie, S., *Homology Theory. An Introduction to Algebraic Topology*. Cambridge University Press, London, England, 1960. (AT)

Hocking, J. G., and Young, G. S., *Topology*. Addison-Wesley, Reading, Mass., 1961. (ST, AT)

Horvath, J., *Topological Vector Spaces and Distributions* (2 Vols.). Addison-Wesley, Reading, Mass., 1966. (a)

Hu, S. T., *Elements of General Topology*. Holden-Day, San Francisco, 1964. (ST)

Hu, S. T., *Cohomology Theory*. Markham, Chicago, 1968. (AT)

Hu, S. T., Homology Theory, *A First Course in Algebraic Topology*. Holden-Day, San Francisco, 1966. (AT)

Hurewicz, W., and Wallman, H., *Dimension Theory*. Princeton University Press, Princeton, 1948. (T)

Husain, T., *Introduction to Topological Groups*. W. B. Saunders, Philadelphia, 1966. (TG, a)

Katznelson, Y., *An Introduction to Harmonic Analysis*. Wiley, N.Y., 1968. (a)

Kelley, J. L., *General Topology*. Van Nostrand, Princeton, 1955. (ST, st)

Kolmogorov, A. N., and Fomin, S. V., *Elements of the Theory of Functions and Functional Analysis*, 2 Vols. (Vol. I trans. by Leo F. Boron, Vol. II trans. by Hyman Kamel and Horace Komm). Graylock, Rochester, N.Y., 1957. (ST, a)

Loomis, L., *Abstract Harmonic Analysis*. Van Nostrand, Princeton, 1953. (a)

Massey, W., *Algebraic Topology. An Introduction*. Harcourt, Brace and World, N.Y., 1967. (AT)

McCarty, G., *Topology, An Introduction, with Application to Topological Groups*. McGraw-Hill, San Francisco, 1967. (ST, TG)

Milnor, J., "Differential Topology". Appears in *Lectures on Modern Mathematics, Vol. II*, ed. T. L. Saaty. Wiley, N.Y., 1964. (T)

Montgomery, D. and Zippin, L., *Topological Transformation Groups*. Interscience, N.Y., 1955. (TG, a)

Munkres, J. R., *Elementary Differential Topology*. Princeton University Press, Princeton, 1966. (T)

Munroe, M. E., *Introduction to Measure and Integration*. Addison-Wesley, Reading, Mass., 1953. (a)

Nachbin, L. N., *The Haar Integral*. Van Nostrand, Princeton, 1965. (TG, a)

Nagata, J. I., *Modern Dimension Theory*. North Holland, Amsterdam, 1965. (T)

Newman, M. H. A., *Elements of the Topology of Plane Sets of Points*. Cambridge University Press, London, England, 1951. (ST)

Paley, H., and Weichsel, P. M., *A First Course in Abstract Algebra*. Holt, Rinehart and Winston, N.Y., 1964. (al)

Pontrjagin, L. S., *Foundation of Combinatorial Topology* (trans. by F. Bagemihl, H. Komm, and W. Seidel). Graylock, Rochester, N.Y., 1952. (AT)

Pontrjagin, L. S., *Topological Groups* (trans. by Emma Lehmer). Princeton University Press, Princeton, 1939. (TG)

Randolph, J. R., *Basic Real and Abstract Analysis*. Academic Press, N.Y., 1968. (a, st)

Reiter, H., *Classical Harmonic Analysis and Locally Compact Groups*. Oxford University Press, London, 1968. (TG, a)

Royden, H. L., *Real Analysis*. Macmillan, N.Y., 1963. (ST, a)

Rudin, W., *Principles of Mathematical Analysis* (2nd Ed.). McGraw-Hill, N.Y., 1964. (st, a)

Simmons, G., *Introduction to Topology and Modern Analysis*. McGraw-Hill, N.Y., 1963. (ST, a)

Singer, I. M., and Thorpe, J. A., *Lecture Notes on Elementary Topology and Geometry*. Scott, Foresman and Co., Glenview, Ill., 1967. (ST, AT, T)

Spanier, E., *Algebraic Topology*. McGraw-Hill, N.Y., 1966. (AT)

Steen, L. A., and Seebach, J. A., Jr., *Counterexamples in Topology*. Holt, Rinehard and Winston, N.Y., 1970. (ST)

Studies in Modern Topology (Vol. 5 of the MAA Studies in Mathematics), ed. P. J. Hilton. Published by the Mathematical Association of America, distributed by Prentice-Hall, Englewood Cliffs, N.J., 1968. (ST, AT, T)

Taylor, A., *Introduction to Functional Analysis*. Wiley, N.Y., 1958. (ST, a)

Thron, W., *Topological Structures*. Holt, Rinehart and Winston, N.Y., 1966. (ST, TG)

Vaidyanathaswamy, R., *Set Topology*. Chelsea, N.Y., 1960. (ST)

Wallace, A. H., *Differential Topology, First Steps*. W. A. Benjamin, N.Y., 1968. (T)

Wallace, A. H., *An Introduction to Algebraic Topology*. Pergamon, N.Y., 1957. (AT)

The World of Mathematics (4 Vols.), ed. James R. Newman. Simon and Schuster, N.Y., 1956. (h)

Whyburn, G. T., *Topological Analysis*. Princeton University Press, Princeton, 1958. (ST, a)

Yosida, K., *Functional Analysis*. Springer-Verlag, N.Y., 1966. (a)

The following four articles appear in Scientific American.

Euler, L., "The Königsberg Bridges", ed. by James R. Newman. July 1953, p. 66–70.
Phillips, A., "Turning a Sphere Inside Out". May 1966, p. 112–120.
Shinbrot, M., "Fixed Point Theorems". Jan. 1966, p. 105–110.
Tucker, A. W., and Bailey, H. S., Jr., "Topology". Jan. 1950, p. 18–24.

The articles by Euler, Shinbrot and Tucker and Bailey also appear in *Mathematics in the Modern World*, W. H. Freeman, San Francisco, 1968.

The following articles appear in recent issues of *The American Mathematical Monthly*.

Bing, R. H., "The Elusive Fixed Point Property". Vol. 76, No. 2, Feb. 1969, p. 119–132.
Comfort, W. W., "A Short Proof of Marczewski's Separability Theorem". Vol. 76, No. 9, Nov. 1969, p. 1041–1042.
Conway, J. B., "The Inadequacy of Sequences". Vol. 76, No. 1, Jan. 1969, p. 68–69.
Howard, E. J., "A Note on Second Category Topological Groups". Vol. 77, No. 5, May 1970, p. 501–502.
Joseph, J. E., "Continuous Functions and Spaces in Which Compact Sets Are Closed". Vol. 76, No. 10, Dec. 1969, p. 1125–1126.
Kannan, R., "Some Results on Fixed Points – II". Vol. 76, No. 4, April 1969, p. 405–408.
Kirch, M. R., "A Class of Spaces in Which Compact Sets Are Finite". Vol. 76, No. 1, Jan. 1969, p. 42.
Kirch, M. R., "A Countable, Connected, Locally Connected Hausdorff Space". Vol. 76, No. 2, Feb. 1969, p. 169–171.
Lorrain, F., "Notes on Topological Spaces with Minimum Neighborhoods". Vol. 76, No. 6, June-July 1969, p. 616–627.
Redheiffer, R., "The Homotopy Theorems of Function Theory". Vol. 76, No. 7, Aug.-Sept. 1969, p. 778–787.
Simon, B., "Some Pictorial Compactifications of the Real Line". Vol. 76, No. 5, May 1969, p. 536–538.
Smale, S., "What Is Global Analysis?" Vol. 76, No. 1, Jan. 1969, p. 4–9.
Thomas, J., "A Regular Space, Not Completely Regular". Vol. 76, No. 2, Feb. 1969, p. 181–182.
Whyburn, G. T., "Dynamic Topology". Vol. 77, No. 6, June-July 1970, p. 556–570.
Williams, R. K., "On Expansive Homeomorphisms". Vol. 76, No. 2, Feb. 1969, p. 176 to 178.

Summary Charts

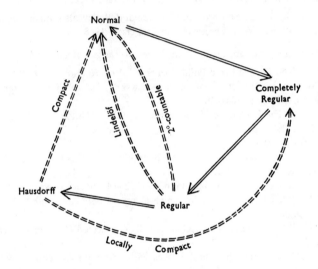

Separation axioms

SUMMARY CHARTS

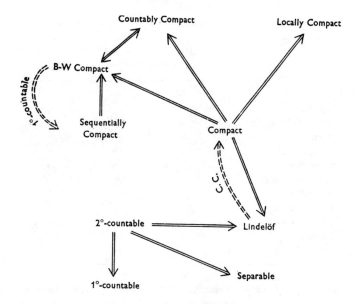

Compactness and covering theorems

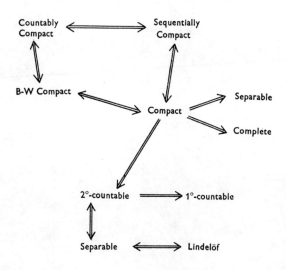

Metric Spaces

Summary table on hereditary and productive properties

	Her.	Prod.	Countably Prod.
Hausdorff	+	+	+
regular	+	+	+
completely regular	+	+	+
normal	−	−	−
compact	−	+	+
connected	−	+	+
path connected	−	+	+
separable	−	−	+
1°-countable	+	−	+
2°-countable	+	−	+
Lindelöf	−	−	−
metrizable	+	−	+

A topological property is called hereditary if it is shared by each subspace. A property is productive if it is shared by any product of spaces having the property. In the first two columns, hereditary and productive properties respectively are indicated by a plus sign. The third column indicates properties which are transmitted to a product of countably many spaces.

Index of counterexamples

Two non-comparable topologies on the same set: Example 3.7.I, p. 17.

A space in which sequences fail to distinguish cluster points: Example 1.6.IV, p. 68.

A non-Hausdorff space in which limits of sequences are unique: Example 1.6.V, p. 89.

A non-regular, Hausdorff space: Example 2.3.V, p. 93.

A product of normal spaces which is not normal: Example 3.3.V, p. 100.

A regular, not completely regular space: Example 4.1.V, p. 110.

A completely regular, non-normal space: Example 4.2.V, p. 115.

A connected, not path connected, space: Example 2.4.VII, p. 154.

A non-normal subspace of a normal space (The Tychonov Plank): Example 1.12.VII, p. 149.

A non-locally compact subspace of a locally compact space: Example 2.6.VII, p. 156.

A countably compact, non-compact space: Example 3.1.VII, p. 163.

A sequentially compact, non-compact space, and a compact, non-sequentially compact space: p. 165 (discussion).

A separable, not $2°$-countable space: Example 5.1.VII, p. 168.

A non-Lindelöf, separable space: Example 5.2.VII, p. 169.

A non-separable, Lindelöf space: Example 5.3.VII, p. 169.

A Lindelöf, not compact, not countably compact, not $2°$-countable space: Example 5.4.VII, p. 170.

A uniformizable, non-metrizable space: Ch. VIII, Problem 27 (f), p. 213.

Symbols and abbreviations

$\in$	denotes element of a set (287)
$\notin$	negation of (287)
$\subset$	denotes subset (288)
$\subsetneq$	denotes proper subset
ϕ	the empty set (287)
$\cup$	union (287, 290)
$\cap$	intersection (287, 290)
$\{x\}$	set consisting of just x (289)
$\{x_1, ..., x_n\}$	set consisting of $x_1, ..., x_n$ (290)
$\mathscr{P}(A)$	power set of A, consisting of all subsets of A (296)
$A \times B$	Cartesian product of A and B (292)
$f: X \to Y$	f is a function on X to Y (292)
$f(x)$, or f_x	value of the function f at x (292)
$f\|A$	f restricted to A (292)
$f \circ g$	composition of f and g (293)
inv (f)	inverse function of f (292)
$f(A)$	$\{f(x)\|x \in A\}$ (292)
$f^{-1}(B)$	$\{x\|f(x) \in B\}$ (293)
1_X	identity map on X, given by $1_X(x) = x$ for each $x \in X$ (293)
$\prod$	Cartesian product (294)
p_β	projection map: $\prod_{\alpha \in A} X_\alpha \to X_\beta$, defined by $p_\beta(f) = f_\beta$ for $f \in \prod_{\alpha \in A} X_\alpha$ (294)
$\prec$	usually denotes a partial order (295)
ω	the set of natural numbers (non-negative integers); equivalently, the first countably infinite ordinal, or set of finite ordinals (299)
Ω	the first uncountable ordinal; equivalently, the set of all countable ordinals (301)
$a \leqq b$	$a \subset b$, for ordinals a and b (300)
$a < b$	$a \subsetneq b$ (alternatively, $a \in b$) for ordinals a and b (300)
$[a, b]$	$\{x\|x$ is an ordinal and $a \leqq x \leqq b\}$
$]a, b[$	$\{x\|x$ is an ordinal and $a < x < b\}$
$]a, b]$	$\{x\|x$ is an ordinal and $a < x \leqq b\}$
$[a, b[$	$\{x\|x$ is an ordinal and $a \leqq x < b\}$

for a, b ordinals

Z	the set of integers
Z^+	the set of positive integers
Q	the set of rational numbers

SYMBOLS AND ABBREVIATIONS

R	the set of real numbers
$[a, b]$	$\{x \mid a \leq x \leq b\}$
$]a, b[$	$\{x \mid a < x < b\}$ for a, b real
$]a, b]$	$\{x \mid a < x \leq b\}$
$[a, b[$	$\{x \mid a \leq x < b\}$
$]-\infty, a]$	$\{x \mid x \leq a\}$
$]-\infty, a[$	$\{x \mid x < a\}$ for a real
$[a, \infty[$	$\{x \mid a \leq x\}$
$]a, \infty[$	$\{x \mid a < x\}$
I	the unit interval $[0, 1]$
$T[S]$	topology generated by S (15)
E^n	n-dimensional Euclidean space, consisting of all n-tuples $(x_1, \ldots, x_n)$, x_i real (1)
ϱ_n	Euclidean metric on E^n, given by $\varrho_n(x, y) = \left(\sum_{i=1}^{n} (x_i - y_i)^2 \right)^{1/2}$ (1)
$\|x\|$	Euclidean norm, defined by $\|x\| = \varrho_n(x, 0)$ for $x \in E^n$
S^n	n-sphere, $E^{n+1} \cap \{x \mid \|x\| = 1\}$ ($n \geq 0$)
B_n	n-ball, $E^n \cap \{x \mid \|x\| \leq 1\}$ ($n \geq 1$)
ϱ	usually denotes a metric
$B_\varrho(x, \varepsilon)$	ϱ-sphere of radius ε about x (21)
$\mathscr{B}_\varrho$	set of all ϱ-spheres $B_\varrho(x, \varepsilon)$ for $x \in X$, $\varepsilon > 0$ (21)
T_ϱ	topology generated by ϱ (B_ϱ is a base for T_ϱ) (21)
(X, ϱ)	used interchangeably with (X, T_ϱ) to denote a metric space (21)
$C([a, b])$	set of real valued, continuous functions on $[a, b]$ (3)
$\mathrm{Int}_T(A)$	set of T-interior points of A (6)
$\mathrm{Bdry}_T(A)$	set of T-boundary points of A (10)
$\mathrm{Der}_T(A)$	set of T-cluster points of A (10)
$\mathrm{Cl}_T(A)$	T-closure of A (10)
$\bar{A}$	closure of A (10)
$\mathrm{Is}_T(A)$	set of T-isolated points of A (10)
$\cong$	homeomorphic to (36)
T_A	relative topology of T on A (40)
$\mathbb{P}$	product topology (46)
$\sim, \approx$	usually denote equivalence relations (302)
$[x]_\sim$	equivalence class of x under an equivalence relation $\sim$ (302)
$A/\sim$	set of equivalence classes generated by an equivalence relation $\sim$ on A (302)
η	usually denotes a natural map (302)
I_f	identification topology determined by f (57)
q	usually denotes a quotient topology
$\mathscr{F}, \mathscr{M}$	usually denote filterbases
$\mathscr{F}_x$	neighborhood filterbase at x (78)
$\mathrm{Comp}_T(x)$	T-component of x (128)
$P\,\mathrm{Comp}_T(x)$	T-path component of x (134)
c.c.	countably compact (161)
B-W compact	Bolzano-Weierstrass compact (164)
Y^X	set of continuous functions on X to Y (175, Prob. 33, 235)

$c(X, Y)$	compact-open topology on Y^X (175, Prob. 33)
G_δ	countable intersection of open sets ⎫
F_σ	countable intersection of closed sets ⎬ (209, Prob. 7)
$\varrho \sim \sigma$	equivalence of metrics ϱ and σ (177)
$\lambda(\mathcal{O})$	Lebesgue number of a cover $\mathcal{O}$ (179)
$H < G$	H is a subgroup of G (304)
$H \triangleleft G$	H is a normal subgroup of G (304)
G	usually denotes a group
e	identity element of a group (303)
x^{-1}	group inverse of x (303)
v	inverse function on G, defined by $v(x) = x^{-1}$ for $x \in G$ (214)
φ	product function on G, defined by $\varphi(x, y) = xy$ (214)
ξ	a function on $G \times G$ to G defined by $\xi(x, y) = xy^{-1}$ (214)
$_xt$	left translation by x (216)
t_x	right translation by x (216)
$\mathscr{I}_x$	inner automorphism determined by x (216)
$G_1 \oplus G_2$	direct sum of G_1 and G_2 (305)
$\simeq$	homotopy equivalence in Y^X (236)
$\cong$	homotopy equivalence for paths (263)
$f \simeq 0$	f is nullhomotopic (241)
Φ, Ψ, Θ	usually denote homotopies
$\{p, q\}$	denotes a simplex with vertices p and q (249)
(p, q)	ordered simplex with vertices p and q (249)
$[p, q]$	cell of the simplex $\{p, q\}$ (249)
$\deg(f)$	degree of f (256)
c_y	constant map: $X \to \{y\}$ (241)
P_{x_0}	set of loops at x_0 (267)
$\pi(X, x_0), \pi_1(X, x_0)$	fundamental group of X at x_0 (267)
$\pi(X), \pi_1(X)$	fundamental group of path connected space X (275)
$\hat{\alpha}$	map induced between $\pi(X, x_0)$ and $\pi(Y, \alpha(x_0))$ by $\alpha : X \to Y$ (276)
$\dot{I}$	$\{0\} \cup \{1\}$ (= boundary of I)
I^n	n-cube, $E^u \cap \{(x_1, ..., x_n) \vert 0 \leq x_i \leq 1\}$ (281)
$\dot{I}^n$	boundary of the n-cube, consisting of $(x_1, ..., x_n)$ in I^n with at least one x_i 0 or 1 (281)
$\pi_n(X, x_0)$	n^{th} homotopy group of X at x_0

Index

Alexander's Theorem 148, 173 (Prob. 16)
Alexandroff, P. S. 138, 151
Alexandroff Compactification 152
Antipodal point 240
Approximate integral 229
Ascoli's Theorem 202
Associativity 303
Automorphism 216
Axiom of Choice 149, 294

Baire Category Theorem 197
Bartle, R. G. 79
Base 12
Base point 262, 266
Base neighborhood 13
Bicompactness 138
Bijection 292
Bolzano-Weierstrass Compactness 138, 164 ff.
Borsuk's Extension Theorem 245
Boundary 10
Boundary point 7
Bounded set 193
Bourbaki, N. 75, 106, 138, 148, 212 (Prob. 27)
Brouwer's Theorem 241, 248, 257, 261 (Prob. 8), 277, 278
Brouwer's Fixed Point Theorem 248

Cantor set 55
Cantor's Theorem 194
Cartan, H. 75, 85 (Prob. 9)
Cartesian product 292, 294
Cauchy filterbase 211 (Prob. 24)
Cauchy sequence 188
Center of a group 232 (Prob. 4)
Closed set 4
Closed map 38 (Prob. 6)
Closure 10
Cluster point 7

Cofinal topology 3
Compact space 139 ff.
Compact Hausdorff space 147
Compact regular space 147
Compact-open topology 175 (Prob. 33)
Complete induction 298
Complete metric space 188 ff.
Completely normal space 117 (Prob. 13)
Completely regular space 109 ff.
Component 128
Composition of functions 293
Cone 61
Connected space 119 ff.
Connectedness
  of an identification space 128
  of a product space 128
  of a subspace 120
Construction by induction 298
Continuity 31
Continuity at a point 28
Continuous vector field 258
Contractible space 219
Contracting map 83 (Prob. 1)
Convergence
  of a filter 85 (Prob. 9)
  of a filterbase 76
  of a net 71
  of a sequence 66
Convex 242
Cover 139
Covering space 283 (Prob. 14)
Countable 301
Countable compactness 138, 161 ff.
Countability
  $1°$- 166 ff.
  $2°$- 161 ff.
Countably infinite 301
Cube 115
Cyclic group 304
Cylinder 59

317

318 INDEX

Deformable 239
Deformation retract 282 (Prob. 10)
Degree 256
DeMorgan's Laws 290
Dense 168
Derived set 10
Diagonalization process 301
Diameter of a set 193
Directed set 70
Direct sum 305
Direction 70
Discrete topology 1
Domain 292
Dugundji, J. 209
Dyadic number 102

Eilenberg's Theorem 247
Element of a set 287
Empty implication 286
Empty set 287
Entourage 212 (Prob. 27)
Equicontinuous 202
Equivalence class 302
Equivalence relation 302
Equivalent metrics 177
Euclidean n-space 1
Euclidean topology 2
Evaluation map 116, 175 (Prob. 33)
Eventually in A 70
Exterior point 6

Filter 85 (Prob. 9)
Filter convergence 85 (Prob. 9)
Filterbase 75 ff.
Filterbase convergence 76
Finite intersection property 143
First category 196
First element 301
First homotopy group 262, 268
Fixed point property 62 (Prob. 4)
Fréchet, M. 86 (Prob. 9), 138
Fréchet filter 86 (Prob. 8)
Free filter 86 (Prob. 9)
Frontier 10
F_σ-set 209 (Prob. 7)
Function 292
Function space 175 (Prob. 33)
Fundamental group 262, 268

Fundamental Theorem of Algebra 272

G_f-set 118 (Prob. 22), 209 (Prob. 7)
Generates a topology 15
Group 303

Haar, A. 228
Haar covering function 229
Haar integral 230, 233 (Prob. 16), 234 (Prob. 17)
Hamel basis 297
Hausdorff space 87 ff.
Hewitt, E. 102
Higher homotopy groups 281
Hille, E. 153
Homeomorphism 35
Homogeneous space 217
Homomorphism 304
Homotopy 235 ff., 263 ff.
Homotopically equivalent spaces 278
Homotopy type 278
Hurewicz, W. xi
Hurwitz, A. 228

Identification map 63 (Prob. 12), 64 (Prob. 14, 15)
Identification topology 56
Identity element of a group 303
Implies 285
Indiscrete topology 1
Induce a topology 17
Induced map 276
Initial point
Injection 131
Inner normal 258
Interior 6
Interior operator 24 (Prob. 13)
Interior point 5
Intermediate Value Theorem 136 Prob. 5)
Intersection 287, 280
Interval 123
Inverse of a function 292
Inverse of a group element 303
Isolated point 7
Isometry 181
Isomorphism 304

Kakutani, S. 149
Kelley, J. L. 148, 149

Kernel 304
Kuratowski closure operator 24 (Prob. 12)

Least upper bound 301
Lebesgue Covering Lemma 179, 180
Lebesgue number 179
Left coset 304
Left translation 216
Left uniformly continuous 226
Lift 269
Limit ordinal 300
Limit point 162
Lindelöf space 169
Local compactness 154 ff.
Locally compact Hausdorff space 159
Locally connected 136 (Prob. 13)
Locally path connected 137 (Prob. 14)

Maximal element 296, 297
Maximum Modulus Theorem 173 (Prob. 22)
McShane, E. J. 75
Metric 20
Metric space 21
Metrizable space 204
Möbius strip 60

Natural map 304
Neighborhood 4
Neighborhood filterbase 78
Neighborhood finite 61 (Prob. 2)
Neighborhood system at x 25 (Prob. 14)
Net 70 ff.
Net convergence 70
Non-limit ordinal 300
Non-vanishing vector field 258
Normal space 97
Normal subgroup 304
Novak, J. 102
Nowhere dense 196
Nullhomotopic 241

One point compactification 152 ff.
Open map 34
Open set 4
Order 17
Order topology 17

Ordinal 300
Ordinal space 19
Outer normal 258

Paracompactness 174 (Prob. 32)
Parallel space 53
Partial order 295
Path 131
Path component 134
Path connectedness 131
  of an identification space 134
  of a product space 133
Peter, F. 228
Point finite 118 (Prob. 18)
Preorder 296
Principle of Mathematical Induction 297, 298
Principle of Transfinite Induction 301
Principle of Uniform Boundedness 201
Product of paths 262
Product topology 45 ff.
  of connected spaces 128
  of compact spaces 148
  of complete metric spaces 210 (Prob. 15)
  of completely regular spaces 114
  of Hausdorff spaces 91
  of locally compact spaces 157
  of metric spaces 204, 205
  of normal spaces 100
  of 1^0-countable spaces 173 (Prob. 24)
  of path-connected spaces 133
  of regular spaces 95
  of separable spaces
  of 2^0-countable spaces 173 (Prob. 24)
Product uniformity 213 (Prob. 27)
Proposition 285
Punctured neighborhood 8

Quotient group 304
Quotient topology 58
Quotient topological group 224

Regular space 93
Regular Lindelöf space 170
Regular 2^0-countable space 171
Relative topology 40
Restriction of a function 292

INDEX

Retract 62 (Prob. 4), 65 (Prob. 15), 117 (Prob. 8), 248
Retraction 62 (Prob. 4), 248, 282 (Prob. 11)
Rice, P. M. 137 (Prob. 15)
Riesz, F. 143
Right coset 304
Right translation 216
Right uniformly continuous 226

Second category 196
Set 287
Set difference 287
Separable 168
Separate 119
Separation 87
Sequence 66
Sequential compactness 138, 165
Sequential convergence 66
Sierpinski topology 1
Sign of a simplex 249
Simplex 249
Slice 53
Slice point 137 (Prob. 15)
Space 1
Starlike 283 (Prob. 16)
Stone, M. H. 106
Subbase 15
Subgroup 304
Subnet 84 (Prob. 5)
Subsequence 165
Subspace 40
sup norm Topology 3
Support 226
Surjection 292
Symmetric neighborhood 218

Terminal point 131
Tietze product topology 45
Tietze-Urysohn Extension Theorem 106
Topology 1
  cofinal 3
  completely regular 109
  generated by a subbase 15
  Hausdorff 87
  identification
  induced by a metric 56

normal 97
order 17
product 46
quotient 58
regular 93
sup norm 3
T_0 219
T_1 224
T_2 87
T_3 93
T_4 97
Topological group 214ff.
Topological invariant 37
Topological space 1
Torus 59
Totally bounded 184, 194
Triadic expansion 56
Triangulation 250
Tucker, A. W. 268
Tychonov, A. xi, 45, 80
Tychonov plank 149
Tychonov product topology 46
Tychonov Theorem 148, 149

Ultrafilter 85 (Prob. 9)
Ultrafilterbase 81
Uniform continuity 179, 225
Uniformity 211 (Prob. 27)
Uniform space 211 (Prob. 27), 232 (Prob. 13)
Uniformly bounded 202
Union 287, 290
Universal net 84 (Prob. 6)
Upper bound 296, 301
Urysohn, P. S. xi, 102, 138
Urysohn function 102
Urysohn's Lemma 104
Urysohn's Metrization Theorem 207

Vector field 258

Weak topology 62 (Prob. 5)
Weierstrass Approximation Theorem 9
Weil, A. 228
Weyl, H. 228

Zorn's Lemma 149, 296